GIS Applications in the Tourism and Hospitality Industry

Somnath Chaudhuri
Maldives National University, Maldives

Nilanjan Ray
Adamas University, India

A volume in the Advances in Hospitality, Tourism, and the Services Industry (AHTSI) Book Series

IGI Global
DISSEMINATOR OF KNOWLEDGE

Published in the United States of America by
IGI Global
Business Science Reference (an imprint of IGI Global)
701 E. Chocolate Avenue
Hershey PA, USA 17033
Tel: 717-533-8845
Fax: 717-533-8661
E-mail: cust@igi-global.com
Web site: http://www.igi-global.com

Library of Congress Cataloging-in-Publication Data

Names: Chaudhuri, Somnath, editor. | Ray, Nilanjan, 1984- editor.
Title: GIS applications in the tourism and hospitality industry / Somnath Chaudhuri and Nilanjan Ray, editors.
Description: Hershey PA : Business Science Reference, [2018]
Identifiers: LCCN 2017035068| ISBN 9781522550884 (hardcover) | ISBN 9781522550891 (ebook)
Subjects: LCSH: Tourism--Management. | Tourism--Marketing. | Hospitality industry--Management. | Hospitality industry--Marketing. | Geographic information systems
Classification: LCC G155.A1 G484 2018 | DDC 910.285--dc23 LC record available at https://lccn.loc.gov/2017035068

This book is published in the IGI Global book series Advances in Hospitality, Tourism, and the Services Industry (AHTSI) (ISSN: 2475-6547; eISSN: 2475-6555)

British Cataloguing in Publication Data
A Cataloguing in Publication record for this book is available from the British Library.

Advances in Hospitality, Tourism, and the Services Industry (AHTSI) Book Series

ISSN:2475-6547
EISSN:2475-6555

Editor-in-Chief: Maximiliano Korstanje, University of Palermo, Argentina

MISSION

Globally, the hospitality, travel, tourism, and services industries generate a significant percentage of revenue and represent a large portion of the business world. Even in tough economic times, these industries thrive as individuals continue to spend on leisure and recreation activities as well as services.

The Advances in Hospitality, Tourism, and the Services Industry (AHTSI) book series offers diverse publications relating to the management, promotion, and profitability of the leisure, recreation, and services industries. Highlighting current research pertaining to various topics within the realm of hospitality, travel, tourism, and services management, the titles found within the AHTSI book series are pertinent to the research and professional needs of managers, business practitioners, researchers, and upper-level students studying in the field.

COVERAGE

- Service Training
- International Tourism
- Service Management
- Casino Management
- Service Design
- Travel Agency Management
- Leisure & Business Travel
- Hotel management
- Sustainable Tourism
- Health and Wellness Tourism

The Advances in Hospitality, Tourism, and the Services Industry (AHTSI) Book Series (ISSN 2475-6547) is published by IGI Global, 701 E. Chocolate Avenue, Hershey, PA 17033-1240, USA, www.igi-global.com. This series is composed of titles available for purchase individually; each title is edited to be contextually exclusive from any other title within the series. For pricing and ordering information please visit http://www.igi-global.com/book-series/advances-hospitality-tourism-services-industry/121014. Postmaster: Send all address changes to above address.

Titles in this Series

For a list of additional titles in this series, please visit:
https://www.igi-global.com/book-series/advances-computer-electrical-engineering/73675

Global Perspectives on Religious Tourism and Pilgrimage
Hatem El-Gohary (Birmingham City University, UK & Cairo University, Egypt) David John Edwards (Birmingham City University, UK) and Riyad Eid (United Arab Emirates Univerity, UAE)
Business Science Reference • ©2018 • 236pp • H/C (ISBN: 9781522527961) • US $185.00

***Legal and Economic Considerations Surrounding Reproductive Tourism Emerging Research** and Opportunities*
Anastasia Paraskou (CROSSBEES Healthcare Consultancy, Cyprus & Swiss Management Center University, Switzerland) and Babu P. George (Fort Hays State University, USA & Swiss Management Center University, Swtzerland)
Business Science Reference • ©2018 • 235pp • H/C (ISBN: 9781522526940) • US $165.00

***Handbook of Research on Technological Developments for Cultural Heritage and eTourism** Applications*
João M. F. Rodrigues (University of Algarve, Portugal) Célia M. Q. Ramos (University of Algarve, Portugal) Pedro J. S. Cardoso (University of Algarve, Portugal) and Cláudia Henriques (University of Algarve, Portugal)
Business Science Reference • ©2018 • 535pp • H/C (ISBN: 9781522529279) • US $295.00

Innovative Perspectives on Tourism Discourse
Magdalena Bielenia-Grajewska (University of Gdansk, Poland) and Enriqueta Cortes de los Rios (Almeria University, Spain)
Business Science Reference • ©2018 • 360pp • H/C (ISBN: 9781522529309) • US $225.00

Virtual Traumascapes and Exploring the Roots of Dark Tourism
Maximiliano Korstanje (University of Palermo, Argentina) and Babu George (Fort Hays State University, USA)
Business Science Reference • ©2018 • 262pp • H/C (ISBN: 9781522527503) • US $195.00

For an entire list of titles in this series, please visit:
https://www.igi-global.com/book-series/advances-computer-electrical-engineering/73675

IGI Global
DISSEMINATOR OF KNOWLEDGE

701 East Chocolate Avenue, Hershey, PA 17033, USA
Tel: 717-533-8845 x100 • Fax: 717-533-8661
E-Mail: cust@igi-global.com • www.igi-global.com

Editorial Advisory Board

Table of Contents

Detailed Table of Contents

Technology has penetrated every aspect of life and it proceeds with the aim of facilitating people's lives. Geographical information systems as a part of the developing technology provide services in every area with different forms. One of these areas is museology. The desire to make visual and written information and cultural heritage that are presented by museums with auditory and interactive experiences has brought together digital technologies and cultural and historical activities. From this point of view, the "Müze Asist" is a remarkable application in terms of the fact that it is a local app and in use in most of the museums in Turkey and the convenience provided for all tourists who want to visit a museum, especially for those who prefer self-guided tours. This chapter may be a guide to its objects and to the next researchers.

Interest of the local population is an important consideration for sustainable development of a region. Researchers recently observed in their studies about the benefits the local population get while being involved in eco-tourism. Though many factors are considered to know how locals are benefited by eco-tourism, an interesting factor to be dealt with utmost care and concern is environmental and ecological issues arising out of these growth factors. All the possible technologies of computer application may be used to maximise the end needs of this specialised subject. One such application is GIS (geographical information system), which can be of great help in this concept. The GIS applications are recently used in the management of parks, facilities management, assessment of visual resources, and new areas identification for new tourism development initiatives.

Understanding how tourists use space in urban environments has practical applications in destination management and planning. It reveals key issues for organizing facilities and essential services, creating new products, impact management, and other purposes related to urban tourism. The geotagged photos shared by Panoramio users, during their visit to the city of Lisbon between 2007 and 2014, allow the authors to incorporate into this chapter a quantitative and geographic reading on the tourist consumption of spaces within the city. In this chapter, they analyze the spatial distribution of tourists and its changes across time considering a period of 8 years. Furthermore, a regression analysis method was carried out in order to find the spatial relations between the observed pattern (geographical agglomeration of tourist́ s photos) and a set of 24 selected independent variables.

The chapter brings out a brief note on the tourist attractions, hotels and lodges, NGOs/travel agencies operating in that region, railway/bus stations, land use profile, etc. in the Sundarban area of West Bengal in conjunction with exploring the potential of ecotourism using GIS and some secondary source data. Moving onto the analysis part, by making use of geo-spatial data, the attributes of ecotourism potential in the Sundarbans has been explored. The author makes use of the Euclidean distance mechanism and principal component analysis to rank the ecotourism sites in Sunderbans (i.e., based on the construction of ecotourism potential index [EPI]). The novelty of the chapter lies in comparing the ranks obtained by constructing the EPI following the principal component analysis and the Euclidean distance function. It needs to be mentioned here that these tourist spots have been selected based on the information collected on the inflow of both domestic and foreign tourists to these spots. The chapter concludes by discussing the future scope of research in this regard.

This chapter proposes and develops a cloud-computing-based SDI model named as TCloud for sharing, analysis, and processing of spatial data particularly in the Temple City of India, Bhubaneswar. The main purpose of TCloud is to integrate all the spatial information such as tourism sites which include various temples, mosques, churches, monuments, lakes, mountains, rivers, forests, etc. TCloud can help the decision maker or planner or common users to get enough information for their further research and studies. It has used open source GIS quantum GIS for the development of spatial database whereas QGIS plugin has been linked with quantum GIS for invoking cloud computing environment. It has also discussed the various spatial overlay analysis in TCloud environment.

The main contribution of this chapter is to critically discuss the benefits of applying geographic information systems (GIS) as a tool for management and promotion of spiritual tourism circuits (STC) and ST destinations. This research-based chapter also examines the extent to which GIS can be used in spiritual tourism management and promotion, proposes a model for the use and benefits of GIS in spiritual tourism management and promotion in India and Pakistan, and proposes GIS connected STC. This chapter identifies the socio-economic and business benefits of applying GIS to spiritual tourism circuits (STC). In this research, the spiritual tourism product is exclusively based on spiritual place, sites, or destinations, which is also called spatial or geographical data.

Geographical data systems (GIS), which enable the creation of entities that are the subject of tourism, tourism planning, the identification of impact areas of tourism values, and the sustainable utilization of tourism entities, are data-collection and decision-making systems that can contribute to tourism planning and the development of tourism Recently, tourists' interests have shifted from sea-, sand-, and sun-focused tourism to alternative tourism types; one of which is rural tourism. With rural tourism, it is expected that economical and socio-cultural opportunities for people who live in rural areas can be improved. The realization of this provides the concept of rural development. Today, in Turkey, 36.40% of the whole population still live in rural areas and agricultural activities are the most important income sources. Therefore, it is possible to investigate rural tourism and rural development in the Turkish context. This chapter deals with suggestions in terms of rural tourism and rural development by using GIS.

Geographical information systems (GIS) are the systems that store location-based data and analyze them. GIS originated from the conventional cartographic techniques of simply drawing maps with a pencil and board. Following the adoption of computer technology, GIS further evolved as a geo-referenced dynamic information system, which can today be considered as a multi-disciplinary instrument that links different disciplines like geography, computer science, remote sensing, civil engineering, statistics, marketing, and other social and behavioral sciences. In this chapter, GIS is defined and its importance and functions are described in detail. Usage of GIS in marketing is explained, especially its development in marketing theory. Also, the value and scope of GIS in tourism management with a view to understand the spread of GIS applications in tourism is explored.

Tourism activities are highly dependent on spatial information. Finding the most interesting travel destinations and attractions and planning a trip are still open research issues to GIScience research applied to the tourism domain. Nowadays, huge amounts of information are available over the world wide web that may be useful in planning a visit to destinations and attractions. However, it is often time consuming for a user to select the most interesting destinations and attractions and plan a trip according to his own preferences. Tourism recommender systems (TRSs) can be used to overcome this information overload problem and to propose items taking into account the user preferences. This chapter reviews related topics in tourism recommender systems including different tourism recommendation approaches and user profile representation methods applied in the tourism domain. The authors illustrate the potential of tourism recommender systems as applied to the tourism domain by the implementation of an illustrative geospatial collaborative recommender system using the Foursquare dataset.

Advancement of location-acquisition technologies with fast development of mobile devices and wireless communication caused a revolution of information. It has been used in location-based social networks (LBSNs), has attracted millions of users to Facebook places, Gowalla, and Foursquare, is an important task to make location recommendations to users, and utilizes user preferences and other information that not only help users explore new places but also make LBSNs more attractive to users. This chapter discusses recommender systems (RS) and its application in different fields like LBSN, big data, and real life. It describes traditional recommendation approaches as well as modern approaches and explains smart community as one of powerful techniques to be used. It also introduces the state-of-art geographical techniques and presents a comparative study of recommendation techniques that can be served as a good guide and a roadmap for research and practice in this area. Finally, the authors discuss measurements and the limitations of RS.

Foreword

This book is a spectacular collection of essays related to the *GIS Applications in the Tourism and Hospitality Industry*. Service industry is a fast-developing industry. It is closely related to many strategies.

There are 12 papers in this collection and all the authors claim responsibility of handling their subjects in a remarkable manner. I'm happy to introduce this marvellous collection to both of you. Through the book, readers will have chance to find crucial and different insights in terms of service sector and GIS applications in service sectors.

In this work author has investigated on storage and retrieval of geospatial data by integrating Hadoop and Cassandra using Prefix-based partitioning and Cassandra's default partitioning algorithm i.e. Murmur3partitioner techniques. Second chapter in this chapter discussed about Geographical Information System applications Utilized in Museums in Turkey within the Scope of the Cultural Heritage Tourism. Third chapter depicts on Geographical Information System in Eco Tourism. Interests of the local population are an important consideration for sustainable development of a region. Researchers recently observed in their studies about the benefits the local population get while being involved in eco-tourism. Chapter 4 focused on Geographical patterns in tourist city. Understanding how tourists use space in urban environments has practical applications in destination management and planning. It reveals key issues for organizing facilities and essential services, creating new products, impact management, and other purposes related to urban tourism. Chapter 5 brings out a brief note on the tourist attractions, hotels and lodges, NGOs/travel agencies operating in that region, railway/bus stations, land use profile, etc. in the Sundarban area of West Bengal in conjunction with exploring the potential of ecotourism using GIS. Chapter 6 proposed a Cloud computing based SDI Model named as TCloud; for sharing, analysis, and processing of spatial data particularly in the Temple City of India, Bhubaneswar. Another chapter based on critical discuss benefits of applying Geographic Information systems (GIS) as a tool for management and promotion of spiritual tourism circuits (STC) and ST destinations. Chapter 7 discussed Geographical Data Systems (GIS), which enable the creation of entities which are the subject of tourism, tourism planning, the identification of impact areas of tourism values and the sustainable utilization of tourism entities. Another chapter focused on The Value and Scope of GIS in Marketing and Tourism Management. Geographical information systems (GIS) are the systems that store location based data and analyze them. Chapter 10 discussed on tourism activities are highly dependent on spatial information. Tourism Recommender Systems (TRSs) can be used to overcome this information overload problem and to propose items taking into account the user preferences. This chapter reviews related topics in tourism recommender systems including different tourism recommendation approaches and user profile representation methods applied in the tourism domain. Another chapter discussed on Recommender systems (RS) and its application in different fields like location based social network (LBSN), Big Data and real-life. We describe traditional recommendation approaches as well as modern approaches and we explain smart community as one of powerful techniques to be used. Chapter 12 discussed on Decision support system (DSS) plays a vital role especially in rural areas to develop rural sector for sustainable development and socio-economic uplifting of the country.

I sincerely think that these issues are of paramount importance and have been pretty well researched and commented in the form of research papers in the book. At the same time, the authors also seek for greater analysis and further study in all these areas of infrastructure development and building of a more vibrant, viable, sustainable, and stable economy in all the growing economies of the world.

I congratulate the editors and the contributors who had put in a lot of effort and time in bringing out this book to address many vital concerns for a better understanding of the issues surrounding our economy.

I hope you'll like this book as much as I do.

Madhusudhan Chakraborty
Adamas University, India

Preface

Geographical information system (GIS) is the information system that provides functions including visual presentations of geographical locations and relevant analysis of geospatial information by processing them in an integrated manner. Both GIS and tourism cross the boundaries of disciplines and application areas. GIS is the inventory for various spatial and non-spatial tourism resources. It provides the information about the tourist location, its topographic condition, natural resources, population density, transport facilities, recreation resource management, socio-economic and cultural factors and other potential conditions that rank the location as a popular tourist destination. It also provides the detailed route from any part of the world to the destination. The online static maps slowly changed to interactive dynamic maps over the world wide web. Many organizations are developing geo-applications to enhance their business operations and transform their business models. The tourism industry is not an exception; because geospatial data and digital maps are the lifeblood of tourism. Tourism by nature involves the transfer of people to places away from home; and all tourism activities take place in certain geographical areas. Indeed, GIS and currently, Web GIS and geoportals are widely adopted in tourism for several purposes, like: trip planning, visitors' management, tracking tourists' behavior, online bookings, location-based services, management and measurement of destinations' carrying capacity, crisis management and for providing localized personalized content to customers. GIS helps in data design and collection, management of database design and application of tourism analysis and problem solving. These fast, cheap, collaborative and multi-layer web map services empower tourists with new trip planning geospatial tools and information. It typically retrieves the data from heterogeneous sources and displays it in the form of a map. The map can be static or interactive. In case of tourism, maps are usually used in selecting the tourist destinations and in planning travel and stay. Potential tourists use maps to navigate on or before travels and for planning about the trip. Interactive online maps provided by Web GIS exploit the two-dimensional capabilities of human vision and represent the information in an effective, precise and "easy to read" manner. At the same time, Web GIS helps in automatic content development and management. Adequate information about a product is one of the key features of promotion and marketing. It is the key driver in tourism marketing also. The creation, aggregation and distribution of creative, innovative and engaging text, imagery and video are at the very core of successful ICT-based tourism marketing using Web GIS. Content-driven marketing is the most cost-efficient and effective form of marketing. However, content development is often the most troublesome component of a tourism marketing campaign. Planning ahead and thinking strategically can create long-term efficiencies in aggregating, developing and distributing content that will be used across all ICT-based tourism marketing platforms. The interactive Web GIS platform solves the problem of content management. Any online user can update online information about any potential tourist destination using Web 2.0 features of Web GIS platforms.

Objectives of this current book are Tourism has developed as a new form of service industry for alternative and sustainable socio-economic development of a country. This service industry has widely implemented Web based Geographical Information System, in order to improve information sharing, monitoring-planning and overall tourism and destination management. The principal objective and mission of the book is to discuss on different pattern of research works related to GIS and Web GIS and their strategic application in tourism and hospitality industry. Besides this, the book will also address contemporary aspects of content-driven tourism management and promotion through Public Participation GIS (PPGIS) like Geospatial Mashups and Geoportals in particular. So the objective of this book is to share research work and will provide a platform for new avenues in GIS-based tourism development, management and promotion and its impact in hospitality industry.

This volume will present a variety of practical application, application tools, models, approaches and strategies that are providing themselves in practice, demonstrating effectiveness with managing diversity and innovation. This volume will also present several visionary proposals for transforming societies, citizens and professionals so all concerned are better prepared to embrace diversity and do their part in creating valuable and necessary innovation that positively impact the global community. The overall objectives and mission of this proposal is to share a different pattern of research work that will provide a platform for new avenues in overall promotional tool for service sector of modern business.

The book is organized in 12 Chapters. A brief description is as follows:

- **Chapter 1:** Now-a-days we are moving towards digitization and making all our devices such as sensors, cameras connected to Internet producing bigdata. This bigdata has variety of data and has paved the way to the emergence of NoSQL databases, like Cassandra for achieving scalability and availability. Hadoop framework has been developed for storing and processing distributed data. In this work we mainly investigated on storage and retrieval of geospatial data by integrating Hadoop and Cassandra using Prefix-based partitioning and Cassandra's default partitioning algorithm i.e Murmur3partitioner .techniques. Geohash value is generated, that acts as a partition key and also helps in effective search. Hence the time taken for retrieving data is optimized. When user requests for Spatial queries like finding nearest locations, searching in Cassandra database starts using both partitioning techniques. A comparison on query response time is made so as to verify which method is more effective. Results showed that Prefix-based partitioning technique is efficient than Murmur3 partitioning technique

- **Chapter 2:** This chapter discussed about Geographical Information System applications Utilized in Museums in Turkey within the Scope of the Cultural Heritage Tourism. Now a day, technology has penetrated every aspect of life and it proceeds with the aim of facilitating people's lives. Geographical Information System as a part of the developing technology provides services in every area with different forms. One of these areas is museology. The desire to make visual and written information and the cultural heritage that are presented by the museums permanent with auditory and interactive experiences has brought together digital technologies and cultural and historical activities. From this point of view, the "MüzeAsist" is a remarkable application in terms of the fact that it is a local app and in use in most of the museums in Turkey and the convenience provided for all tourists who want to visit a museum, especially for those who prefer self-guided tours. This work may be a guide to its objects and to the next researchers.
- **Chapter 3:** This present chapter depicts on Geographical Information System in Eco Tourism. Interests of the local population are an important consideration for sustainable development of a region. Researchers recently observed in their studies about the benefits the local population get while being involved in eco-tourism. Though many factors are considered to know how locals are benefited by eco-tourism, an interesting factor to be dealt with utmost care and concern is environmental and ecological issues arising out of these growth factors. All the possible technologies of computer application may be used to maximise the end needs of this specialised subject. One such application is GIS (Geographical Information System), which can be of great help in this concept. The GIS applications are recently used in the management of parks, facilities management, assessment of visual resources, and new areas identification for new tourism development initiatives.
- **Chapter 4:** This present study focused on Geographical patterns in tourist city. Understanding how tourists use space in urban environments has practical applications in destination management and planning. It reveals

key issues for organizing facilities and essential services, creating new products, impact management, and other purposes related to urban tourism. The geotagged photos shared by Panoramio users, during their visit to the city of Lisbon between 2011 and 2015, allow us to incorporate in this presentation a quantitative and geographic reading on the tourist consumption of these spaces: which are the most sought city elements by tourists? what is its location in the urban context? In this chapter, we analyze the spatial distribution of tourists in Lisbon based on data from 'Panoramio' social network between 2011 and 2015. Furthermore, a regression analysis method was carried out, in order to find the spatial relations between the observed pattern (geographical agglomeration of tourist´s photos) and a set of 24 selected independent variables.

- **Chapter 5:** The study brings out a brief note on the tourist attractions, hotels and lodges, NGOs/travel agencies operating in that region, railway/ bus stations, land use profile, etc. in the Sundarban area of West Bengal in conjunction with exploring the potential of ecotourism using GIS and some secondary source data. Moving onto the analysis part, by making use of geo-spatial data, the attributes of ecotourism potential in the Sundarbans has been explored. The author makes use of the Euclidean distance mechanism and Principal Component analysis to rank the ecotourism sites in Sunderbans, i.e. based on the construction of Ecotourism Potential Index (EPI). The novelty of the paper lies in comparing the ranks obtained by constructing the EPI following the Principal Component analysis and the Euclidean distance function. It needs to be mentioned here that these tourist spots have been selected based on the information collected on the inflow of both domestic and foreign tourists to these spots. The paper concludes by discussing the future scope of research in this regard.
- **Chapter 6:** The present chapter proposed and developed a Cloud computing based SDI Model named as TCloud; for sharing, analysis, and processing of spatial data particularly in the Temple City of India, Bhubaneswar. The main purpose of TCloud is to integrate all the spatial information such as tourism sites which includes various Temples, Mosques, Churches, Monuments, Lakes, Mountains, Rivers, Forests etc. TCloud can help the decision maker or planner or common users to get enough information for their further research and studies. It has used open source GIS Quantum GIS for the development of spatial database whereas QGIS Plugin has been linked with Quantum GIS for invoking cloud computing environment. It has also discussed the various spatial overlay analysis in TCloud environment.

- **Chapter 7:** The main contribution of this chapter is to critically discuss benefits of applying Geographic Information systems (GIS) as a tool for management and promotion of spiritual tourism circuits (STC) and ST destinations. This research based chapter also examines the extent to which GIS can be used in spiritual tourism management and promotion, and to proposes a model for the use and benefits of GIS in spiritual tourism management and promotion in India and Pakistan, and propose GIS connected STC. This paper identifies the socio-economic and business benefits of applying GIS to Spiritual Tourism Circuits (STC). In this research the spiritual tourism product will be exclusively based on spiritual place, sites or destinations, which is also called as spatial or geographical data. Since it could be better aligned with GIS, rather people or events, which is a part of non-spatial or attribute data.
- **Chapter 8:** Geographical Data Systems (GIS), which enable the creation of entities which are the subject of tourism, tourism planning, the identification of impact areas of tourism values and the sustainable utilization of tourism entities, are data collection and decision making systems which can contribute the tourism planning and the development of tourism Recently, tourists' interests have shifted from sea, sand and sun focused tourism to alternative tourism types one of which is rural tourism. With rural tourism, it is expected that economical and socio-cultural opportunities for people who live in rural areas can be improved. The realization of this provides the concept of rural development Today, in Turkey 36.40% of the whole population still live in rural areas and agricultural activities are the most important income sources. Therefore, it is possible to investigate rural tourism and rural development in Turkey context. This study deals with suggestions in terms of rural tourism and rural development by using GIS.
- **Chapter 9:** This chapter focused on The Value and Scope of GIS in Marketing and Tourism Management. Geographical information systems (GIS) are the systems that store location based data and analyze them. GIS originated from the conventional cartographic techniques of simply drawing maps with a pencil and board. Following the adoption of computer technology, GIS further evolved as a geo-referenced dynamic information system, which can today be considered as a multi-disciplinary instrument that links different disciplines like geography, computer science, remote sensing, civil engineering, statistics, marketing and other social and behavioral sciences. In this study, GIS is defined and its importance and functions are described in details. Usage of GIS in marketing area is explained especially its development in marketing theory. Also it is delved into the value and scope of GIS in tourism management with a view to understand the spread of GIS applications in tourism.

- **Chapter 10:** Tourism activities are highly dependent on spatial information. Finding the most interested travel destinations and attractions and planning a trip are still open research issues to GIScience research applied to the tourism domain. Nowadays, huge amounts of information are available over the World Wide Web that may be useful to plan a visit to destinations and attractions. However, it is often time consuming for a user to select the most interested destinations and attractions and plan a trip according to his own preferences. Tourism Recommender Systems (TRSs) can be used to overcome this information overload problem and to propose items taking into account the user preferences. This chapter reviews related topics in tourism recommender systems including different tourism recommendation approaches and user profile representation methods applied in the tourism domain. We illustrate the potential of tourism recommender systems as applied to the tourism domain by the implementation of an illustrative geospatial collaborative recommender system using the Foursquare dataset.
- **Chapter 11:** In this paper, we discuss Recommender systems (RS) and its application in different fields like location based social network (LBSN), Big Data and real-life. We describe traditional recommendation approaches as well as modern approaches and we explain smart community as one of powerful techniques to be used. We also introduce the state-of-art geographical techniques and present a comprehensive survey of recommendation techniques, which can be served as a good guide and a roadmap for research and practice in this area. Finally, we discuss measurements and the limitations of RS
- **Chapter 12:** Decision support system (DSS) plays a vital role especially in rural areas to develop rural sector for sustainable development and socio economic uplifting of the country. To take appropriate decisions and to develop village economy, decision support system is useful for the mandal revenue officer, collector, Surpanch and different administrators. It deals both spatial and non-spatial data at village level and comprises various ancillary information including mandal maps and village-wise information of Anantapur and Kadapa districts of Andhra Pradesh such as number of houses, male and female, SC/ST/OBC/general (or) OC population, literate and illiterate population, total working and non working population. Datasets are collected from district collector office, mandal revenue officer (MRO) and inserted in GIS database. This book chapter makes an attempt to inbuilt features of various decisions at Anantapur and Kadapa districts by integrating various layers of information both spatial and attribute at village level.

Acknowledgment

First and foremost, I would like to thank my parents Sri Nirmalendu Ray and Smt. Rina Ray for their unending inspiration and standing beside me throughout my career and for writing this book. I would like to thank to my wife Trina Ray for providing me moral support and inspiration as well as the necessary time and resource toward the applications and maintaining databases. I would like to especially thank to my Doctoral Supervisor and my teacher Dr. Dillip Kumar Das, Asst. Professor and Head of the Department, Department of Tourism, The University of Burdwan India to guide me and develop the applications for this project and for giving me the freedom to manage my projects and provide the necessary time and resource toward the applications and databases.

I acknowledge my indebtedness to all the members of editorial advisory board and technical reviewers of this volume. I am also grateful to all the authors whose valued contributions have enriched the volume. I wish to thank the officials at IGI Global Publications for their invaluable efforts, great support and valuable advice for this project towards successful publication of this book.

I wish to acknowledge my sincere thanks to my co-editor and friend Dr. Nilanjan Ray who have stretched his helping hand and supported me through every step of this challenge. This work would not have been possible without the guidance and moral support of him. I am grateful to my parents and sister for "everything", their guidance and love are always with me in whatever I pursue. I wish to thank Panacea Yamin for her motivation and inspiration to improve my knowledge and move my career forward.

I am indebted to all the contributing authors and officials at the IGI Global Publications with whom I have had the pleasure to work during the project. The collaborative effort from all helped the train to reach its destination.

Nilanjan Ray
Adamas University, India

Somnath Chaudhuri
Maldives National University, Maldives

Chapter 1
Framework for GeoSpatial Query Processing by Integrating Cassandra With Hadoop

S. Vasavi
V. R. Siddhartha Engineering College, India

Mallela Padma Priya
V. R. Siddhartha Engineering College, India

Anu A. Gokhale
Illinois State University, USA

ABSTRACT

We are moving towards digitization and making all our devices, such as sensors and cameras, connected to internet, producing bigdata. This bigdata has variety of data and has paved the way to the emergence of NoSQL databases, like Cassandra, for achieving scalability and availability. Hadoop framework has been developed for storing and processing distributed data. In this chapter, the authors investigated the storage and retrieval of geospatial data by integrating Hadoop and Cassandra using prefix-based partitioning and Cassandra's default partitioning algorithm (i.e., Murmur3partitioner) techniques. Geohash value is generated, which acts as a partition key and also helps in effective search. Hence, the time taken for retrieving data is optimized. When users request spatial queries like finding nearest locations, searching in Cassandra database starts using both partitioning techniques. A comparison on query response time is made so as to verify which method is more effective. Results show the prefix-based partitioning technique is more efficient than Murmur3 partitioning technique.

DOI: 10.4018/978-1-5225-5088-4.ch001

1. INTRODUCTION

Companies that use big data for business challenges can gain advantage by integrating Cassandra with Hadoop. Hadoop distributed file system framework can process voluminous of data generated from various sources. Out of various NoSQL databases, Cassandra supports linear scalability and high availability for ensuring fault tolerance. As such, when integrated, Cassandra and Hadoop together increase the processing capabilities to manage big data efficiently. *G*eospatial data helps in identifying the geographic location of an object, its features and boundaries on Earth. Such data can be analyzed to serve various purposes such as tourism, health care, geo marketing and intelligent transportation system. There are two data types for spatial data, vector and raster. Both data types stores object reference as latitude and longitude (Vertices/paths or grid cells) as shown in figure 1. Raster data includes remote sensing, photogrammetric, and vector data includes Geographical Positioning System (GPS), digitizing. Cassandra integration with Hadoop helps in Querying spatial data efficiently by reducing the query response time.

Raster data can be represented at its original resolution and form without generalization. But the location of each vertex needs to be stored explicitly. Advantage of Vector data is the geographic location of each cell is implied by its position in the cell matrix. Disadvantage is it is difficult to adequately represent linear features depending on the cell resolution. The following figure 2 and figure 3 presents example for vector data and raster data.

Figure 1. Longitude and Latitudes of Earth (Geohash and its format, 2016)

Figure 2. Geospatial Vector data type
(Ruslan Bobov, 2017)

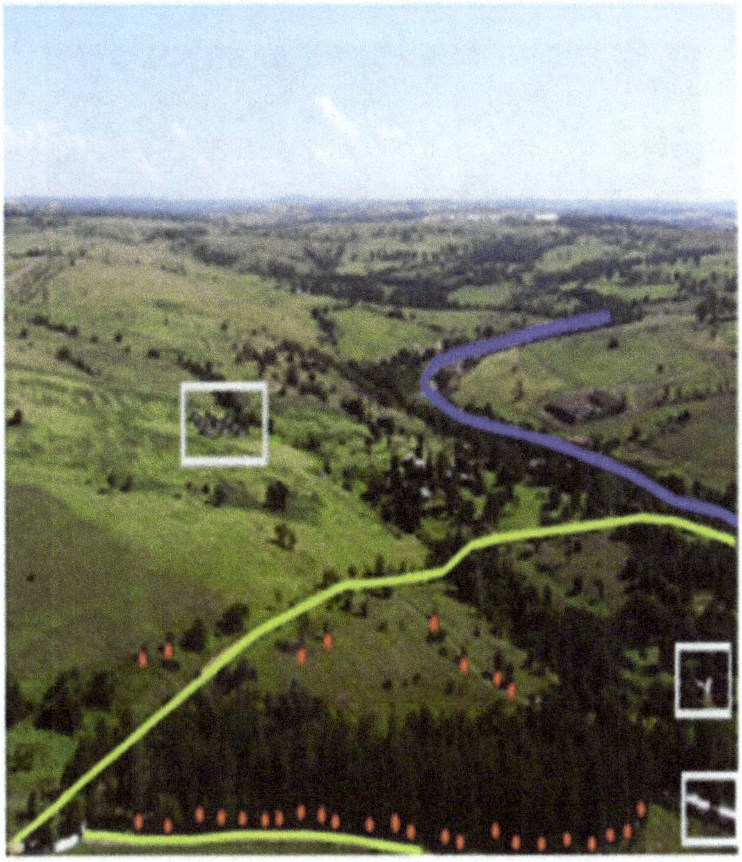

Motivation

Traditional databases (Relational database) are suitable for storing and querying structured data that guarantees ACID properties. With the emergence of internet, large amount of unstructured data is being produced. NoSQL databases, that guarantees CAP properties are suitable for storing such unstructured data. Dynamo, MongoDB, BigTable, HBase, Cassandra are designed to handle the data storage and processing with less response time. Even though MongoDB suits for complex queries such as social networking applications where we have to be optimise for latency, HBase and Cassandra when integrated with Hadoop are equivalently good in such a scenario. Cassandra DB and its query language CQL supports queries such as indexing, search libraries but not spatial queries. Even though some works

Figure 3. Geospatial Raster data type
(Ruslan Bobov, 2017)

are reported for labeling and retrieving Cassandra database, are not efficient. This chapter aims at adding the functionality of spatial querying for Cassandra database by integrating Cassandra with Hadoop.

Problem Statement

Every location on this Globe is an intersection of Latitude and Longitude co-ordinates as shown in Figure 4. Hence this point of representation is used to find a location on map. There are many different kinds of spatial operations such as reclassifying maps, spatial joins, raster analysis, Topological overlay, Measuring distance and connectivity, Characterizing neighbourhood (Berry, 1987). Proximity analysis

Figure 4. Geographic coordinate system (What are Longitudes and Latitudes, 2017)

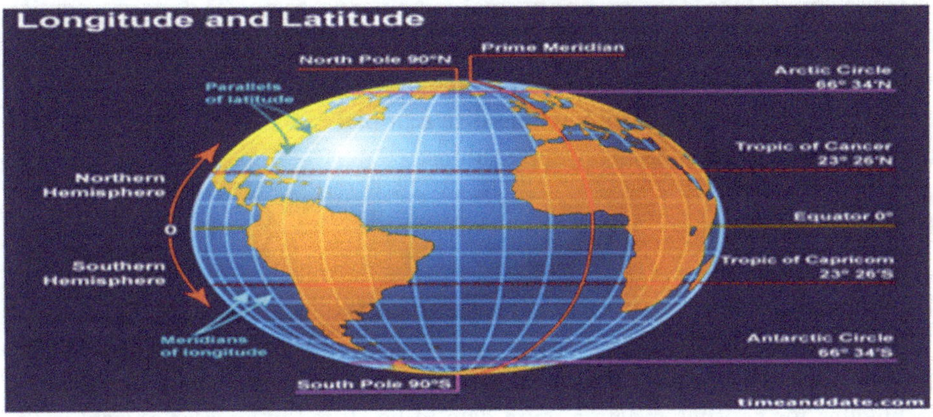

queries such as "which parcels are within 100 meter of the subway", contiguity analysis queries such as "which states share a border with Coorg", Neighbourhood analysis queries such as "calculate an output value at a location from the values at nearby locations". Other GIS functions include zooming and panning, reordering layers, and selecting features. Most commonly used operations are to find the nearest locations of a specified source location. But finding these locations based on latitude-longitude co-ordinate values are really a bit difficult task, especially when dealing with high precision values.

Geohashing technique can be used to overcome this problem. It takes latitude-longitude pair as input and produces a geohash value, whose length is based on precision value specified. Another major pitfall is searching the entire database sequentially for a required destination using this geohash value may not deliver efficient results as expected. Thus parallel processing must be done to get rid of this issue. To achieve this, Cassandra is integrated with Hadoop which is an efficient distributed parallel processing framework by using MapReduce programming model.

BACKGROUND

Humans interact with the Earth system to understand the functioning of its components in ways that have alarming and potentially irreversible consequences for Earth's systems. The outcome of their research activities has many societal impacts in our day- to-day life such as climate prediction, agricultural activities, and terrestrial vegetation. Google Maps and Google Earth are some applications of this kind. Geographic data is usually large in size because of spatiotemporal associations. As such spatial queries are compute intensive because of geometric computations and execution might take long time on a normal computer processor. The query execution time can be reduced when run in a distributed manner, because in a distributed system different parts can be executed at the same time thereby reducing the overall execution time. Now-a-days searching different places based on current location have become important in many applications.

Geospatial Data

Geographical Information System (GIS) data has explicit location information such as road network, GPS data, geo-tagged satellite image (Orlandini, 2017). Geospatial data are characterized by their location properties. As such query processing techniques for geospatial data require data to be stored in simpler objects in the form of rectangles, polygons. The following figure 5 presents pie chart indicating components of geographic data. Geospatial data specifies where it is, attribute data specifies what it is and metadata specifies both geospatial and attribute data (Intro to advanced GIS and a review of basic GIS, 2017).

The following figure 6 presents geospatial data challenges and opportunities.

Query Processing Techniques

Query processing on non geospatial data requires data to be stored either as centralized or in distributed form. Preference based query processing technique executes the query basing on the user supplied preferences. Approximate query processing techniques are suitable for spatial data. Continuous query processing techniques are suitable for stream data. Adaptive query processing techniques uses runtime input so as to change the execution pattern for reduced execution time and efficient CPU utilization. Query based approximation techniques for spatial data can be classified into various groups namely query relaxation, preference based queries and approximate query

Figure 5. Pie chart indicating components of geographic data (Intro to advanced GIS and a review of basic GIS, 2017).

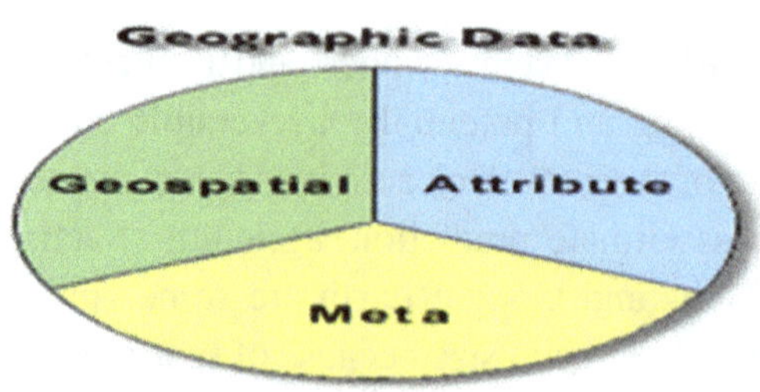

Figure 6. Small things of Big (Geo Spatial) Data (Raghavendran, 2017)

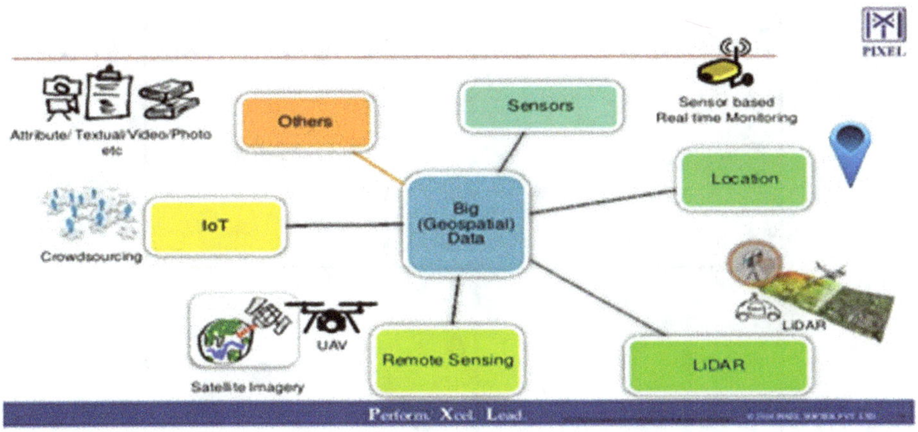

processing (Catania & Migliorini, 2012). Query relaxation and preference based queries improves the quality of the result inspite of data heterogeneity and limited input knowledge. Approximate query processing is aimed at improving query performance. Top-k queries are yet another set of operators that outputs the result based on ranking function or index structure. Spatial query processing techniques can be classified as techniques that find k nearest neighbor spatial joins, and range queries, closest pairs, distance join. Following are the definitions for spatial range join and k-NN join operators on two dataset, say Q and D (Tang et al., 2016).

Definition 1: Spatial Range Search – Range (q, D)

Suppose q is the spatial range area for objects such as circle or rectangle and let D is the dataset, then range(q, D) finds all the tuples from D that may overlap the spatial range defined by q.

Definition 2: Spatial Range Join - Q ⋈sj D

Let Q and D are the two datasets, Q ⋈sj D, combines each object say q ∈ Q with its range search results from D, Q ⋈sj D= {(q, o)|q ∈ Q, o ∈ range(q, D)}.

If spatial data is partitioned, then figure 7 and figure 8 presents two spatial join operators, namely range and k-NN joins, to combine two datasets, say D and Q, with respect to a spatial relationship. For each point q∈ Q, a spatial range join as shown in figure 7 returns data points in D that are inside the radius of the circle centred at q. In contrast, a k-NN join as shown in figure 8 returns the k nearest-neighbors from the dataset D for each query point q ∈ Q (Tang et.al., 2016).

Figure 7. Spatial Range join (Tang et al., 2016)

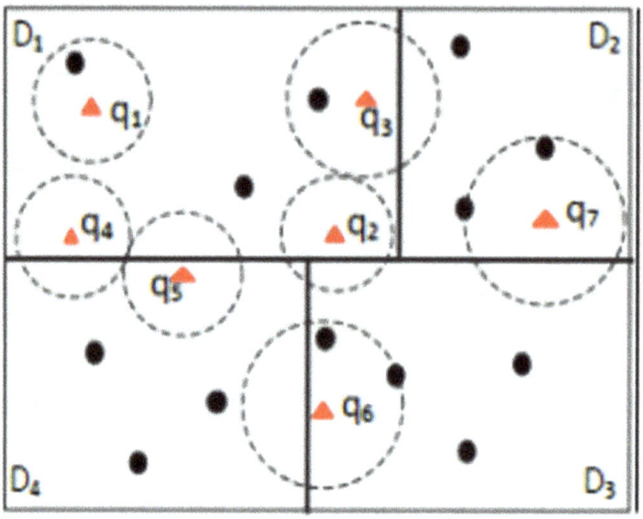

Figure 8. k-NN join(Tang et al., 2016)

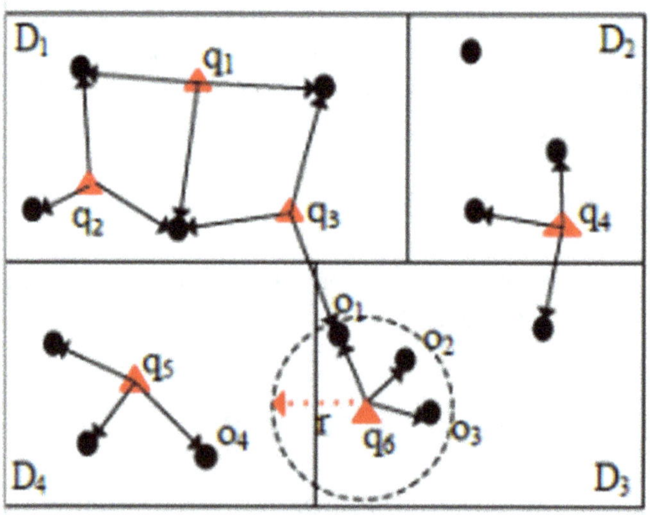

Definition 3: kNN Search - kNN(q,D):

Suppose q is the query tuple, D is the dataset, k is the integer, kNN(q,D), returns the output set $\{o | o \in D \text{ and } \forall s \in D \text{ and } s \neq o, \|o, q\| \leq \|s, q\|\}$, where the number of output objects from D, $|kNN(q, D)| = k$.

Definition 4: kNN Join – Q ⋈knn D

For a parameter k, kNN join of Q and D computes each object q ∈ Q with its k nearest neighbors from D. Q ⋈knn D= {(q, o)|∀q ∈ Q, ∀o ∈ kNN(q, D)}.

Applications of Geospatial Data

Following are the various applications of geospatial data (What are geospatial technologies, 2017; Hodroj 2016):

1. **Remote Sensing:** This data is collected using specialized equipment such as airborne camera, radar systems, sensor platforms, satellite image provider software.
2. **Geographic Information Systems (GIS):** This data is useful for monitoring disasters, extracting mineral deposits, classifying land and water covers, natural calamities, location analysis, nearest neighbors and guiding users.
3. **Global Positioning System (GPS):** This data is gathered using navigation systems or satellites, that gives precise address in terms of latitude and longitude.
4. **Web Mapping:** Internet Mapping Technologies such as Microsoft Virtual Earth, Mapquest, virtual globe Google Earth, NASA world wind, Yahoo Maps, SkylineGlobe, Celestia, EarthSlot, GEpro, Google Maps provides virtual environment to view location and to perform spatial analysis.
5. **Financial Services:** Risk can be assessed by using geospatial data such as demographic characteristics past transactional data. Such data also helps in fraud detection in insurance claims, medical bill claims, credit card transaction bills and income tax evasion.
6. **Retail Marketing:** After performing customer segmentation, demographic data of customers such as address, mobile phone, life style data is used to find out customer churn analysis.
7. **Intelligent Transportation:** Police authorities and other public can know the information on accidents, crimes and build a predictive model so as to predict probability of a crime/ accident occurring at a particular place.

Spatial Query Processing Architectures

The following figure 9 presents system design and architecture for Geospatial data query processing.

GeoSpatial data is stored in a SQL Server 7.0 database. ArcGIS software is used to update spatial data and for further spatial analysis. It is also used to plot maps as needed. CadClient is used to perform data exchange between geo database and

Figure 9. System design and architecture for Geo spatial query processing (Benkirane, Kettani, 2017)

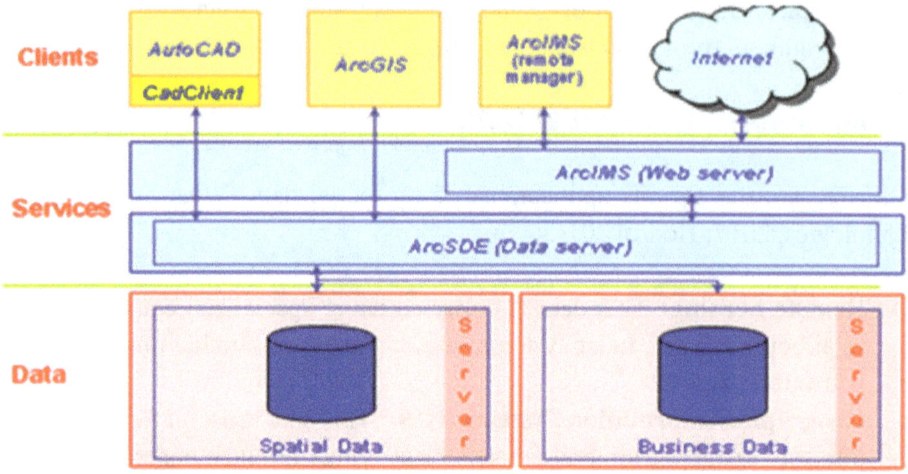

AutoCAD. ArcIMS software is used to deploy spatial information over the web. ArcSDE will hold all the pieces together. The following figure 10 presents three layer architecture for geo spatial data that supports data model, query language, query processing, file organization and indices, query optimization (Shekhar & Chawla, 2003).

The following figure 11 presents Hadoop-GIS (Hive) architecture that includes data partitioning, data storage, query language, query translation, and query engine.

Figure 10. Three-layer architecture for geospatial data query processing (Shekhar & Chawla, 2003)

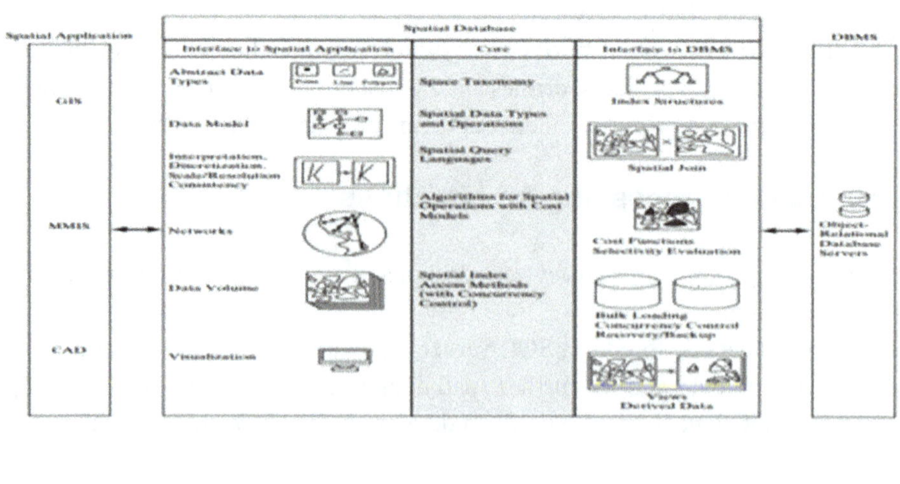

Figure 11. Architecture overview of Hadoop-GIS (Hive)[Ablimit Aji etal., (2013)]

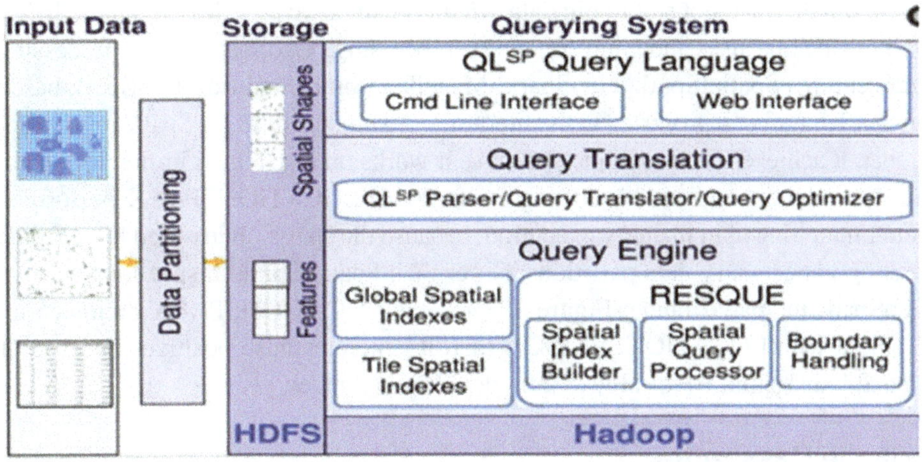

The query engine consists of index building, query processing and boundary handling on top of Hadoop.

Hive is used for feature queries on top of MapReduce. HiveQL is used for spatial data processing. But this architecture will be difficult as we were unable to do traditional MapReduce job. The architecture shown in figure 12, SpatialHadoop is directly integrated into the Hadoop making. It is more efficient than Hadoop-GIS.

Figure 12. Architectural view of GeoSpark (Lenka et al., 2016).

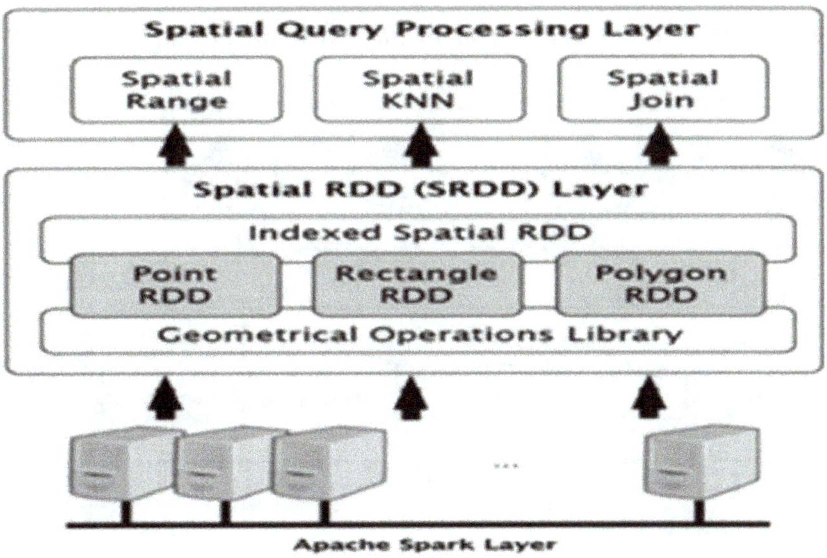

Problem with using Hive for geospatial data processing is that, Hive will do cross join that will not suit to big data . In this chapter we propose Cassandra based framework that uses indexing for query processing. GeoSpatial data processing architectures SpatialSpark, GeoSpark, Magellan, Simba and LocationSpark based on Spark have been proposed in the literature. Advantage of using Spark framework is that, it achieves fault tolerance because it works on distributed in-memory data. In experiments conducted by (Tang et al., 2016) LOCATIONSPARK performed better than Magellan in query execution, because the spatial index used by it avoids visiting unnecessary data partitions, better than SpatialSpark in speed, faster than GeoSpark in spatial range. Figure 13 presents LOCATIONSPARK architecture. Problem with LOCATIONSPARK is that, it generates false positives because of embedding Spatial Bloom Filter into global spatial index

Apache Cassandra

Cassandra is one of the open source frameworks for distributed database management of NoSQL systems, developed at Facebook and built on Amazon's Dynamo and Google's BigTable. It is consistent, scalable and fault tolerant column oriented database. It can accommodate variety of data and supports ACID properties. It can work on commodity hardware with quick response time. Figure 14 presents physical deployment of Cassandra cluster. Nodes of the cluster contain some chunks of data and are shown as a ring. Data is partitioned based on primary key with two parts: partition key based on hash function and second part will store information on clustering and sorting the data for a given partition.

Figure 13. LOCATIONSPARK system architecture (Tang et al., 2016)

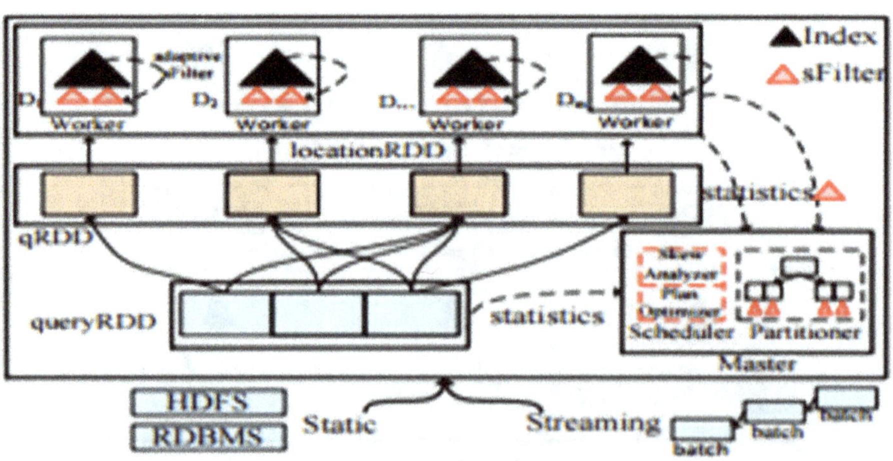

Figure 14. Physical deployment of Cassandra Cluster (Lakhshman & Malik, 2010).

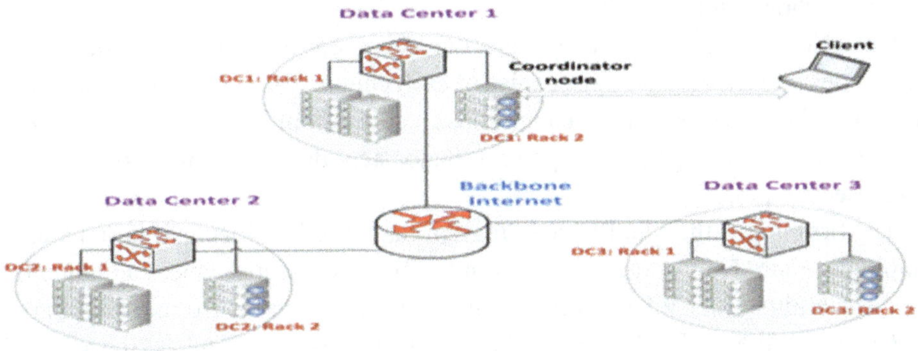

Apache Hadoop

Apache Hadoop is a open-source library for distributed computation for large-scale data sets. It's a large-scale batch processing system as shown in figure 15 that is primarily meant for big data analysis using Hadoop Distributed File System (HDFS) and MapReduce programming. HDFS supports scalability over commodity hardware and MapReduce implementation for parallel processing on the data nodes of a given job. The dataset is divided into fixed size blocks called chunks and is stored in multiple copies for achieving fault tolerance and high availability. Figure 16 presents how voluminous of big data is distributed over multiple nodes and job execution is done in parallel. Job tracker and task tracker automatically controls task

Figure 15. Architecture of Hadoop ecosystem (Dittrich, Quian´eRuiz, 2012)

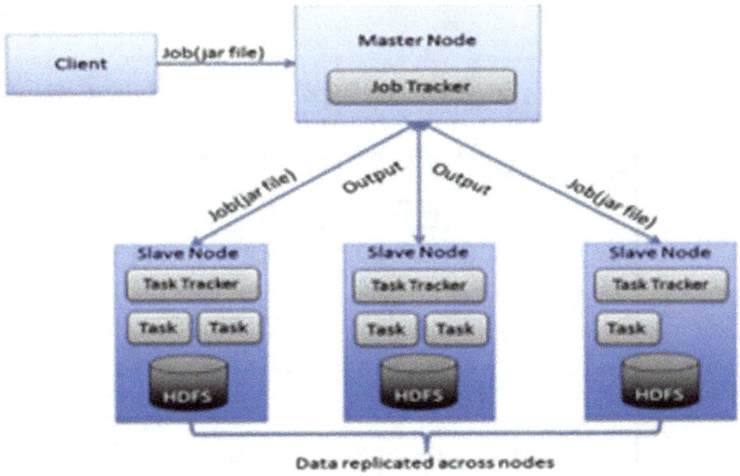

scheduling and execution. MapReduce controls data distribution so as to decrease latency as shown in figure 17.

NameNode of each cluster manages the filesystem namespace and metadata. DataNodes of each cluster stores copies of each chunk. If there is a problem to any of the DataNode, copies of the chunks are created automatically, so as to prevent data loss. The following figure 18 presents major steps during the process of running a MapReduce job.

The first step is the MapReduce Program that specifies Jobclient to run a user MapReduce job. Now a message is given to Jobtracker that produces a unique ID for the submitted job. When all the required resources are loaded into HDFS, the JobClient tells the JobTracker to start the job. The JobTracker calculates the required number of mappers so as to maximize throughput. JobTracker receives messages from TaskTrackers periodically on the status of job execution. If TaskTracker is failed, then it appoints another TaskTracker and resumes the job execution.

Figure 16. Architecture of Hadoop MapReduce framework (Dittrich, Quian´eRuiz, 2012)

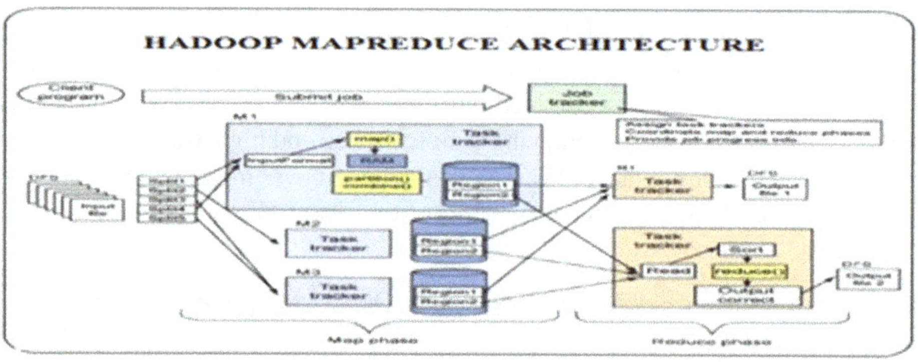

Figure 17. Various nodes in MapReduce (Dittrich, Quian´eRuiz, 2012)

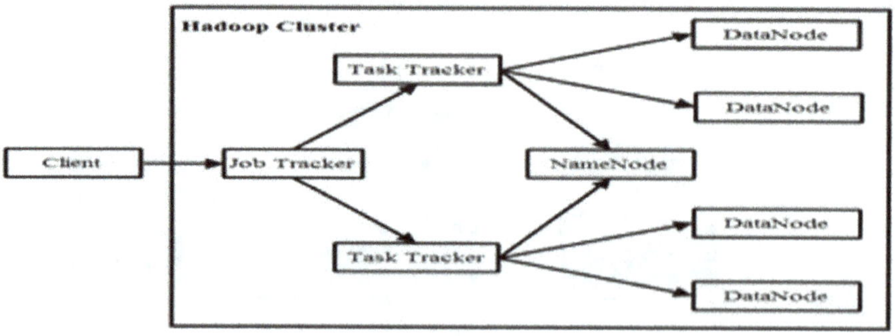

Figure 18. Execution flow in MapReduce (Mingyao, 2015)

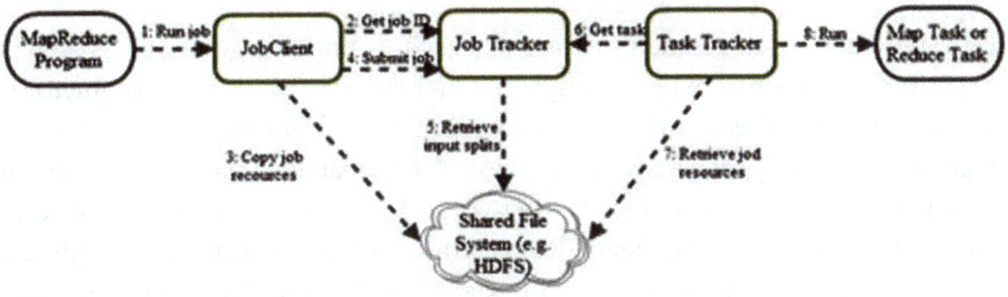

Integrating Cassandra With Hadoop

Hadoop is a distributed system for processing variety of data and Cassandra aims at high availability. By combining both Hadoop and Cassandra, we can process spatial operations such as joins, indexing, geohashing on massive amounts of geospatial data effectively. The following figure 19 presents MapReduce process in a Cassandra/Hadoop cluster.

This architecture will have Hadoop cluster over the Cassandra nodes. Each Cassandra node will have Hadoop TaskTracker(TT) and Data Node. When client submits the job, JobTracker/ResourceManager(JT/RM) initiates the task execution by first interacting with NameNode(NM). JT/RM queries Cassandra for DataNodes information, further sends a MapReduce job request to the Task Trackers/Node Managers (TT/NM). Cassandra nodes run a TT process on each node for processing Map and Reduce task.

Figure 19. MapReduce process in Cassandra/Hadoop cluster (Hadoop Support, 2017)

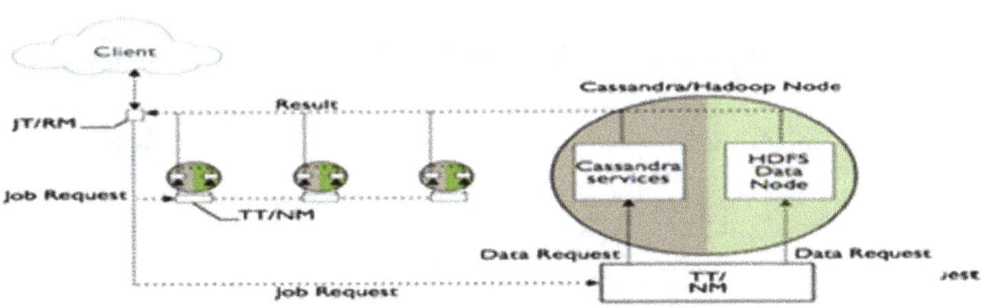

Geospatial Query

Geospatial query uses geometry data types such as points, circles and polygons and perform queries that exploit the spatial relationship between these geometries. Geospatial data defines a particular location in terms of latitude and longitude i.e geography of locations or geospatial database. A Geospatial co-ordinate system can be imagined to have latitude representing vertical axis and longitude representing horizontal axis and elevation (height from earth's reference point). Such coordinate system maps locations on earth to numeric. The following figure 20 presents locations of gas stations on earth converted to numeric.

Geospatial data can be applied in a variety of applications as mentioned in section 1.3.3. The following paragraph presents examples of geospatial queries ranging from point query, box query. Radius query and polygon query.

1. Geofencing query allows raising alerts based on the defined coordinates of a geographic area. The following figure 21 presents Geofencing query that

Figure 20. Geography locations of gas stations on earth to numeric (Geospatial Queries, 2017)

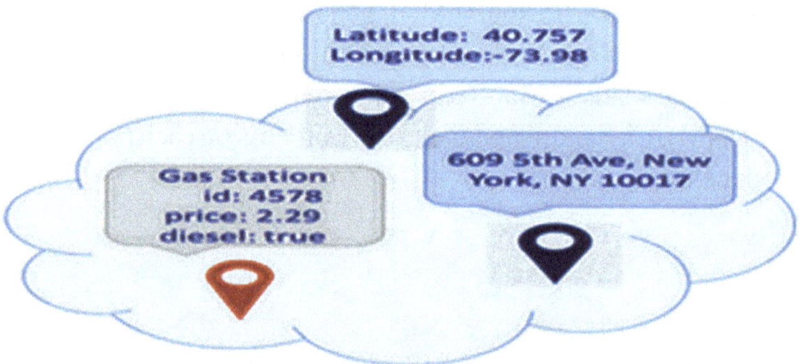

Figure 21. Geofencing query to define virtual borders (Geospatial Queries, 2017)

defines boundaries for arriving, dwelling and leaving. They are the virtual barriers that help to identify intruders entering at the borders of the country, thefts and also in video surveillance applications.

2. Distance query returns all Point Of Interest POI such as shopping malls, hospitals, whose distance is calculated using a distance metric (Euclidean, Minkowski, Manhattan) from a given position those are less than a particular threshold. Figure 22 shows inputs and corresponding output for distance query.
3. k-Nearest Neighbor Query returns *k* nearest POI which are the closest to a certain position and are ordered by proximity. Figure 23 shows inputs to k-NN query, where k-NN is bounded within a distance.
4. Range Query returns all POI within a certain space range such as polygon. Figure 24 shows inputs to range query to find cells that belong to a polygon.

Figure 22. Distance query (Niyizamwiyitira & Lundberg, 2016)

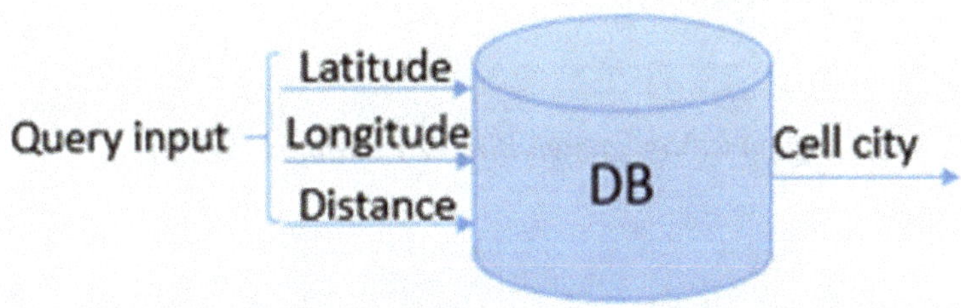

Figure 23. K-Nearest neighbor query (Niyizamwiyitira & Lundberg, 2016)

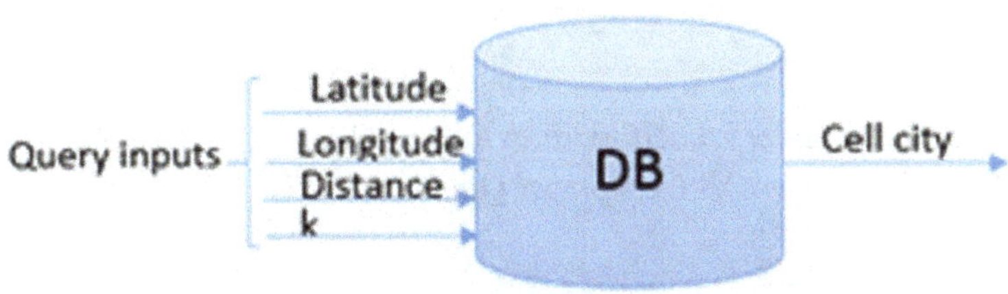

Figure 24. Range Query (Niyizamwiyitira & Lundberg, 2016)

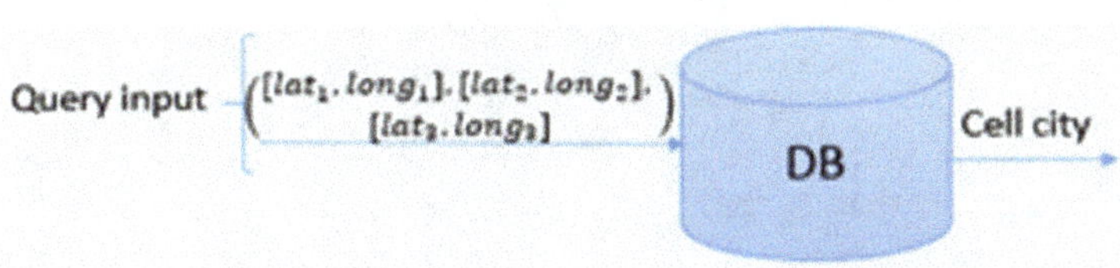

5. Region query finds region which is more likely to be passed by a user at certain time throughout a given time interval. Figure 25 shows inputs to region query to find cell city that is the most visited at certain time.
6. Point query outputs locations that match a particular point. Initially a geospatial point needs to be constructed.
7. Bounded box query where query is specified for some range: start_range and end_range. For example we want to find out medical pharmaceutical shops that are opened 24X7 that are opened round the clock.
8. Radius query outputs all points within a region or a region containing points or points with a specified radius. Such queries help in tracking humans, vehicles within a region. The following figure 26 presents geospatial query for finding the historical sites within a 5-mile radius of the Thomas Jefferson Library.
9. Polygon query searches any point within a polygon defined by n-sides. The following figure 27 presents output of a polygon query for finding restaurants corresponding to locations defined.

Figure 25. Region Query (Niyizamwiyitira & Lundberg, 2016)

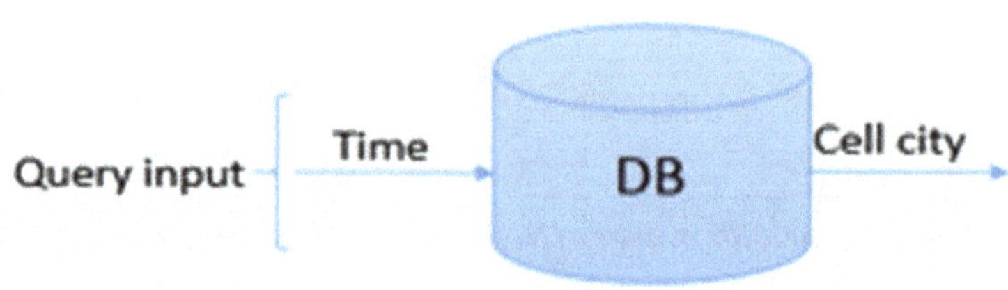

Figure 26. Geospatial query that illustrates Radius Query type (GeoFile: Everything in the Radius with MongoDB Geospatial Queries, 2017)

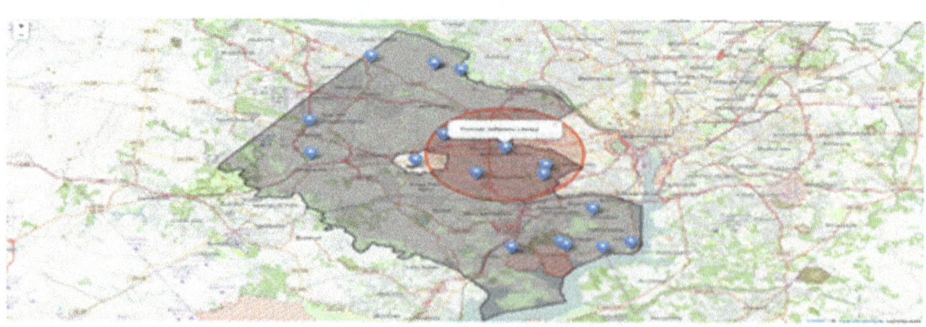

Figure 27. Geospatial query that illustrates Polygon Query type (Find Restaurants with Geospatial Queries, 2017)

Geohashing

Geohashing is a way to express a location using latitude and longitude or alpha numeric string. Geographic Coordinate System (GCS) uses a 3D spherical surface to define locations on the earth. It forms a grid like structure called graticule. Equator will have graticule(0,0) to define 0 latitude and 0 longitude. The geohash identifies a particular grid cell defined by latitude and longitude. Figure 28 presents how the world is divided into two halves “0”and “1”.

If suppose our location in question is in the half “0”. Divide portion “0” into two halves again as “0”and “1”, making two bit precision “00”and “01”as shown in figure 29.

Figure 28. Subdividing the earth surface up to one bit (Geohash and its format, 2016)

Figure 29. Subdividing the earth surface up to two bits (Geohash and its format, 2016)

A greater precision will give more exact location. That is portion "00"can further be divided as "000"and "001" and the process continues until we come across with exact location as shown in figure 30.

As the process of subdividing is repeated, Geohash of 32 bit looks like: 00111 10001011001011010100011110110110101

Each 5 bits are converted to one character as shown in Table 1.

00111 10001 01100 10110 10100 01111 01101 10101

Finally geohash value is **"7jdqngep"**

It is clearly illustrated that greater precision will give longer strings as shown in figure 31.

Figure 30. Subdividing the earth surface up to three bits (Geohash and its format, 2016)

Table 1. Base-32 Character Map (Geohash and its format, 2016)

Decimal	0	1	2	3	4	5	6	7	8	9	10	11	12	13	14	15
Base 32	0	1	2	3	4	5	6	7	8	9	B	c	d	E	f	g
Decimal	16	17	18	19	20	21	22	23	24	25	26	27	28	29	30	31
Base 32	h	j	k	m	N	p	q	r	s	t	u	v	w	x	y	z

Figure 31. Moveable type Scripts: Enter latitude, longitude & precision to obtain Geohash (Movable Type Scripts: Geohashes, 2017)

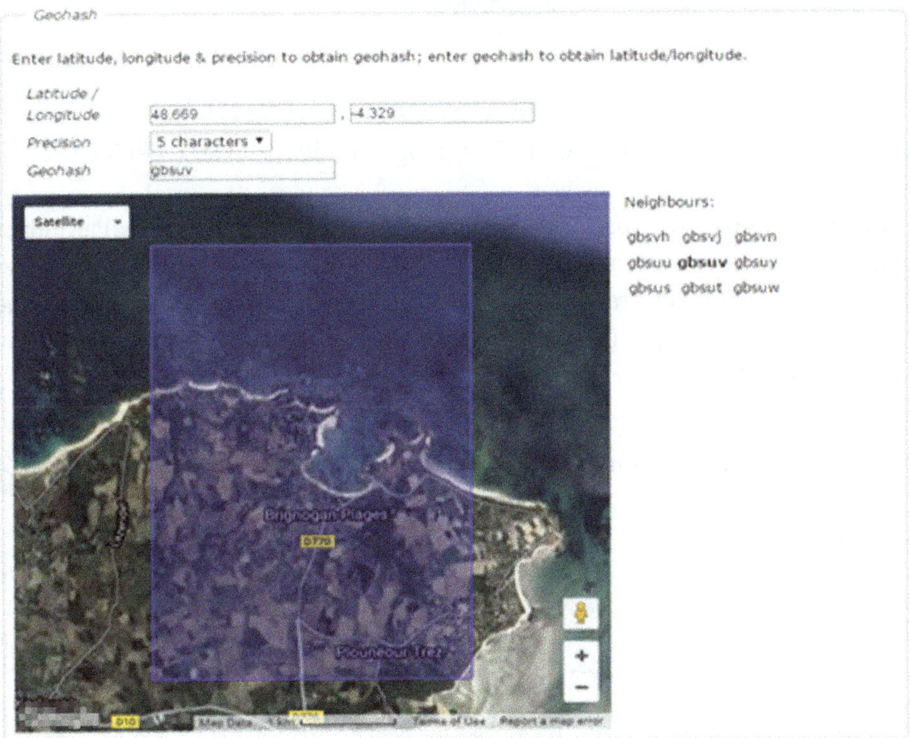

Data Partitioning in Cassandra

Cassandra distributed database runs on multiple nodes. Partitioning schemes such as Random Partitioner (MD5 based distribution of data), ByteOrdered Partitioner (key bytes based data distribution), Murmur3Partitioner (MurmurHash values based data distribution) will determine in which node data should be stored. Cassandra data storage is shown in shown in figure 32 such as a circular ring. The ring will have ranges that are equal to the number of nodes, where each node is responsible

Figure 32. Four node cluster partitioned with Token ranges (DataStax Apache Cassandra Documentation, 2016)

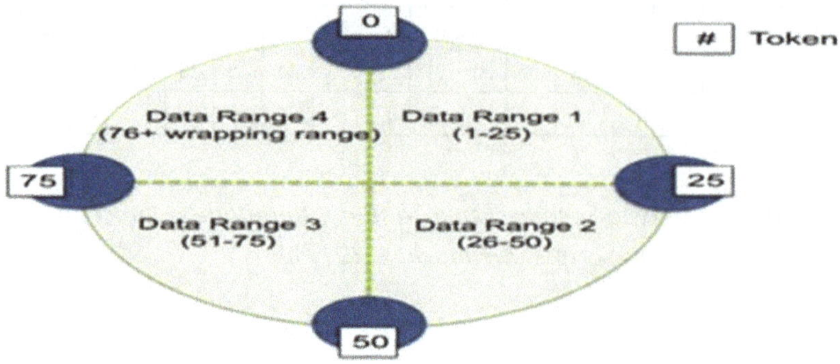

for other ranges of data. A node is first assigned with a token (position) before joining the ring.

Cassandra uses random partitioning technique specifically ByteOrderedPartitioner (that in turn uses consistent partitioning technique) for assigning data to cluster nodes. MD5 hash value (range is from 0 to 2^127) is generated for each of the record. A simple example for random partitioner is shown in figure 33[DataStax Apache Cassandra Documentation (2016)]

Figure 33. Random partitioner (DataStax Apache Cassandra Documentation, 2016)

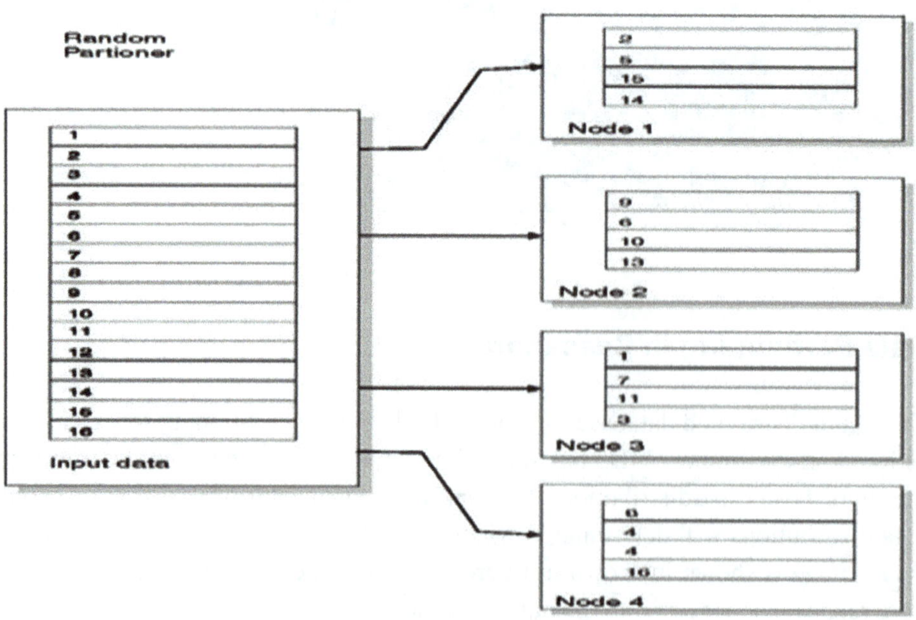

There are three types of Byte Ordered Partitioner (Bop). In the First type of partitioner (ByteOrderedPartitioner), instead of using encoded strings, keys(hexadecimal representation) are stored in the order of their raw bytes. Second type of partitioner (OrderPreservingPartitioner) UTF-8 encoded value of the keys is used to store the data. Last type of partitioner(CollatingOrderPreservingPartitioner) storage of row keys is based on United States English locale (EN_US). Latest versions of Cassandra usc Murmur3Partitioner that in turn uses Murmurhash function and virtual nodes for storing the data. Hash function generates a 64-bit hash value (in the range -2^63 to +2^63 -1) as partition key. Murmur3 partitioner also uses consistent hashing that manages distribution of data based on partition key across the cluster.

The main objective of this chapter is to propose a novel geo spatial computing system "GeoHadoop Architecture" for efficient processing of spatial queries in a distributed in-memory environment that takes advantage of integrating Cassandra with Hadoop for geo spatial querying by using prefixing, spatial indexing and partitioning techniques. This objective can be achieved with the following methodology which is summarized as follows:

1. Understanding architecture and Data Model of Cassandra.
2. Collection of Geographical datasets according to the requirements.
3. Partitioning of Cassandra data using Murmur3Partitioning Technique.
4. Apply spatial query using Cassandra Query Language (CQL).
5. Process the same data using Hadoop MapReduce framework.
6. Apply "Prefix-based partitioning algorithm"
7. Compare the performance evaluation values between Murmur3Partitioning technique and Prefix-based partitioning algorithm.

LITERATURE REVIEW

Centralized systems have more response time and slower data access. Since 1980's there is a huge demand for decentralized databases. Spatial query processing has been used in many GIS based applications where significant conclusions can be made from end user location information. But spatial data is voluminous and requires scalable techniques, distributed computing platforms and technologies for efficient processing of spatial queries. Even though techniques given by Hadoop MapReduce, Hadoop Distributed File System (HDFS) and NoSQL databases for spatial queries have been proposed, but requires modification of systems or frameworks to implement indexing techniques (Lee, Ganti, Srivatsa, Liu, 2014). Spatial query processing techniques as mentioned in the previous section, can be classified into two categories basing on the amount of data selected: huge selection such as k-NN queries and low data selection such as k-NN join.

Work reported in Zhang, Han, Liu, Wang, and Feng (2009) describes MapReduce and showed how spatial queries can be expressed without explicitly addressing any of the details of parallelization. Their experimentation on sample spatial queries proved that MapReduce is also appropriate for small scale clusters and computing intensive applications.

R-Tree based approach is proposed in Liao, Han, and Fang, (2010). It constructs hierarchical index structures to organize datasets in dimensional space so as to improve the query performance on HDFS. Authors concluded that when data is stored in HDFS where index structures, buffer management are supported, query response time can be reduced.

In another work by Liu, Han, Zhong, Han, and He (2009), summarized that even though HDFS can handle big data, could not perform well for large amounts of small files. Because of this, web based applications could not be benefited from Hadoop. An approach is proposed to optimize the I/O performance of small files in HDFS environment. This is done by combining small files into a single large file and to create index for each file. Experiment results showed that their approach achieved good performance. Their work also focused on methods for parallelizing task computation, and distributing the data over clusters.

Finding nearest location is one of the important spatial queries that find the nearest location(s) from the user location. Existing methods for finding nearest location such as zip code method (zip code of the places are matched and those locations whose zipcode belong to the same region or which are nearer are extracted), k-d tree method(a binary tree that finds nearest locations in log(n) time), polygon method(boundaries are defined on the map and queries are performed on points belonging to this polygon), Voronoi diagram (Voronoi cells are defined as shown in figure 34(a) and based on seed point as shown in figure 34(b) nearest locations

Figure 34. (a) Voronoi cells (Lee, 1982) (b) Nearest location using seed point (Lee, 1982)

are identified), could not process large data and not find nearest location with high precision (Dubey, Agrawalan, 2015). Work reported by (Lee, 1982) is on finding nearest locations using Voronoi diagrams in O(nlogn) time.

There are other spatial partitioning techniques such as grid, quad-tree(spatial indexing for RDBMS), STR/STR+, Z-curve, Hilbert curve. In the work reported by Moniruzzaman AB, Hossain SA(2013), most of the NoSQL databases such as Cassandra databases are not supporting geospatial indexes and hence spatial queries. In the work reported by Brahim, Drira, Filali, and Hamdi (2016) handles spatial data retrieval within Cassandra NoSQL database. A framework is proposed and implemented for spatial queries. In the work by Fox, Eichelberger, Hughes, and Lyon (2013) proposed a spatio-temporal index structure for achieving horizontal scalability of NoSQL databases.

DRAWBACKS OF EXISTING METHODS

The main drawback of existing system is time complexity for processing large datasets. Our work mainly focuses on reducing the query response time by using integration framework and implementing various effective techniques. Also random partitioning technique poses several drawbacks such as uneven load balancing, sequential writes does not distribute data across the cluster. As such murmur3 partitioning and prefix based partitioning are used in our framework. We also made an attempt for using multiple partitioners to determine which partitioner is performing well after using geohashing technique to improve geo spatial query processing.

Proposed System

This section presents the detailed functionality of the proposed system to perform geospatial querying in the GeoHadoop Architecture. In this architecture, Cassandra is integrated with Hadoop MapReduce framework as shown in figure 35 so that query response time for spatial data analysis could be optimized.

Methodology for Processing Spatial Queries

Spatıal Data Storage

- Geospatial data is collected from various external devices or sensing systems. Collection of data can also be through Event messages or current geo location of end users. This data is in the form of <Latitude, Longitude> pair for

Figure 35. Architecture for processing Spatial queries by integrating Cassandra with Hadoop MapReduce framework.

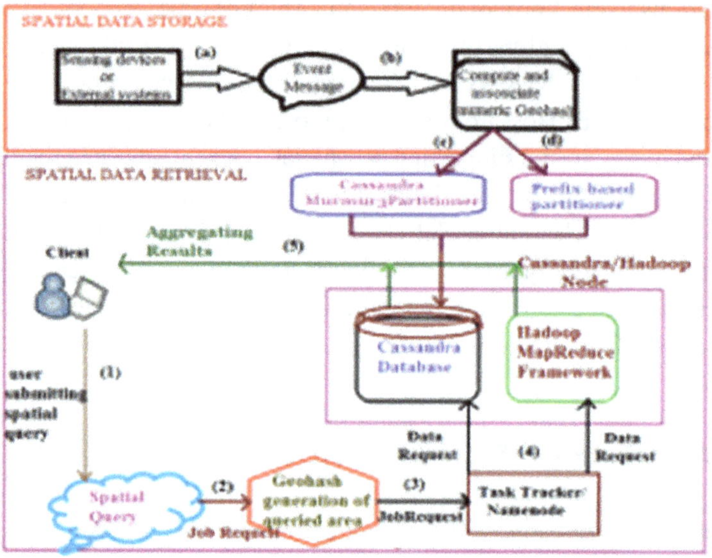

different locations, where spatial queries are processed on this pair. A dataset is constructed with this pair of data.

- Geohashing technique (that converts the longitude, latitude pair into a single value which is represented in binary format) is applied for each tuple in the dataset which takes <Latitude, Longitude> pair as input and produces "Geohash" value as output. Thus a new dataset is formed by adding an additional column containing Geohash values. Data is stored according to Cassandra schema i.e columnar database schema.

Spatial Querying

- User submits spatial query.
- Compute Geohash value for the queried area using Geohashing technique. MapReduce framework is used for parallel implementation of the input query. The input query logic is executed in parallel over the independent chunks and over different mappers.
- Hadoop MapReduce framework performs Map and Reduce task as described in the algorithms of section 3.4. First data is retrieved from Cassandra database for input to map function, after it performs reduce function it writes back to the Cassandra database.
- Finally the output result is delivered to the end user.

Hadoop MapReduce Framework

Data Processing using Hadoop framework is divided into 2 phases:

- Mapper phase: Performs mapping of required attributes using Map() function.
- Reducer phase: Reduces the number of entries using Reduce() function and returns required tuples.

Here both Mapper and Reducer uses "Prefix Based Matching Algorithm" as shown in figure 36.

Prefix-Based Partitioning Algorithm

For spatial indexing, a hierarchical spatial data structure, called Geohash (Moniruzzaman, Hossain, 2013), which is a geocoding system for latitude and longitude. A Geohash code, represented as a string, denotes a rectangle (bounding box) on the earth. It provides a spatial hierarchy and it can reduce the precision (i.e., represent a bigger rectangle) by removing characters from the end of the string. In other words, the longer the Geohash code, the smaller the bounding box represented by the code. Another property of Geohash is that, two places with a long common Geohash prefix are close to each other. Similarly, nearby places usually share a similar prefix. However, it is not always guaranteed that, two close places share a long common prefix.

Mapper job mainly takes Geohash, Place_name as <key,value > pair and produces Place_name and its prefix matched count as shown in figure 37.

The output of Mapper phase is taken as input to Reducer phase as shown in figure 38. Basing on the length of prefix matched, each entry in datanode is compared

Figure 36. Phases of Hadoop MapReduce framework for Spatial search

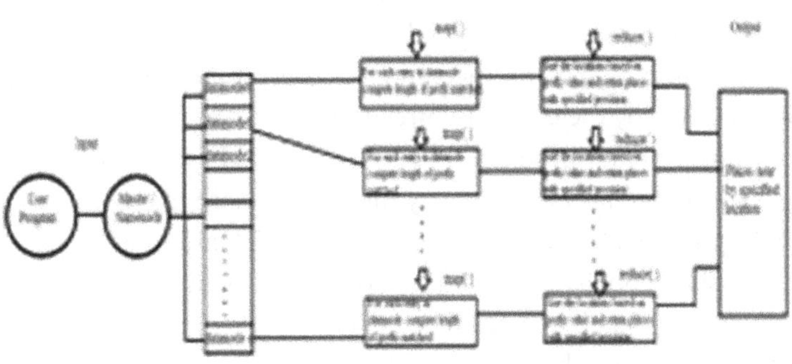

Figure 37. Mapper Task

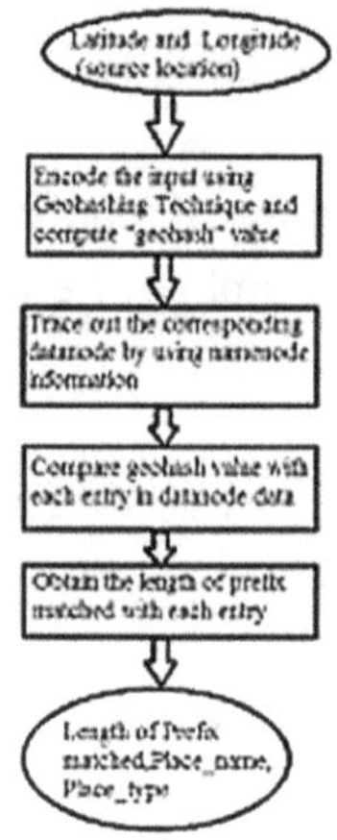

Figure 38. Reducer task

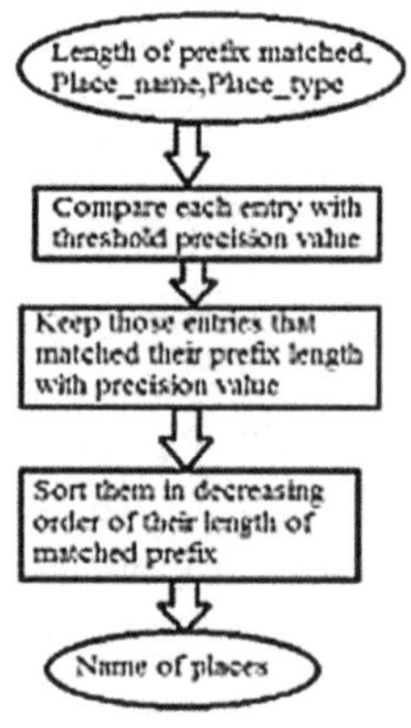

whether it is greater than or equal to the precision value. If this condition is satisfied then entry is stored otherwise discarded. Finally the result set is sorted in decreasing order of their length of prefix matched and displays all the places that are nearer to the location specified by the client. Figure 39 presents the flowchart for processing spatial queries using data presented in Cassandra database. Firstly it receives the spatial queries from the user through any interface. These queries are submitted to the Query interpreter which interprets the syntax of CQL queries. After approving that queries are syntactically correct, query optimization is performed for reducing query response time. Finally the matched data is retrieved from Cassandra database containing Spatial/Geographical data.

Figure 39. Spatial query processing

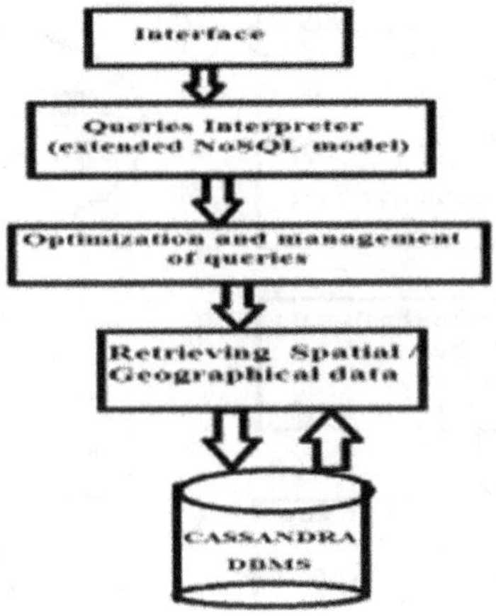

Murmur3 Partitioning

Figure 40(a) and Figure 40(b) shows Mapper and Reducer phases that can be performed using Murmur3Partitioner technique. During Mapper phase, geohash value is calculated using Geohashing technique for the required place based on its latitude and longitude. At the backend Cassandra is used as a storage database and it uses Murmur3Partitioning technique to partition the data. Hence data is distributed across multiple datacenters (or) nodes equally. As a result, data is fetched based on its Murmur3 hash value corresponding to geohash value. During reducer phase, data such as length of prefix matched, Place name and Place type is collected and sorted in decreasing order to display nearby places.

Implementation & Results

Dataset Used

Data Requirements which involves collection of required data, is the first step for the database analytics projects. Partitioning techniques on Geographical dataset of India is performed which contains about 50000 tuples of data available at http://www.latlong.net/country/india-102.html. The dataset is taken in the form of .csv format, which contains Place_name along with its Latitude and Longitude values. The Place_type is also considered for processing the dataset. Figure 41 presents a snapshot of database. These tables are in denormalized form, since denormalization is best practice for modelling NoSQL columnar databases. Table 2 presents list of

Figure 40. (a) Mapper phase (b) Reducer phase

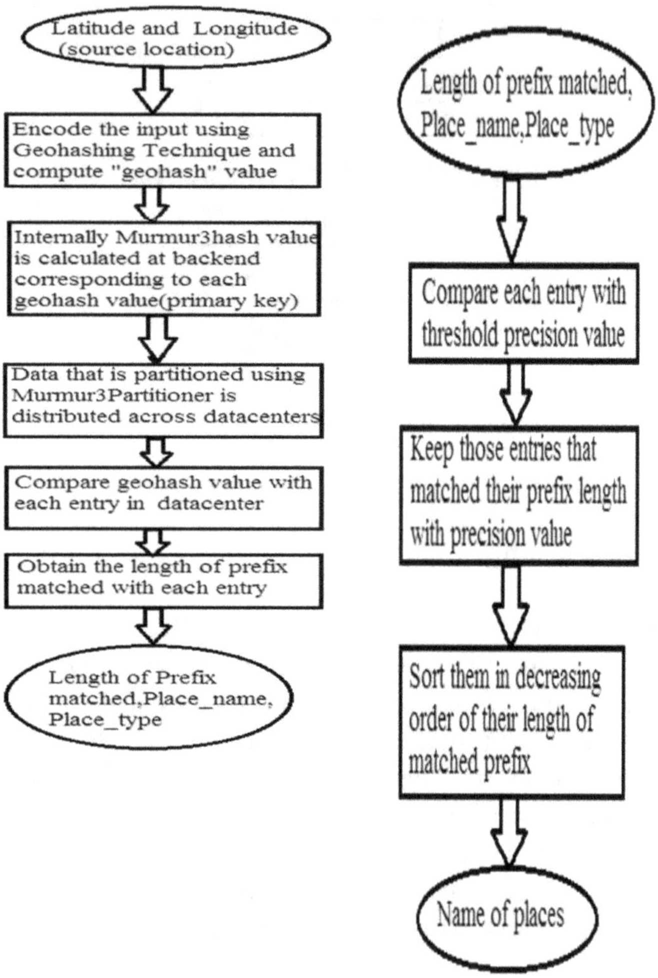

attributes along with data types. Place_Name represents the name of the location. Place_type refers to the type of the place like Village, Town, Garden, Hospital etc. Latitude and Longitude columns represent the position of location on the globe.

Figure 42 shows the conceptual data model representation of World Map entities whereas Figure 43 presents Physical data model representation of the database.

Software Requirements

- Operating Systems: Windows 8.1 and above, Ubuntu 10.0.14+.
- Java versions 1.7 and above.

Figure 41. Geographical Dataset of India (Bobov, 2017)

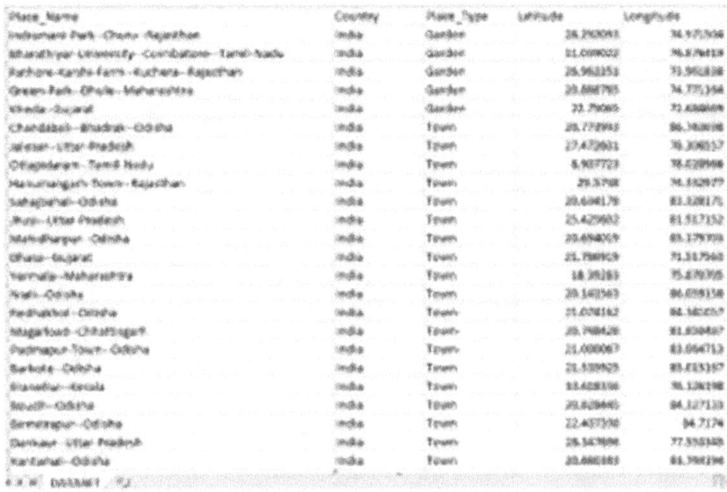

Table 2. List of Attributes with data types

Attribute	Datatype (CQL)
Place_Name	Text
Country	Text
Place_type	Text
Latitude	Decimal
Longitude	Decimal

- Hadoop versions 2.6 and above.
- Cassandra versions 3.0 and above.
- Maven version 3.0 and above.
- Eclipse IDE

All Hadoop and Cassandra projects support Linux operating system. Therefore, install any Linux flavored OS. As a groundwork we implemented Java and Cassandra Installation, and Hadoop 2.7.3 is installed and integrated with Cassandra environment

Implementation Overview

1. Initially a dataset containing places with their <latitude, longitude> is collected and stored in .csv format.

Figure 42. ER (Entity-Relationship) diagram for World Map dataset

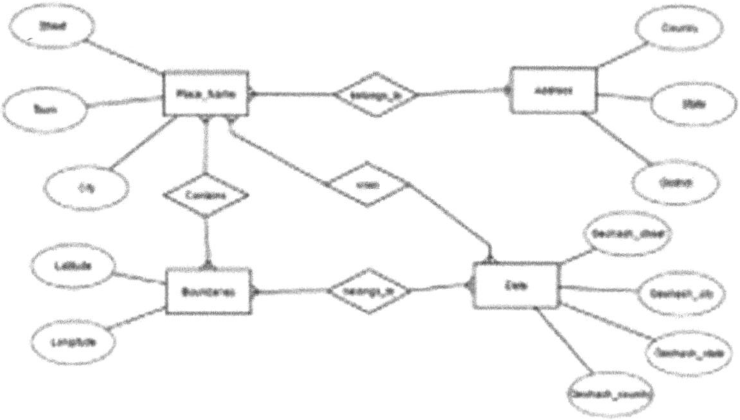

2. As shown in figure 44, a new dataset is formed by adding an additional column named "Geohash" which contains geohash values computed using Geohashing technique.
3. Connect to the Hadoop and Cassandra server
4. MapReduce executes in parallel, so our large data will be processed in parallel altogether. All the work is carried out with these two functions only. Map function will take the input in the form of <key, value> pair and provides a new <key, value> pair as output. This output then goes to reducer which then processes this data and gives the final output.
    - Here <key, value> pair will be <Place_Name, geohash>
    - Long stands for longitude
    - Lat stands for latitude
    - Place stands for name of the location
    - Type stands for category of the place

Now every location has its own unique (longitude, latitude) pair, so we will represent the location by its (longitude, latitude) pair in our implementation. Name of the places will be required at the time of displaying the names in the output. Type tells that what is the category of the place like there can be so many categories school, library, hospital, pubs, colleges, temples, museums etc.

Figure 45 depicts calculation of Murmur3hash values for corresponding geohash value for the given dataset.

Figure 43. Physical data model representation on World Map dataset

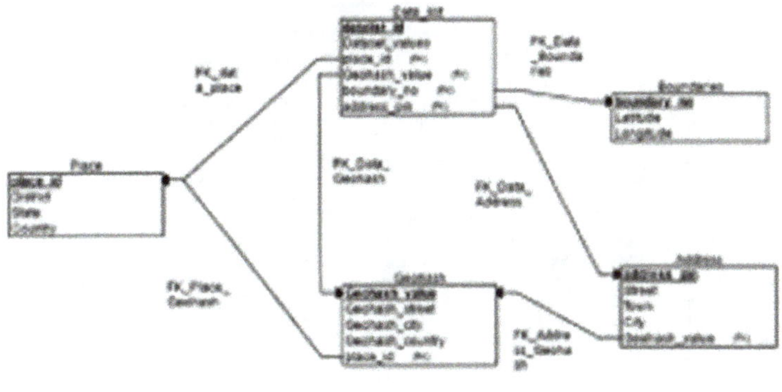

Figure 44. Modified dataset with geohash values

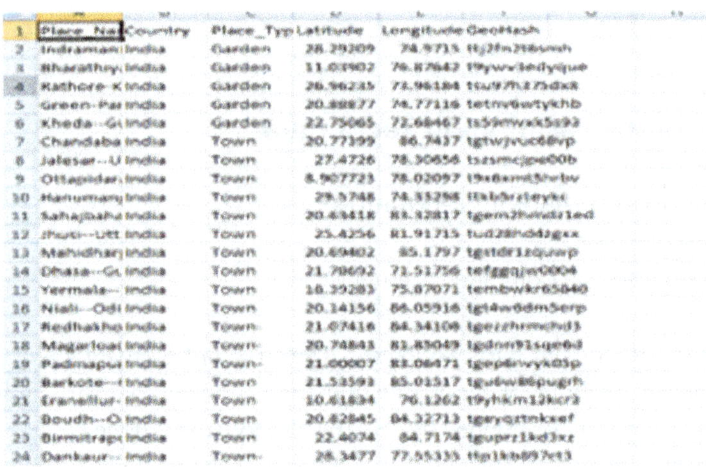

Algorithms

As explained in the earlier sections. integrating Cassandra with Hadoop increases the performance efficiency, the Map reducer tasks can be performed by using various Partitioning techniques. The default partitioning technique for Cassandra is Murmur3partitioner. But considering the query response time criteria, Prefix based partitioning technique gives faster response when compared to the Murmur3Partitioning algorithm as proved in the next section 3.5 and is shown in figure 46. The reason behind this is, Prefix based partitioning will partition the data based on prefix of geohash value, which makes spatial search easier.

Figure 45. Murmur3Hash values for corresponding Geohash value

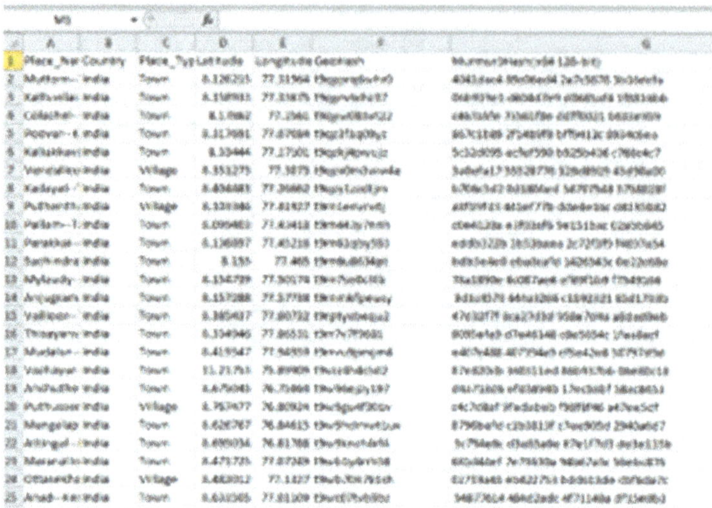

Using Murmur3 Partitioning Technique

Algorithm for Geohashing Technique

Procedure Geohashing (Longitude, Latitude)

1. The world is divided into two halves. The first half is given value 0 and the other half is given the value 1. The division is done with a vertical line.
2. Each division is again divided into two parts. Division is done with a horizontal line. Each divided portion is again assigned the numbers 0 and 1. "0" is assigned to the first half and "1" is assigned to the other half.
3. This is done until a very long series of bits is obtained.
4. This bitwise series is converted into the base-32 format.
5. In whichever region the location is falling, the code of that region is the geohash code of the location.

Algorithm for Mapper

```
Procedure Map(place_type,            /*input key*/
          Murmurhashvalue,                /*input value*/
          Place_type              /*output key*/
          Prefix_match        /*output value*/
```

```
          Geohash          /*geohash value for latitude
longitude co-ordinates*/
)
{
 For each column in dataset
Murmuhash = values[6];    /* Murmurhash value of each entry in
datacenter */
hash=values[5]; /* Geohash value of corresponding MurmurHash
value*/
if(Murmurhashvalue !=Murmurhash)
   For (i=0; i< geohash.length; i++)
              If(geohash.charAt[i]==hash.charAt[i])
                          Count++;
           Context.write(place_name, count)
  }
Output of Mapper:  Key-value pair of   place_name with its
prefix Matched.
Algorithm For Reducer
Reducer (Place_Name                /* key */
               Count                        /*value*/
               Output Place_Name   /*output*/
              )
{
  For every column in dataset
               If(count>=5)         /*  assume precision for
matching prefix be 5 */
                    Context.write(Place_name)
}
Output: Places nearby the given Place_Type.
```

The drawback of the above specified method is that Cassandra database will store data randomly according to its 64-bit token generated internally. As a result the required queried data must be searched on all nodes for finding nearest neighbours. Hence the efficiency of Query response time may be reduced.

Using Prefix Based Partitioning Method

Partitioning Method

```
1)        Consider 32 datanodes, where each node contains
values starting with each character in table .
              Node-0: geohash values starting with '0'
              Node-1: geohash values starting with '1'
              Node-2: geohash values starting with '2'
              Node-3: geohash values starting with '3'
                  .                       .
                  .                       .
              Node-z: geohash values starting with 'z'
2)        Consider a node array and assign node starting
values to that particular array
For example:
                 Node[0]='0';
                 Node[1]='1';
                 Node[2]='2';
                  ….
                 Node[31]='z';
Algorithm For Mapper
Procedure Map(place_type,          /*input key*/
                             Geohash,             /*input
value*/
                             dataNode            /* node
[corresponding geohash value letter ]*/
                           Place_type            /*output key*/
                          Prefix_match       /*output value*/
                     )
{
  For each row in node
Hashvalue= node[ i];
if(geohash !=hashvalue)
      For (i=0; i< geohash.length; i++)
              If(geohash.charAt[i]==hashvalue.charAt[i])
                                         Count++;
                 Context.write(place_name, count)
   }
Output:  Key-Value pair of   place_name with its prefix
```

```
matched.
Algorithm For Reducer
Procedure Reducer (Place_name      /* key */
                                     Count              /*value*/
                                   Place_name    /*output*/
                                 )
{
For every row in dataNode
         If(count>=5)        /* assume precision for matching
prefix be 5 */
          Context.write(place_name)
 }
Output: Prints corresponding place_name
```

Performance Evaluation

This section presents results of query processing both using murmur and prefix based partitioning methods. Sample queries for finding nearby places of specified locations are shown as follows:

- **Query 1:** Find places nearby Muttom Town in TamilNadu
- **Query 2:** Find nearest Places of Kandla
- **Query 3:** Find all the places nearer to Jairampur in ArunachalPradesh

For each query we are going to give nearest locations based on the precision values. If the user wants only the nearest places then we will consider higher precisions as threshold values. But if user specifies to find nearer places then the threshold value can be reduced. Hence considering all the possibilities we have calculated the query response time for both precisions 2 and 3 (according to our dataset values). As prefixes matching with precision-2 will be more, the time taken to retrieve data from database will be more compared to higher precision values. Here we have plotted a graph as shown in figure 46 for more clear explanation considering precision values on x-axis and Time (in milliseconds) taken for querying on y-axis.

From the figure 46, it is very clear that the performance of Prefix-based partitioning technique is more comparatively. The reason behind this is splitting of data into different datanodes and performing MapReduce functionality reduces the number of searches as well as increases the efficiency which in turn reduces the query response time. On an average Murmur3 partitioning took 238 milli seconds where as prefix based partition took 103.5 milli seconds.

Figure 46. Performance evaluation of Murmur3Partitioner and Prefix-based Partitioner

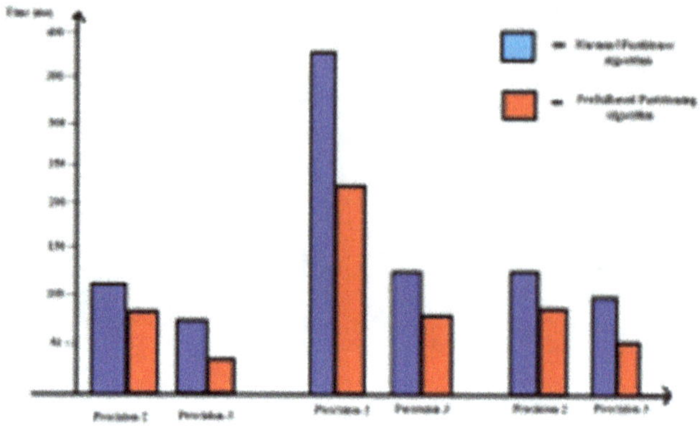

CONCLUSION AND FUTURE WORK

Finally Cassandra works as an effective storage system providing scalability, high performance, and wide applicability. Cassandra can support a very high update throughput while delivering low latency. Murmur3partitoner plays a major role in providing better performance than other databases because of its 32 or 128-bit hash values. But considering Geospatial search use case another user defined partitioning technique called Prefix-based Partitioning is giving better results comparatively. And from our work we can conclude that Cassandra can also support for spatial queries with user defined functions. Hence when voluminous amount of data is to be processed and analyzed, Cassandra may play a best role in deriving expected results as fast as possible within seconds. Even though Hadoop itself provides HDFS for storing data, we preferred another database Cassandra externally, reason being Cassandra is suitable for online applications where as Hadoop is suitable for batch oriented analytical tasks (Hadoop Vs. Cassandra, 2017). İntegrating Hadoop and Cassandra made us to process spatial query data stored in Cassandra without having to move data into Hadoop environment. It took some time for us to create the dataset, because Hadoop can store any formatted data where as Cassandra needs structured data. Geohash helped to distribute data and also enabled to improve query performance.

In Future we are planning to set up configurations for Hadoop as well as Cassandra at Multi-node clusters which can produce best results for processing large datasets especially Geographical data. We can also propose some other effective searching technique for finding nearest places as Geohash search fails in some edge cases like when same point belongs to different domains. Finding nearest locations can also be used as an application for various organizations like Telecom industry to implant cell towers, etc.

REFERENCES

Aji, A., Wang, F., Vo, H., Lee, R., Liu, Q., Zhang, X., & Saltz, J. (2013, August). Hadoop GIS. *Proceedings of the VLDB Endowment International Conference on Very Large Data Bases*, *6*(11), 1009–1020. doi:10.14778/2536222.2536227 PMID:24187650

Ben Brahim, M., Drira, W., Filali, F., & Hamdi, N. (2016). Spatial data extension for Cassandra NoSQL database. *Journal of Big Data*, *3*(1), 11. doi:10.1186/s40537-016-0045-4

Benkirane & Kettani. (n.d.). Retrieved from www.aui.ma/personal/~D.Kettani/courses/gis/GDB-benkirane.ppt

Berry, J. K. (1987). Fundamental operations in computer-assisted map analysis. *International Journal of GIS*, *1*, 119–136.

Bobov. (n.d.). *Spatial data Visualization*. Retrieved from https://portal.opengeospatial.org/files/?artifact_id=73214

Cassandra. (2017). *Contrasting Hadoop & Apache Cassandra.* Retrieved April 12, 2017 from https://www.datastax.com/nosql-databases/nosql-cassandra-and-hadoop

DataStax Apache Cassandra Documentation. (2016). Retrieved from http://www.odbms.org/wp-content/uploads/2013/11/cassandra10.pdf

Dittrich, Arnulfo, & Ruiz. (2012). Efficient Big Data Processing in Hadoop MapReduce. *Proceedings of the VLDB Endowment, 5*(12).

Dubey, N. K. (2015). Efficient Approach To Find Nearest Location Using Geohashing On Hadoop And Pig. *International Journal Of Engineering Research-Online*, *3*(3), 771–777.

Find Restaurants with Geospatial Queries. (n.d.). Retrieved from https://docs.mongodb.com/manual/tutorial/geospatial-tutorial/

Fox, Eichelberger, Hughes, & Lyon. (2013). *Spatio-temporal Indexing in Non-relational Distributed Databases*. Commonwealth Computer Research, Inc.

GeoFile. (n.d.). *Everything in the Radius with MongoDB Geospatial Queries*. Retrieved from https://www.compose.com/articles/geofile-everything-in-the-radius-with-mongodb-geospatial-queries/

Geohash and its format. (2016). Retrieved from http://geohash.org/site/tips.html

Geospatial Queries. (n.d.). Retrieved from https://docs.gigaspaces.com/xap110/query-geospatial.html

Hadoop Support. (2017). Retrieved from https://docs.datastax.com/en/cassandra/2.1/cassandra/configuration/configHadoop.html

Intro to advanced GIS and a review of basic GIS. (n.d.). Retrieved from www.utsa.edu/lrsg/Teaching/EES6513-09/L1_Intro.ppt

Lakhshman, A., & Malik, P. (2010). Cassandra: A decentralized structured storage system. *ACM SIGOPS Operating Syst Rev.*, *44*(2), 35–40. doi:10.1145/1773912.1773922

Lee. (1982). On -Nearest Neighbor Voronoi Diagrams in the Plane. *IEEE Trans. Computers*.

Lee, K., Ganti, R. K., Srivatsa, M., & Liu, L. (2014). Efficient Spatial Query Processing for Big Data. *ACM SIGSPATIAL '14*.

Lenka, Barik, Gupta, Ali, Rath, & Dubey. (2016). *Comparative Analysis of SpatialHadoop and GeoSpark for Geospatial Big Data Analytics*. Cornell University Library.

Liao, H., Han, J., & Fang, J. (2010). Multi-dimensional Index on Hadoop Distributed File System. *Proceedings of IEEE Fifth International Conference on Networking, Architecture, and Storage*, 240-249.

Liu, X., Han, J., Zhong, Y., Han, C., & He, X. (2009). Implementing WebGIS on Hadoop: A case study of improving small file I/O performance on HDFS. *Proceedings of IEEE International Conference on Cluster Computing and Workshops*, 1-8.

Belussi, Catania, & Migliorini. (2012). Approximate queries for spatial data. In *Advanced Query Processing: Vol. 1. Issues and Trends*. Springer Science & Business Media.

Moniruzzaman, A. B., & Hossain, S. A. (2013). Nosql database: New era of databases for big data analytics—classification, characteristics and comparison. *Int J Database Theor Appl.*, *6*(4), 1–13.

Niyizamwiyitira & Lundberg. (2016). Performance evaluation of trajectory queries on multiprocessor and cluster. *Proceedings of CNSA*, 145-163.

Orlandini, R. (n.d.). *Geospatial Data*. Retrieved from http://researchguides.library.yorku.ca/content.php?pid=245987&sid=2176375

Raghavendran. (n.d.). *Small things of Big (Geo Spatial) Data*. Retrieved from https://www.slideshare.net/raaags/small-things-of-big-geospatial-data

Scripts, M. T. (n.d.). *Geohashes*. Retrieved from http://www.movable-type.co.uk/scripts/geohash.html

Shekhar, S., & Sanjay, C. S. D. (2003). *A Tour. Prentice Hall.*

Tang, Yu, Aref, Mahmood, Malluhi, & Ouzzani. (2016). *In-memory Distributed Spatial Query Processing and Optimization*. Purdue technical report 2016.

What are geospatial technologies. (n.d.). Retrieved from https://www.aaas.org/content/what-are-geospatial-technologies

What are Longitudes and Latitudes. (n.d.). Retrieved from https://www.timeanddate.com/geography/longitude-latitude.html

Zhang, Han, Liu, Wang, & Feng. (2009). Spatial Queries Evaluation with MapReduce. *Proceedings of GCC '09.*

Zhou. (2015). *Analysis and performance test of MapReduce, Spark, Phoenix, Disco, Mars DeveloperWorks*. IBM.

Chapter 2
Geographical Information System Applications Utilized in Museums in Turkey Within the Scope of the Cultural Heritage Tourism:
A Case Study of Mobile Application of Müze Asist

Aysu Altaş
Aksaray University, Turkey

ABSTRACT

Technology has penetrated every aspect of life and it proceeds with the aim of facilitating people's lives. Geographical information systems as a part of the developing technology provide services in every area with different forms. One of these areas is museology. The desire to make visual and written information and cultural heritage that are presented by museums with auditory and interactive experiences has brought together digital technologies and cultural and historical activities. From this point of view, the "Müze Asist" is a remarkable application in terms of the fact that it is a local app and in use in most of the museums in Turkey and the convenience provided for all tourists who want to visit a museum, especially for those who prefer self-guided tours. This chapter may be a guide to its objects and to the next researchers.

DOI: 10.4018/978-1-5225-5088-4.ch002

INTRODUCTION

Cultural heritage tourism is type of tourism concentrated on visiting artifacts, historical buildings, monuments and art objects inherited from various cultures, which could not be manufactured once again and which reflect characteristics of a certain society. Geographical Information System (GIS) is a tool allowing today's users to access and handle information conveniently and efficiently. Owing to the advantages of the GIS, everyday various new GIS applications are introduced and find themselves a new application area such that millions of applications increase progressively and serve millions of individuals for different purposes. Museum is one of these areas. Museums are the places first come to mind as areas for exhibition of inherited artifacts within the cultural heritage and they are frequently visited by tourists. On the other hand, tourists participated in cultural heritage tourism do not always tend to travel in a tourist groups leaded by a tourist guide. There are also individuals who plan their personal travel route, organize their accommodation, dining needs, and who prefer to move independently along their travel. These individuals are capable of performing operations such as navigation, buying travel tickets, hotel reservation by means of smart phones, tablets and mobile PCs. The only necessity for these operations is internet. The *"Müze Asist"* application was designed for all tourists and especially for solitary travelers enjoying independency in their journey, which could operate on smart phones, tablets and PCs of individuals. This application functions as a travel guide accessible from the digital environment, which provide information, details, images, videos with voice explanations of artifacts, especially for museums, other cultural and art establishments. This research on the "Müze Asist" one of the geographical information system applications used in museums in Turkey within the scope of cultural heritage tourism, has been prepared by analyzing the content of the web site of the "Müze Asist" application, using the "Müze Asist" application on a smart phone with an Android operating system and making a literature review. It is aimed to make a contribution to the existing literature.

CONCEPT OF CULTURAL HERITAGE

Culture; is defined as the whole of material and spiritual values created in the process of historical and social development and the means showing the measures of the sovereignty over the natural and social environment of man, which is used to create and transmit those values to the next generation (Turkish Language Society, n.d.). According to another definition, culture is the way of life and thought that constitutes its identity with its own art, beliefs, customs, understanding, and behaviors that make a society different from other societies and continue to change in time. It is the whole of the material and spiritual values that give society a sense of solidarity, and unity, that maintain order in it and make society develop an identity (Republic of Turkey Ministry of Culture and Tourism, n.d.). Cultural heritage, on the other hand, can be described as all of the documents that reflect the characteristics of a society and the past of a society. Cultural heritage is divided into two as tangible cultural heritage and intangible cultural heritage. Tangible cultural heritage involves tangible artefacts such as archaeological excavations, ancient cities and museums, while intangible cultural heritage involves food culture, traditions, dance, music and so on.

United Nations Educational, Scientific and Cultural Organization (UNESCO) bring forward the World Heritage Concept with the aim of introducing cultural and natural sites with universal values that are accepted as common heritage of mankind to the world, creating consciousness to preserve the universal heritage in society and providing necessary cooperation for the survival of the destroyed cultural and natural values which have been corrupted by various causes. From this point forth, at the 17th Session of the 16th General Conference, organized by UNESCO in Paris from 17 October to 21 November 1972, this case was decided to be an international convention. On November 16, 1972 the "Convention Concerning the Protection of the World's Cultural and Natural Heritage" was adopted and it was a historic step to protect the cultural and natural assets of all the world's nations on an international platform and to transmit them to future generations. With this convention, the assets that have been delivered by different cultures and civilizations throughout the history of mankind and that provide insights into different phases, stages, wealth and differences of this common past have been protected with the understanding that "one's loss is a loss to all human memory". Again in this convention, UNESCO pointed out that the cultural and natural heritage is under threat of extinction not only due to the traditional deterioration but also by the changing social and economic conditions. UNESCO underlined that the deterioration or destruction of any part of the cultural and natural heritage constitutes a nonrecoverable impoverishment for the heritage of all nations in the world. UNESCO has laid the foundations for an effective system for protecting cultural and natural heritage in exceptional value and transferring it to future generations in accordance with modern-scientific methods, taking into account the fact that the conservation of the world heritage which the whole humanity possesses by the contract is again the task of the international community. The convention called upon the state parties to make an inventory of all cultural and natural heritage assets within their borders and to define, preserve, protect, expose and/or present them and transmit them to future generations in the best possible way by taking all necessary legal, administrative, scientific and technical measures and the states parties have taken on responsibilities in this respect (Directorate of Istanbul Historical Areas, n.d.).

In accordance with the convention, all cultural and natural properties that are guaranteed to be protected by the country and which is of "outstanding universal value" are called "World Heritage", and the list of these properties is called the World Heritage List. The World Heritage Committee, the World Heritage List, the World Heritage List in Danger, the World Heritage Fund were established under the convention, and the World Heritage Center was established in the following years to serve as the secretariat for the Committee. The World Heritage Committee evaluates the applications made by the state parties with the help of the institutions with international experts whether or not they meet the criteria and classifies the assets into three groups: Cultural Heritage, Natural Heritage, and Mixed Heritage. By the year of 2016, there are 1052 cultural and natural sites around the world registered in the UNESCO World Heritage List, of which 814 are cultural, 203 are natural and 35 are mixed (cultural/natural) sites. With the annual World Heritage Committee meetings, these numbers are increasing every year (Republic of Turkey Ministry of Culture and Tourism, General Directorate of Cultural Heritage and Museums, n.d.).

According to the data of the year 2017, 17 items from the UNESCO World Heritage List were obtained from Turkey which is the subject of this study. These are; Göreme National Park and the Rock Sites of Cappadocia, Great Mosque and Hospital of Divriği, Historic Areas of Istanbul, Hattusha: the Hittite Capital, Nemrut Dağ, Hierapolis-Pamukkale, Xanthos-Letoon, City of Safranbolu, Archaeological Site of Troy, Selimiye Mosque and its Social Complex, Neolithic Site of Çatalhöyük, Bursa and Cumalıkızık: the Birth of the Ottoman Empire, Pergamon and its Multi-Layered Cultural Landscape, Diyarbakır Fortress and Hevsel Gardens Cultural Landscape, Ephesus, Archaeologicl Site of Ani and Aphrodisias. Göreme National Park and the Rock Sites of Cappadocia and Hierapolis-Pamukkale among these places are classified as the Mixed sites, and the remaining places are classified as the Cultural sites (UNESCO, 2017). On the other hand, there are many other cultural heritages in Turkey that are not listed on the heritage list by UNESCO.

Cultural Heritage Tourism

Cultural heritage tourism consists of historical architectural monuments, artistic monuments and archeological constructions, limited understanding of cultural heritage and conservation, as well as the sum of all the intangible and tangible assets that are formed as a result of the acceptance of the structures made in modern times by the society (Karapınar & Barakazı, 2017). The idea of managing cultural heritage assets as products for tourism consumption is relatively new, for cultural tourism professionals and scholars have been advocating this idea only since the late 1990s. This idea centres in the goal of providing an enriching experience to satisfy tourists' needs (Ho & McKercher, 2004). Although cultural heritage has been the focus of tourism since the end of the 1990s, the countries have been striving hard to be in UNESCO's list, tourism professionals competes to promote and to earn revenue by marketing the places on the lists, since the first day of publication of the Cultural Heritage List, which has been published by UNESCO since 1972. The features of cultural heritage tourism, which is very popular nowadays are (Albayrak, 2012: 63):

- The focus of cultural heritage tourism is the properties inherited from various cultures such as historical buildings, monuments, and works of art.
- Products subject to cultural heritage can not be reproduced. Because the subject products are works that reflect the characteristics of a society. For this reason, it is not possible to recreate these works. On the other hand, the cultural heritage is unique and endemic that is why they have characteristics that can not be imitated.
- Cultural heritage tourism, as well as cultural tourism, allows people to bridge the past and the future, solve identity problems, gain knowledge of history and sense of belonging, and communicate between generations.
- Cultural heritage tourism is not a tourism based on time. On the other hand, it is a type of tourism that can be integrated with other types of tourism.
- One of the most important features of cultural heritage tourism is to use the past heritage for tourism purposes and to allow for continuous consideration and preservation of the works which are accepted as heritage due to the income from it.
- Due to cultural heritage tourism, the economic conditions of the local people living in the region are developing as a result of tourism activities. On the other side, depending on the economic development of the local people, their living conditions are improving and they perceive the values and problems of the environment they live in better.
- Cultural heritage tourism transforms these areas into protected and used attractions by contributing to raising awareness of people about heritage characteristics, rather than having monuments and their surroundings become abandoned ruins in urban life.
- It enables social relations and cultural exchanges between local people and tourists.

Museums come to mind first when many features of cultural heritage tourism are considered. Because a museum is a place to exhibit cultural heritage and art works by various cultures that are the focus of cultural heritage tourism. In this context, it would be appropriate to define museology and to talk about the historical development of museology in Turkey, before the topic of Müze Asist, one of the geographical information system applications used in museums in Turkey.

MUSEUM CONCEPT AND HISTORICAL DEVELOPMENT OF MUSEUMS IN TURKEY

The word of museum originated from the *Mouseion* in Greek mythology; and it refers the place in which "Muses" live or the place dedicated to "Muses". Muses are fairies of inspiration who are daughters of God *Zeus* and *Mnemosyn*, Goddess of memory. They are nine daughters each of which represents an individual creative profession to protect and inspire them (Özkasım, 2015). In terms of today's reference, museum, in the simplest way, refers a place or building in which artifacts and scientific objects or the things useful for art and scientific practices for the purpose of public exhibition (Turkish Language Society, n.d.). Based on another definition, museum is a non-profit and open institution which collects, preserves, investigates, inherits and displays the tangible or intangible inheritance of human beings and their environment for examination, education and entertainment purposes in line with serving and developing society (International Council of Museums, n.d.).

Museology phenomenon which plays essential role in preservation of cultural values and inheriting them to the next generations emerged in Turkey with collection of objects in parallel to whole world. In the official website of the Republic of Turkey Ministry of Culture and Tourism, General Directorate of the Cultural Assets and Museums, museology in Turkey was explained as below:

In the Seljukian Period (13th Century), the practice of formal placing objects from previous period around the periphery of the castle walls surrounding the Konya Cairn whose remaining are not present today was considered as the preliminary traces of Turkish museology. Then it is known that late Hittite objects were collected around the Castle in Kahramanmaraş City in the Dulkadir Principality period. In the Ottoman Empire period, valuable heirloom objects, presents and trophies captured during the wars were kept safe in the treasury sections of the palace. As a result of numbers of valuable goods such as sacred relics of Islam to Ottoman Empire afterwards of the conquest of Istanbul, especially transfer of caliphate to Turks after eastern campaign of Yavuz Sultan Selim, part of the palace was reorganized in the form of a treasury house in the period of Sultan Mehmet and these developments resulted in a rich collection at the empire. These collections and rich Anatolian history paved the way of Turkish Museology.

Indeed, foundation of the Turkish Museology relies on Ancient Object Collection which constitutes the ground of the Istanbul Museum of Archeology. Upon Sultan Abdulmecit's encounter with Eastern Roman Inscriptions during his trip to Yalova in 1845, he decided to transfer them to Istanbul and stored in Aya Irini which had been utilized as an arsenal by the date. The museum was organized in two sections one of which constituted foundation of the Harbiye Military Museum. In 1869, the first museum of the empire was established under the title of "Empire Museums". In the very same year, Edward Goold, one of teachers of the Galatasaray High School, was assigned as museum chair. Additionally, a general circular was communicated to the city governors to transfer all historical artifacts to the museum by preserving them from any harm. As a result of the number of objects collected at the museum populated, the museum was moved to *Çinili Köşk* which commissioned in 1880. Afterward of movement of the museum, Osman Hamdi Bey who has significant place in Turkish museology history was assigned as museum chair upon death of previous museum chair, Anton Dethier.

In 1887, as a result of the excavations implemented in Sayda, a group of tombs including the Alexander's Tomb were exposed and transferred to Istanbul. Upon this development, an initiative was taken to build a new museum building. Architect Alexandre Vallaury was assigned to design the building of the museum the first building designed solely to serve as museum building in the country. The museum was commissioned in 1891. Shortly after opening of the museum, a library, photograph laboratory and model workshop were established. In order to exhibit historical artifacts explored during excavations, two buildings were annexed to the museum in 1903 and in 1907. Again, in this period, some initiatives were introduced to establish museums in Anatolian cities beside Istanbul. New museums were established in Konya and Bursa in 1902 and 1904, respectively. By consulting foreign archeologists and specialists, registry and museum collections, catalogues and exhibitions were developed. In the period of 1912-1914, Gustav Mendel's catalogue with title of "Catalogues de Sculptures Grecgues, Romaines et Byzantines" in three volumes about the stone artifacts were the prominent resource promoting Turkish museums to the world.

Efforts regarding establishment of an art museum in Turkey initiated in the 19th century in Turkey. In parallel, an inventory collection was commenced but opening plans of the museum failed unfortunately. In this regard, the first art museums emerged after proclamation of the Republic of Turkey; Istanbul Art and Sculpture Museum was founded by Ataturk in 1937. In the republic period, The Directorate of Culture under the Ministry of National Education was put in charge about state museums; then, the General Directorate of Ancient Arts and Museums in 1944. Then in 1924, based on the council of ministers decision, it was adopted to open the Topkapı Palace to public visitors together with its current collection of artifacts. As Atatürk requested Hittite Museum to be founded in Ankara, *Kurşun Han* and *Mahmut Paşa Bedesten* Manors were renewed by the Ministry of National Education to assign for the purpose of establishing the Ankara Archeology Museum. Then, it was reorganized and renamed as the Anatolian Civilizations Museum subject to the variety of objects in the museum in 1967.

The first museum building in the republic period was the Ankara Ethnography Museum; while its construction was started in 1925, the building was commissioned in 1930. Subject to the law enacted in 1925, madrasahs and dervish lodges were banned and their valuable goods were exhibited in the Ankara Ethnography Museum; and illustrations of ceremonial and daily public were exhibited by using original goods. On the other hand, Konya Mevlevi Tomb was not closed and instead it was reorganized as a museum with its collection of goods along the instructions of Atatürk. In addition to the aforesaid museums in the early years of the republic, new museums were opened across Anatolia; and special interest was paid to allow them to proliferate around the country.

Today, there are totally 370 museums in Turkey; whereas 187 were under administration of Ministry of Culture and Tourism, 183 were private museums subject to supervising of the ministry. As number of museums increases every day, they are not only the places where artifacts are exhibited or stored, rather they have become educational and cultural establishments for social and cultural events, national and international conferences, seminars for public education and contribute in promotion of the country (Republic of Turkey Ministry of Culture and Tourism, n.d.).

Today, Turkey has a total of 370 museums and cultural heritage sites with millions of visitors every year. The number of museums, the number of artifacts and the number of museum visitors between 2000 and 2015 of the Republic of Turkey in Table 1 also support this statement.

According to Table 1, number of artifacts in the inventory of museums under the Republic of Turkey, Ministry of Culture and Tourism increased to 3 million and

Table 1. Number of Museums, Works and Visitors to the Museums and Ruins in Turkey

Year	Number of Museums	Number of Works	Total Visitor	Paying Entrance Fee	Free
2005	156	2 790 566	19 663 014	15 212 404	4 450 610
2006	175	2 866 303	16 086 050	10 375 980	5 710 070
2007	165	2 767 149	18 023 618	12 111 952	5 911 666
2008	159	2 989 749	22 736 238	15 823 171	6 913 067
2009	183	3 044 197	23 000 957	16 504 935	6 496 022
2010	185	3 096 599	25 107 203	17 209 442	7 897 761
2011	189	3 149 982	28 459 311	21 984 444	6 474 867
2012	188	3 177 446	28 781 308	19 211 461	9 569 847
2013	187	3 174 867	29 566 691	23 081 525	6 485 166
2014	192	3 217 173	29 799 790	23 692 984	6 106 806
2015	***193***	***3 235 113***	***28 122 934***	***19 699 239***	***8 423 695***

Resource: (Turkish Statistical Institute, 2016)

235 thousand 113 pieces by 0.6% in 2015 with respect to 2014; number of visitors decreased to 28 million and 122 thousand 934 by 5.6% in 2015 with respect to 2014. Again, number and proportion of the visitors who admitted into the museums and archeological sites under the ministry with admission fee decreased from 23 million and 692 thousands 984 and 79.5% in 2014 to 19 million and 699 thousand 239 and 70% in 2015. Again, number and proportion of visitors who admitted into museums and archeological sites under the ministry free of charge increased from 6 million and 106 thousands 806 and 20.5% in 2014 to 8 million and 423 thousands 695 and 30% (Turkish Statistical Institute, 2016).

According to the statistics published by the Ministry of Culture and Tourism, Directorate of Cultural Assets and Museums, the most frequently visited museum in Turkey in 2015 was İstanbul Ayasofya Museum with 3,466,638 visitors. This museum was followed by Topkapı Palace Museum with 3,252,524 visitors and Konya Mevlana Museum with 2,337,850 visitors, İstanbul Topkapı Palace - Seraglio (877,331 visitors), İstanbul Archeology Museums (411,797 visitors), Ankara Republic Museum (381,294 visitors), Antalya Demre Museum (364,529 visitors), Nevşehir Hacıbektaş Museum (363,457 visitors), Ankara Anatolian Civilizations Museum (272,814 visitors) and İstanbul Kariye Museum (233,000 visitors) (Republic of Turkey Ministry of Culture and Tourism, n.d).

Nowadays, museology in Turkey come to the point of virtual museology and using geographical information systems in the museums. It is possible to use many applications on smartphones or tablets that make museum visits more efficient in today's digital era thanks to the developing technology. These applications are now regarded as indispensable for contemporary museum management.

Geographical Information Systems

Developments in information technologies are mostly experienced in the field of software and hardware. After having the required processor speed and disk space, commercial software increased rapidly, which is inspired by CAD software and able to use numerical and alphanumeric databases, in which data sets are processed, queries are made, new data is created, and positional analyzes. The same developments have also occurred in the field of remote sensing. Thus, Geographic Information System (GIS) has become an essential tool for all scientific disciplines (Coşgun, 2015: 1-2). The fact that Geographical Information System is important for all disciplines makes it difficult to make a definite description of it. Because each discipline has made a definition by setting priorities related to itself. For example, Bensghir and Akay (2007) emphasized that GIS has different definitions in terms of technological and managerial aspects. From a technological point of view, GIS is defined as a system that stores, processes, transforms, displays, and collects actual spatial data; it is evaluated as the integration of spatial data into the decision support system from the administrative point of view. In addition, other reasons for not being able to make a common definition of GIS are the commercial expectations from the system and the rapid developments in GIS technology (Bensghir & Akay, 2007: 5-6). As a result, GIS has gone a long way since 1912 when the first studies forming GIS are executed, and now it has become a must for today's technology. GIS, which has many uses in different disciplines, has been studied through a museum application in this study.

Müze Asist

The Müze Asist application; is a mobile guide and virtual tour application that can be used in museums, exhibitions, galleries, fairs, and any other area that needs contactless interaction. It is a free application which is always accessible in the digital environment. It provides professional guidance it works on users' mobile phones, tablets (GooglePlay-AppStore) and their computers, with explanations, detail visuals, videos with multi language support and audio presentations of works in cultural and artistic areas, especially in museums. The Müze Asist application which includes all museums wherever it is in Turkey and which is spreading rapidly among the museums, is a digital travel guide that can be accessed by a single application in every destination.

The Müze Asist is designed to enable visitors to benefit from museums, culture and art centers, where societies maintain and demonstrate their historical and cultural background, and pass on their values to future generations in the most favourable and lasting way. The application is 100% native work as a result of R&D studies and it is developed for museums, running on mobile platforms. The application provides visual, auditory and interactivity-based contactless interaction with the electronic module specially designed for it. The application aims to make visual and audible presentations in every language and to bring the museology together with the latest technology congruously via the interactive working system.

The communication between the application and the visitor is based on the Bluetooth ® Low Energy (BLE) Smart Hub technology, which keeps off all distraction during the visit and also is the only protocol that can be applied according to the safety rules in the museum. The system is based on contactless positioning technology for the museums via the mobile application. This positioning structure is based on the principle that the mobile application should be uploaded by visitors, or the information should be presented to visitors in a way that informs the users visually and audibly, in the direction of the information received from the electronic devices located near the artifacts. In this way, application users will have a useful structure that will provide them with flexible and very different functions on smart devices instead of static structures such as voice tour programs prepared in advance.

These innovations combine with Bluetooth® Smart Ready technology to provide a continuous and contact-free experience.

The Müze Asist makes it easy to update museum contents, audio and visual data dynamically as well as to offer language options and also it is possible to use gamification techniques in the application to increase memorability. In the light of all these, it is possible to list the features of the Müze Asist as (Yalçın, 2015; Catalog of Müze Asist, n.d.):

- Find the nearest museum to the visitor's current location,
- Show the museum on the map and get audio and video directions based on the desired route,
- Connect to the website of the museum to visit,
- Save your favorite museums,
- Share the desired museum on social networking sites,
- Contact the museum with the registered phone number,
- List all the activities of the museums,
- See all notifications from the museums.

Various photographs were used during the study to better understand the operating principle of the system. For example, Figure 1 shows an example, the Müze Asist marking all museums on the map from a selected city in Turkey.

The Müze Asist also have other features than those listed above. These are:

- Thanks to the Müze Asist's multilingual feature (Turkish, English, Arabic, and Russian), visits to museums accompanied by the narration of professional guides become a more satisfying experience.
- The Müze Asist runs as a digital archive for museums, exhibitions, and galleries, where they can hold audio, visual or written information.
- Thanks to that the application's synchronized with the web portal, the museums have easier management and accessibility. Studies based on social media and viral promotion for the publicity of artifacts and museums are also created within the content management system.
- Through the application, the museums are categorized as "History, Art, Culture, Science, Nature, and Children". Thus, the users will have the chance to choose between the museums according to their interests through the application.
- Thanks to the Müze Asist's Virtual Tour feature, when users approach the exhibited artifacts, they can reach audio, visual and written explanations without contact. However, it is necessary to have a MultiGuide that works with IBEACON devices in the museum. With the contact-free operation, a

Figure 1.

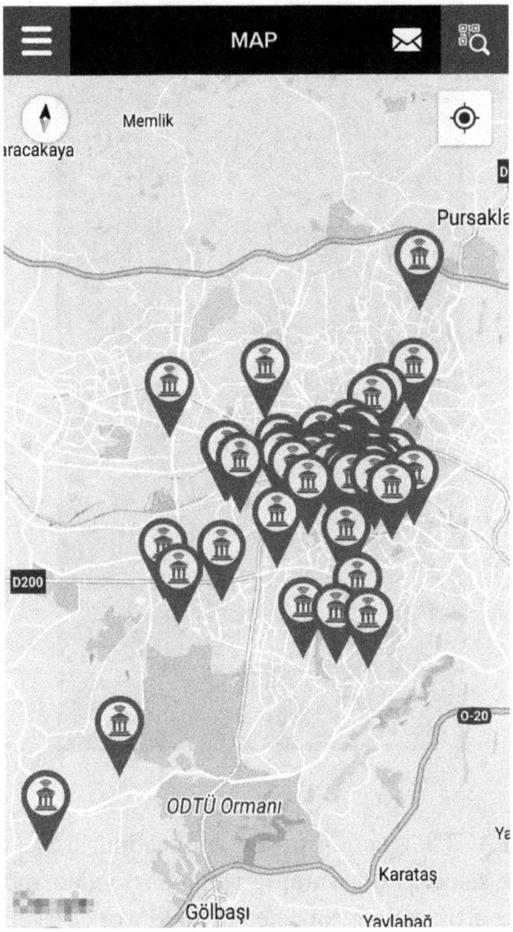

virtual tour can be done with audio, picture, video and text support. To start the virtual tour, the key in Figure 2 should be turned on. All these are executed by the control software that is given to the museum and gallery authorities and working on the web (Yalçın, 2015; Catalog of the Müze Asist, n.d.):

- In the "PLACES" section of the application, the display areas of the selected museum (Kitchen, Hammam, Harem, etc.) are located separately. These places are gathered by field studies.
- In the "CATEGORIES" section of the application, assorted and classified states of the artifacts belonging to the selected museum (Study Rooms, Dining Rooms, Bed Rooms, etc.) are included.

Figure 2.

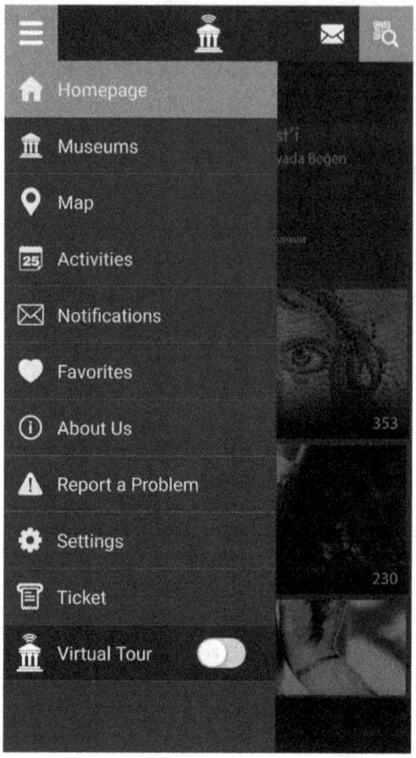

- In the "ARTIFACTS" section of the application, it is possible to find the artifacts of the selected museum, photographs, video and audio files of the artifacts. These artifacts are collected by field works.
- An advanced statistical system of the Müze Asist Control Panel analyzes the visitor information of the active museums and presents them as easy-to-read data to the user. This information includes parameters such as the view frequency of artifacts, the average duration of visit, and visitor density. The statistics can be generated by the system for both general users and individual users. Figure 3 will be helpful to build an idea of the advanced statistical system.

On the other hand, museum assists are thought to be superior in all respects compared to other exhibition technologies used in museums. Table 2 shows the state of exhibition technologies used in the museums and their superiority to each other.

In addition to the information in Table 3, the following innovations have been added to the Müze Asist thanks to innovations in the field of Bluetooth 4.0 (Smart).

Figure 3.

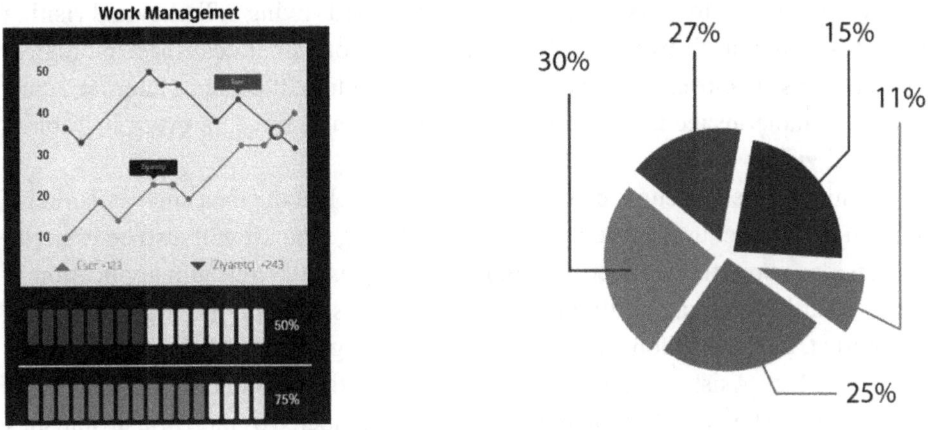

Table 2. Exhibition Technologies Used in the Museums

FEATURES	Audio Guide	Manual Entry	QR Code	NFC	Augmented Reality	Müze Asist
Language Option	x	x	x	x	x	✓
Platform Independence	x	x	x		x	✓
Security	x	x		x		✓
Electronic Guide		x	x	x	x	✓
Interactivity		x	x	x	x	✓
Random Access		x	x	x	x	✓
Easy Update		x	x	x	x	✓
Remote Access						✓
Uninterrupted Experience						✓
Non-contact Interaction						✓

Table 3. Innovations about the Müze Assist

Device	BR/EDR (Classic Bluetooth) Support	BLE (Bluetooth Low Energy) Support
Pre- Bluetooth 4.0	Yes	No
4x Single-Mode (Bluetooth Smart)	No	Yes
4x Dual-Mode (Bluetooth Smart Ready	Yes	Yes

It is thought that Museum Assist will be successful and widely used in all museums by especially Turkish citizens, but also tourists living in Turkey and visiting museums within the framework of cultural heritage tourism. Because technological developments are closely followed in Turkey. In addition, the rate of internet, smart phone and tablet usage is also very high. In addition to the data given in Table 1 on annual visitor

statistics of the museums in Turkey, when you look at the data such as Turkey's population, internet usage rate and smart phone usage rate, it will also be easier to have an idea about the number of people who may use the museum application in Turkey, that is potential cultural heritage tourists. As a result, the higher usage of the internet, the smartphone, and the tablet is, the higher the probability of people using the Müze Asist will be. In this context, statistical results of the Household Information Systems Technologies Usage Survey conducted by the Turkish Statistics Institute in 2015 were summarized below (Turkish Statistical Institute, 2016):

- Proportion of internet user population in Turkey is 55.9%; and population of Turkey is 78 million 741 thousand 53 as of December 31st, 2015. That is, there are about 44 million internet users. Computer and internet user rates among 16-74 aged population were 54.8% and 55.95 as of April 2015, respectively. Whereas these rates were 64% and 65.8% for males, 45.6% and 46.1% for females.
- In Turkey, seven out of ten houses have internet access. Again, proportion of houses with internet access in Turkey was 69.5% as of April 2015. The houses with broadband internet connection are 67.8%. Accordingly, whereas 37.4% of overall households has broadband connection (ADSL, cable internet, fiber etc.), 58.7% has mobile broadband connection. Narrowband was used for internet access by 2.7% of houses.
- In Turkey, whereas there is a cell or smart phones in 96.8% of overall houses, there is a land line phone in 29.6% of them. In the same period, while there is table-top PC in 25.2% of household, there is laptop PC in 43.2% of household; and there is smart TV with internet connection in 20.9%.
- In terms of internet usage purpose in Turkey, 80.9% of internet users in the first quarter of 2015 were using internet for having personal profile on the social media, sending messages, or sharing photo or content; 70.2% of users followed online news, newspaper or magazine, 66.3% searched for medical information, 62.1% used personal blogs for their texts, images, photograph, video, and music contents; 59.4% searched for market research for goods and services.

- In the first quarter of 2015 in Turkey, 87.1% of internet users from the age group of 16-74 were from home; 42.5% were from workplace; 37.7% were from houses of relatives and friends; 29.2% were from wireless areas such as malls, airports and etc.; and 10.6% were from internet café. Usage of mobile devices for internet access outside house work place has increased. While 74.4% of individuals used cell and smart phones for wireless connection outside their house and work place, 28.9% used mobile PC (laptop, net-book, tablet etc.).
- In the twelve-month period between April 2014 and March 2015 in Turkey, 53.2% of users used internet to communicate with public institutions for personal purposes or to access a public service. Accessing information in the pubic websites was most frequent reason for internet usage by 50.5%. Whereas the proportion of the individuals who used internet for online shopping was 33.1% in 2015, this rate was 30.8% in 2014. Of these individuals who made shopping on the internet along the concerned period, 57.4% purchased apparel and sport goods; 27% travel tickets and rent a car etc.; 25.5% household goods (furniture, toy, electric appliances), 22.4% electronic devices (Cell phones, camera, radio, TV, DVD player etc.); 18.4% books, magazine, newspaper (e-book inc.). In terms of the internet usage frequency along the last quarter of 2015, proportion of the regular user profile who use internet daily or minimum once a week was 94.2%.

As it could be understood from the results of the survey conducted by the Turkish Statistic Institution, which concerned Household Information Technology Usage in 2015, usage of technology and internet is remarkably high in Turkey. But it is also notable that there was no sub-titles of culture, art or museum under the option of "having information" given for internet usage purposes.

CONCLUSION

The museums, which protect historical and cultural wealth, constitute one of the most important fields of activity in terms of tourism and cultural promotion. Museology is based on firstly gathering the important objects about the human and human environment from the cultural, artistic or scientific point of view, making of their collections, and then the preservation and exhibition of these collections. A museum is a public domain where information is shared with people. The effort of collecting, accumulating and protecting different objects or knowledge is, of course, an important level in terms of humanity (Karabıyık, 2007). Thus, the museums which collect, research, protect and present the cultural heritage of society for the education of the people are accepted as one of the basic institutions that should be present in a contemporary society.

Thanks to the developing technology, the era we are in is called “Digital Age” and museums such as all institutions have not been left indifferent to technological developments. Museum administrations are now making maximum use of technological developments and striving to keep up with the times. The geographical information system provides a great deal of contribution to the new ways of the museum administration. This has made it possible for people traveling within cultural heritage tourism to access the museums and to get maximum benefit from them. The Müze Asist which is 100% native work constructed as a result of R&D studies, is an application based on geographical information systems and equipped all kinds of details for tourists’ museum visits. Although it just started up in Turkey, it is expected that it will be used nationwide in a short period of time.

The Müze Asist aims to increase the memorability of cultural heritage by supporting presentations with technology in today’s world where the information circulates very fast globally. The Müze Asist is a system that can respond to questions like how many museums are located in Turkey or where the nearest museum is located with a single click. People can access general information such as entrance fees, visiting days and hours, open addresses, or get directions to each museum in Turkey, share the museum on social media platforms, add to favorites thanks to the Müze Asist. The museum managements can organize their current information thanks to the easy-to-use control panel and they can inform the art-lovers about the events instantly. With this system, individual tourists who do not want to participate in guided tours can plan their own trips on the map and visit the cultural heritage points in Turkey in the exhibition environments. The Müze Asist aims to provide more useful museum visits for the visitors with developed systems aggregate. At the same time, all the museums are presented with information to visitors in order to make museum visits more frequently.

Technological developments are followed closely in Turkey, where the use of the Internet, smartphones, and tablets are very high. For this reason, thanks to the Müze Asist and the many applications like it, it is expected that museum visits will be increased, the whole world heritage, especially the ones in Turkey, will be recognized and these practices will be further developed.

REFERENCES

Albayrak, A. (2012). *Alternatif turizm*. Ankara: Detay Yayıncılık.

Catalog of Müze Asist. (n.d.). Retrieved July 15, 2017, from http://www.muzeasist.com

Çoşgun, S. (2015). *Coğrafi bilgi sistemleri kullanılarak ekolojik faktörlerin bazı asli ağaç türlerinin konumsal dağılımına etkilerinin incelenmesi* (Unpublished master thesis). University of Süleyman Demirel, Isparta, Turkey.

Directorate of Istanbul Historical Areas. (n.d.). *World Heritage Areas*. Retrieved from http://www.alanbaskanligi.gov.tr/unesco_dunya_miras_alanlar%C4%B1.html

Ho, P. S. Y., & McKercher, B. (2004). Managing heritage resources as tourism products. *Asia Pacific Journal of Tourism Research*, *9*(3), 255–266. doi:10.1080/1094166042000290655

International Council of Museums. (n.d.). Retrieved January 19, 2017, from http://icom.museum/the-vision/museum-definition

Karabıyık, A. (2007). *Çağdaş sanat müzeciliği kapsamında Türkiye'deki müzecilik hareketlerine bir bakış* (Unpublished master thesis). University of Atatürk, Erzurum, Turkey.

Karapınar, E., & Barakazı, M. (2017). Evaluation of cultural heritage tourism as a sustainable tourism: Gobeklitepe ruins. *Güncel Turizm Araştırmaları Dergisi*, *1*(1), 5–18.

Kaya Bensghir, T., & Akay, A. (2007). *Yerel yönetimlerde coğrafi bilgi sistemleri Türkiye uygulamaları*. Ankara: Public Administration Institute for Turkey and Middle East Publication.

Özkasım, H. (2015). Müze ve müzecilik. In Altınsapan & Küçükhasköylü (Eds.), Müzecilik ve sergileme (pp. 2-21). Ankara: Anadolu Üniversitesi Yayınları.

Republic of Turkey Ministry of Culture and Tourism. (n.d.). Retrieved July 04, 2017, from http://www.kultur.gov.tr/TR,96254/kultur.html

Republic of Turkey Ministry of Culture and Tourism, General Directorate of the Cultural Assets and Museums. (n.d.). Retrieved July 05, 2017, from http://www.kulturvarliklari.gov.tr/TR,44423/dunya-miras-listesi.html

Turkish Language Society. (n.d.). Retrieved July 04, 2017, from http://www.tdk.gov.tr/index.php?option=com_gts&arama=gts&guid=TDK.GTS.595b81c56284d9.35935636

Turkish Statistical Institute. (2016). *Kültürel miras 2015.* Retrieved January 31, 2017, from http://www.tuik.gov.tr/PreHaberBultenleri.do?id=21550

United Nations Educational, Scientific and Cultural Organization (UNESCO). (2017). Retrieved July 07, 2017, from http://whc.unesco.org/en/list

Yalçın, C. (2015). *Müze assist.* Retrieved July 14, 2017, from http://www.muzeasist.com/kklavuz.pdf

KEY TERMS AND DEFINITIONS

Collection: Group of objects gathered and classified based on natural, historical, scientific, artistic, and cultural values of humanity.

Destination: Destination is a city, town, or other area that is dependent to a significant extent on the revenues accruing from tourism, or a country, state, region, city, or town that is marketed or markets itself as a place for tourists to visit.

Digital Age: The information age (also known as the computer age, digital age, or new media age) is a period in human history characterized by the shift from traditional industry that the industrial revolution brought through industrialization to an economy based on information computerization.

Exhibition: This is the work of organization of museum collections under certain construct and logic so as to represent the relevant time interval or theme deliberately. Exhibitions are classified in three categories: persistent, temporary, and mobile.

Mobile Application: Mobile application is a software application designed to run on a mobile device such as a smartphone or tablet computer.

Chapter 3
Geographical Information System in Eco-Tourism

Ishrat Nasreen Wahab
Garden City University, India

Sachin Soonthodu
Garden City University, India

ABSTRACT

Interest of the local population is an important consideration for sustainable development of a region. Researchers recently observed in their studies about the benefits the local population get while being involved in eco-tourism. Though many factors are considered to know how locals are benefited by eco-tourism, an interesting factor to be dealt with utmost care and concern is environmental and ecological issues arising out of these growth factors. All the possible technologies of computer application may be used to maximise the end needs of this specialised subject. One such application is GIS (geographical information system), which can be of great help in this concept. The GIS applications are recently used in the management of parks, facilities management, assessment of visual resources, and new areas identification for new tourism development initiatives.

INTRODUCTION

According to the World Tourism Barometer of the United Nations World Tourism Organisation (UNWTO), Foreign Tourist Arrivals or FTAs in the World Tourism Market is growing at an average rate of 4% each year and had reached a massive number of 1.2 billion in the year 2015. In fact, the demand for international tourism

DOI: 10.4018/978-1-5225-5088-4.ch003

remained robust in 2016 despite many challenges. International tourist arrival grew by 3.9% to reach a total of 1,235 million, and some 46 million more tourists travelled internationally compare to 2015 (www.unwto.org). The phenomenon of tourism has shown as astonishing growth in the recent years. The widespread of this domain has turned it into a tool for development by contributing towards employment generation in various sectors, adding to the growth of the economy, as well as providing opportunities to and supporting the development of many MSMEs. Other positive impacts of tourism would include cultural exchange, environmental protection, improvement in the standard of living, creating a better image of the destinations and of course spread of the message of peace and harmony throughout the world.

With more and more places being developed as tourist sites/destinations, proper planning, development and operations have become critical since damage to the environment, community and economy is a price one cannot afford to pay. Thus, with the evolution of ethical travel, concepts such as sustainable tourism, green tourism, responsible tourism, eco-tourism, etc. are gaining momentum. These concepts provide a continual process of improvement and one which applies equally to tourism in cities, resorts, rural and coastal areas, mountains, and protected areas. It can apply to all forms of business and leisure tourism. These forms of tourism aim at achieving the goal of development of a destination without harming the environment or the surrounding. Multiple case studies have showcased the problems faces when tourism is not eco-friendly, which causes severe adverse environmental impacts caused by infrastructure development and measures taken to meet tourists' demands, such as energy consumed and water resources provided, waste disposal and so on. Therefore, a proper management system in the sector of tourism development is crucial (Ulmasova, Sabrina, 2007)

GIS, being a tool for analysing and mapping of data, serves very effectively for tourism planners to predict various scenarios, trace the dynamics, and make appropriate decisions in tourism planning and implementation. It plays an important role in assessing the proper land use for tourism development, and also helps the stakeholders in use of spatial decision support systems in the tourism business.

What distinguishes GIS from other information systems are their abilities to represent, storing, managing, analysing, and visualising spatial and non-spatial data in an integrated environment. GIS, as spatial information systems, have found many possible applications in various fields, including business. It can also be a valuable tool for investigating specific questions that pertain to tourism development including location, the condition of the area, trends and changes, routeing to and through the site, and patterns associated with resource use. (Dye, Andrew, 2007)

OBJECTIVES OF THE STUDY

- To understand the concept of Geographic Information System and its relevance in tourism industry
- To analyse the role of Geographical Information System from Eco-tourism perspective
- To suggest various measures for effective application of Geographic Information System in Eco-tourism

REVIEW OF LITERATURE

Geographic Information System is the popular tool of mapping the earth and its topology. It is the technology to capture, store, manipulate, present and compare geographical data. The term 'Geographical Information system' was introduced by Roger Tomlinson in the year 1968 who is considered as the father of GIS. The Geographic Information System, in general, is a computer mapping system, which uses various digital data creation and output methods (Chang, K., 2007)

According to Star, Jeffrey et al., (1990), "A geographic information system (GIS) is an information system which is designed to work with data referenced by spatial or geographic coordinates. In other words, a GIS is both a database system with specific capabilities for spatially-referenced data, as well as a set of operations for working with data . . . In a sense, a GIS may be thought of as a higher-order map."

From this definition Geographic Information System can be understood as, a) It is a database system which captures, stores and exhibits data. b) It is associated with digital geographic coordination and c) It helps in the comprehensive decision-making process by integrating the technology

Tourism is a generic industry which is highly dependent on geography and natural resources. Forests, Mountains, water resources and the greenery, etc., are the major source of tourist attraction. Today sustainable practices and development are the major challenges of the tourism industry. This can be effectively attained through practising eco-tourism activities. Developing and managing eco-tourism sites have become a most concerned issue today. Eco-tourism sites are encroached, over-utilised and engraved due to human activity. Eventually, government and other social institutions turning up to safeguard the ecology of which, technology is used in its entire possible means. Geographic Information system is one such technology which can be harnessed for the better and effective utilisation of sustainable eco-tourism.

Wei, Wei (2011) in his research article explains how Geographical Information System can be used for tourism management. The article discusses the collaboration of Global Information system and technology to construct a database for tourism management. As the information technology continuously changes there is the need of upgrading Geographical Information System accordingly this can be redressed by combing the expert system of technology with Geographical Information System for the better future.

Shahrak, Fataneh et al., (2015), refer Makhdoum Ecological model in implementing Geographical Information System. A comparative analysis of the model has shown how Geographical Information Model helps in identifying and developing various ecological sites of Sistan and Balouchestan Province. The research also provides statistical data about the ecological site development in this area and various criteria for locating and specifying the eco-tourism zones which helps the local governments to enforce policy implementation programme in this area. Thus the complete research primarily built on Geographical Information System mechanism.

Mennecke, Brian E. et al., (1998) in their research study had an attempt of introducing Geographical Information System in a broad way. The special analysis of the study emphasises on the features, functions applicability and the opportunity for further research in GIS. Further, the study discusses the social applicability and impact of GIS in a broad way.

Most of the social, economic and physical problems are associated with water management. Miroslav Rusko, Roman Chovanec, Dana Roskova 2010) make an attempt at using GIS technology and Digital Elevation Model Data (DEM) in the conservation of water. The study identifies that designing an intelligent model for water, waste water and storm water through GIS model is an effective tool for water conservation and ecological balance. Thus GIS applications have the ability to integrate all spatial data to the decision-making process in water management.

Tourism is an activity oriented industry. One of the major problems of the tourism industry is that service providers don't have the kinds of information which are necessary for their decisions assertiveness regarding the territory planning. Sousa, Paulo R. et al., (2008) suggest that GIS has the ability to exhibit social, cultural, ethnographic and geographic diversity which has the ability to enhancing the system use and economic revenue of the agents involved. This results in tourism demand management through that balanced regional development of a nation. T. Turk, M. U. Gumusay believes that GIS planning and implementation helps the end user to initiate and implement optimum planning for the industry.

According to Anshu, Anshu (2017) application Digital Elevation Model as a part of GIS for the identification of eco-tourism sites has the greater impact on potential development. This will also help in marking and developing the basic infrastructure facilities such as road connectivity, transport network to the eco-tourism sites. Thus the application of GIS contributes to the greater socio-economic development of a region.

Many researches have focused on eco-tourism and GIS. There is larger scope for identification of the potential areas for eco-tourism in enhancing the socio-economic status of the indigenous communities using GIS (Wanyonyi, Jeremiah Waswa, 2016). Identification of eco-tourism exploitable area and implementation of eco-tourism model in the sites such as national parks, wild life sanctuaries and Game reserve etc benefits the sustainable objectivity. GIS also helps in deciding the suitability of a destination for eco-tourism project implementation.

Google earth the technology which is completely based on GIS, which gives a virtual presentation of the topography of the Earth (Henry, Andrew, 2009). Eventually, the browser can have the comparative analysis of future Earth through that the scientists can predict the possible changes of earth and frame necessary action plans to save the earth from possible dangers. GIS also has the database on the past-earth which provides more weightage to the comparison. Web based tourist maps supported by GIS assist the visitors during their travel which reduces the dependency on individuals (Agus et al., 2013). The tourist will also have a comprehensive tourist guide in his/her pocket to assist him/her round the clock.

Eco-tourism is also concerned with the geographical area development factors. Hence Multiple Criteria Evaluation methods are used for land suitability analysis or fulfil specific objectives based on various features that exist in the selected area. Thus Multiple Criteria Evaluation (MCE) and GIS have a major role in Eco-conservation.MCE method helps in solving the conflict between conservation and recreation which has the potential objective of minimising the environmental impact (Mohd, Zulaikha Hana et al., 2012). this also shows how an area can be selected on a priority basis for different activities.

Climatic conditions play a major role in Eco-tourism activities (Kumari et al. 2010), fragile ecosystem (Mohd, Zulaikha Hana et al., 2012), and historical resources, natural area and traditional culture (Geremew et al., 2015) also influence eco-tourism activities.

GIS and Remote sensing (RS) is used to detect the potential eco-tourism sites based on the environmental indicators such as environmental resilience, ecological fragility and eco-tourism attractivity (Kumari et al. 2010).

Creation of unified tourist products and environmental data system helps both the tourists and the scientists to use the data base easier and to predict the possible results (Cekule, Marita)

A good portion of eco-tourists are professional travellers who always look for technological convenience. Implementation of GIS in Elbasan region has improved the tourism competitiveness of Albania both in the Domestic and International market. (Shyti, Bederiana et al., 2012).

The tropical developing countries such as Thailand have implemented sustainable eco-tourism development, protected area management based on GIS. GIS plays an evident role in the process of documenting natural conditions, developing and documenting the suitable resources for tourism, exposing conflicts, and identification of cause-effect relationships (Rannveig, O. et al., 2009) and also provides a digital data bank for long-term and beneficial monitoring of sites (Mazaher, M, 2010)

GEOGRAPHICAL INFORMATION SYSTEM IN TOURISM

Geographical information system (GIS) is a technological application of geography, mathematics, statistics, management, survey and mapping. In simple it is imagery representation of the earth's topography. GIS plays a vital role in the rational decision-making process related to geography and it is widely used in land management, urban planning disaster management, weather forecasting and other physiographic commercial activities. GIS through its hardware and software systems collects, stores, manages transforms and displays the data. Google mapping is one of the examples for the GIS data management system.

GIS is an analytical tool which is basically originated from Computer Aided Design (CAD) which creates maps at different scales in different positions by using new input information and displays it in various formats.

Tourism is the major source of revenue generation for many economies around the globe. The governments and the travel organisations adopt various promotional strategies to attract more number of tourist in-flow to their respective nations. Destination improvements, swift connectivity, bulk information management, price offers are some of the strategies that the service providers offer to attract the potential customers.

Quick communication and extensive geographical information are one among such services offered by the service providers. Today Geographic Information System (GIS) plays a crucial role in the international tourist market. Web based GIS system is the popular technology that the tourists prefer today. Digital multi media mapping and mobile mapping are the latest trends in GIS system.

GIS is the database application which uses geo-references as the primary source of storing and accessing information. This helps the traveller to get accurate information about the site that they intend to visit. Eventually, GIS integrates the technology by analysing Ariel photographs and satellite images through that drafts the map. This helps the traveller to decide his accessibility to a destination. Besides these, GIS powerful tool of decision making which can be used in the research process. Thus the GIS process includes,

- Data Input
- Data Storage

- Data Transformation and
- Data Output

According to Jovanovic, Verka (2008) in tourism, GIS is used to provide,

- A digital map base for printed maps
- Digital files for Internet mapping
- Digital files for mobile mapping
- Attractions map
- Website with interactive mapping

Apart from these the GIS also creates and outputs special database about acquiring graphical maps covering a particular region, digitalizing the paper maps, the creation of topography and real world coordination of data. To visualise the data, vector graphic shapes, textual information and images are integrated. A multimedia presentation of data is also available for the users to make the information more accurate and attractive. Thus the tourist will have a unique experience of browsing a travel destination to make his holidays hassle-free. This map will show the distinctive attributes of the places to be visited. Traveller will have a more convenient itinerary. The GIS is proximately used in travel connectivity network analysis, identification of hotels and restaurants, distance calculation, identify the destination, destination planning and so on. One of the noted uses of GIS is Eco-tourism development and decision making. GIS will also help the government to offer quality service through continuous evaluation and experimenting tourism resources. There are varieties of usage one can extract by applying the GIS. The great feature of the GIS is the integration of its all components like software, hardware, capturing of data, managing and analysing and displaying geographically identified information (Environmental Systems Research Institute [ESRI], 2010). Many uses of GIS which includes planning for environmental sustainability, infrastructure selling, management of the property, planning emergencies, market analysis, navigation systems, etc.

GEOGRAPHICAL INFORMATION SYSTEM IN ECO-TOURISM

Eco tourism is the process of visiting the exotic natural tourist sites with the objective of biodiversity conservation. The International Ecotourism Society defines ecotourism as "responsible travel to natural areas that conserves the environment, sustains the well-being of local people, and involves interpretation and education".

Tourism is a recreational activity and it is a human-environment struggle. One of the negative consequences of tourism is that it threatens the biodiversity and natural habitats. Urbanisation and commercialisation due to intensive tourist flow are a threat to the natural environment. Eco-tourism is a growing concern which has the fundamental objective of visiting a destination without harming the natural environment.

GIS can be effectively implemented in identifying the eco-tourism sites and forecasting a development model into the region. GIS Applications such as Multimedia Imagery system serves in identifying the boundaries of eco-hubs, protected areas etc. and to make swift decisions during emergencies. Canadian forest management department has successfully adopted GIS in terms of real time decision making and realistic modelling of forest fire danger and behaviour prediction. A real time calculation of the consequences of the forest fire is possible through the GIS system.

IDENTIFICATION OF POTENTIAL ECO-TOURISM SITES

The fundamental objective of eco-tourism is to protect the natural and cultural resources for future and also to generate the revenue. There is larger scope in the identification of wild life distribution, ecological value, eco tourism attractivity and diversity through the GIS (Kumari et al., 2010). Land use or forest cover mapping based on GIS gives a clear account of natural vegetation and biodiversity. GIS illustrates mountain, hill, terraces etc and gives information about the abilities of these sites.

GIS is also an effective tool in designing and monitoring sustainable tourism development. Balanced regional growth can be achieved by adopting the GIS system in tourism. A study on Application of GIS on Sundarbans reserve forest (Rahman, Md. Azizur, 2010) shows that the density of forest has gone down because of the excessive human activities and the climatic change. GIS study proposed to create a buffer zone and implementing eco-tourism plan in this region by which the GIS helps in planning and decision-making process.

GIS applications are found in areas involved in the following kinds of work (Cetin, Mehmet, 2016):

1. Processing and interpretation of digital images
2. Identification of species, habitats, and biodiversity
3. Detection, analysis, and evaluation of landscape structures (biotope, land use, ecological areas, etc.)
4. Protecting and monitoring of detected valuable ecological areas
5. Creation of maps of these studies

Eco-tourism development is a sensitive issue which requires proper planning and implementation strategy. GIS implementation can solve the complex issues of concrete bio-diversity, life and diversity of deep marine etc. To plan eco-tourism activities understanding and internalising natural and biodiversity is essential which the GIS can identify easily. Thus, resource identification and inventory resource management for tourism activities in an eco-tourism hub is possible through GIS application.

Visual Analysis

GIS provides visual analysis tools to analyse the impact on the earth virtually. 3D output of a destination can provide useful information about the resources planning and also plan for future hypothetically.

Implementation Strategy: The Competitive Advantages

Local administrative units and the related organisations act as the procurement of knowledge system. These administrative authorities produce the data and use specific management and distribution system through applications such as GIS. GIS accepts, internalises and output the result which further used for various purposes such as planning and policy making, product knowledge etc.

Figure 1. Conceptual Role of Tourism Knowledge System (Kim, 2002)

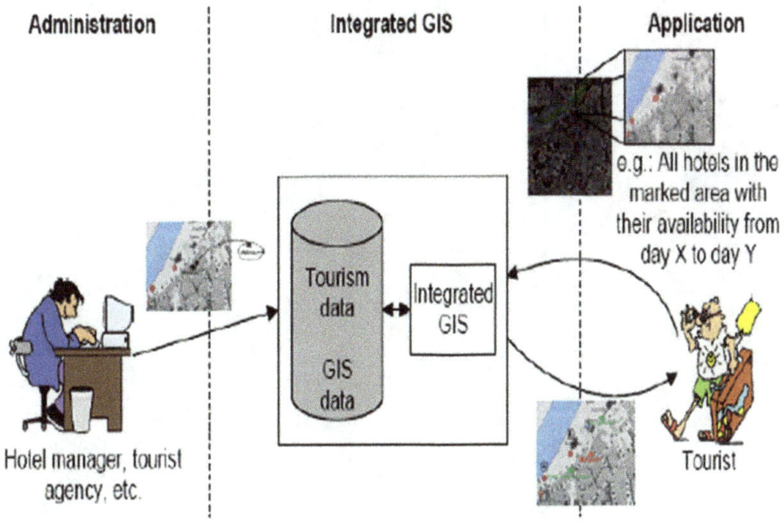

Above image shows the potential site for high eco-tourism suitable area. Based on the GIS the policy makers can judge the location for eco-tourism activities.

Virtual tour of eco-tourism associated with images, video presentation and audios help the tourists to have a better understanding of the site. Eco-tourism is a highly competitive sector which requires maximum utilisation of available technical resources to attract the niche segment. Government and travel organisations need to identify the eco-tourism buffer zone and develop it as a prime tourist hub which shall not deteriorate ecology.

While considering the eco-tourism development feasibility experts have identified five criteria and nine factors based on GIS based layers to determine the suitable area for eco-tourism development. It includes landscape/naturalness (visibility, land use/cover), wildlife (reservation/protection, species diversity), topography (elevation, slope), accessibility (proximity to cultural sites, distance from roads) and community characteristics (settlement size). This criterion helps in prioritising the developmental model and application of GIS for the better presentation of eco-tourism hubs. For instance, it is useful for the hikers and trekkers to decide the season and time of visit, elevation, vegetation and the route-mapping.

Brihat Bangalore Mahanagara Palike (BBMP) has adopted GIS mapping to divide the whole city into various sub divisions (B1). This helps in easy accessibility to the various locations in and around the city and also for the smooth administrative

Figure 2. GIS map showing eco-tourism potentials in Kwale Country (Wanyonyi, Jeremiah Waswa, 2016).

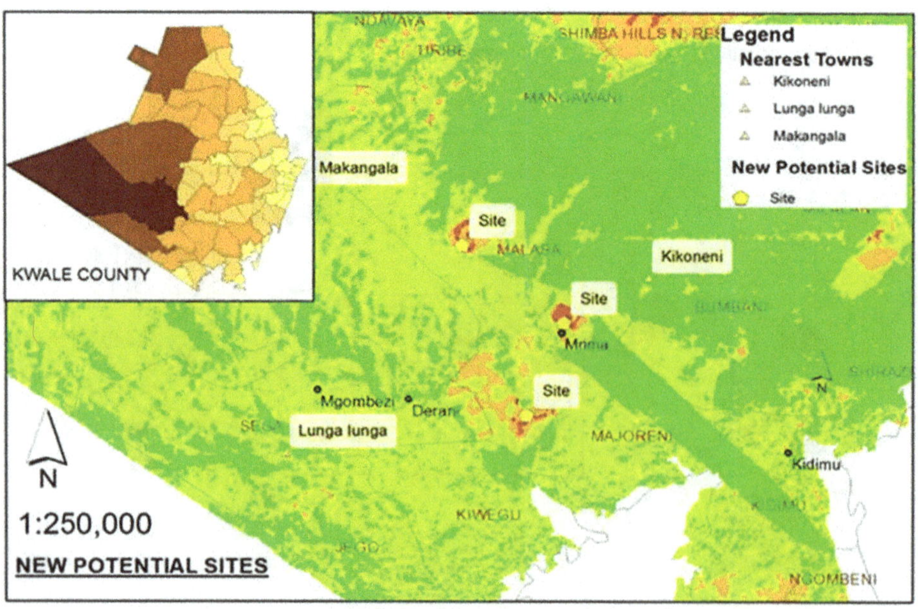

Figure 3. B1. Bangalore South Zone on GIS Mapping (Courtesy: BBMP)

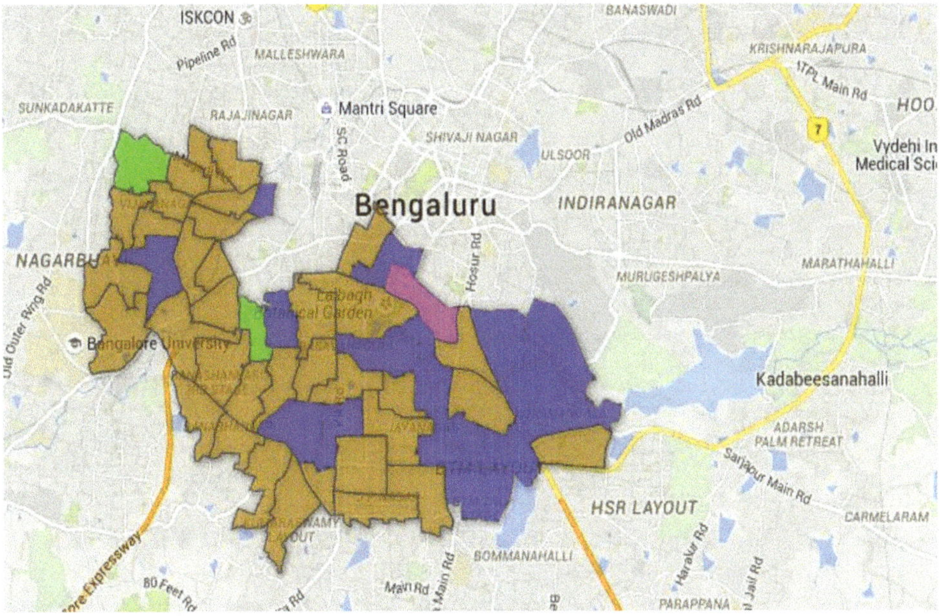

mechanism. BBMP has also created a GIS property identification map (B1) which is the watch guard for revenue collection for the administration of this huge metropolitan city.

Tourism Web GIS through Bhuvan facilitates the users to select the city/tourism places of their interest and find the places of interest, search nearby places along with routeing. Department of tourism of Karnataka also initiated GIS mapping to locate the tourist attractions in and around Hampi, the UNESCO heritage site of Karnataka (H1)

SUGGESTIONS

- Involvement of tourists, local people and the local community in eco-tourism activities. Bringing these segments under GIS application criteria.
- There is larger scope for improvement of mobile-based GIS application in eco-tourism developments
- Integration of regional and national eco-tourism organisations to develop a suitable eco tourism hub
- Technological advancement for pop-up – helpline for the browsers and tourists

Figure 4. B2. BBMP GIS mapping for revenue collection

Figure 5. H1. Ariel view of Hampi (Courtesy: DOT, Karnataka)

- Creation of spatial database for ecological awareness and resource sustainability
- Creating and developing GIS application and linking individual end-user to the eco-tourism institutions/geological departments, through that attaining individualistic attention towards ecological conservation.
- Use of remote sensing and GIS in forecasting the physical and environmental changes and to take necessary corrective measure.
- The implication of GIS in high priority area such as forest conservation zones, forest fire zones, coastal bio diversity zones etc.
- Linking GIS with social network sites for better implementation of GIS objectives.
- Establishment of GIS-Ecotourism regulatory authority to frame policy and implementation programmes.
- The introduction of Multiple menu GIS – which offer various eco-tourism products such as eco-resorts, jungle lodges, forest walks and safaris, authentic food etc.

CONCLUSION

GIS is technology based geographic application system which plays a vital role in global socioeconomic and environmental sustainability. Eco-tourism is a sustainable developmental practice where the GIS can be implemented to fulfil its objectives. Effective implementation of GIS and involving various stakeholders in strategic management can boost the sustainable tourism across the world. This would result in balanced ecological growth and attain economic stability across the globe.

REFERENCES

Sudarma, A., Piarsa, N., & Buana, W. (2013). Design and Implementation of Geographic Information System on Tourism Guide Using Web-Based Google Maps. *International Journal of Computer Science Issues*, *10*(1-2).

Henry, A. (2009). *Using Google Earth for Internet GIS. Institute of Geography School of GeoSciences*. University of Edinburgh.

Anshu, A. (2017). *Geographic Information System and Ecotourism Sites Identification of Jamui District*. Bihar, India: Academic Press.

Shyti, B., & Kushi, E. (2012). The impact of GIS application in the tourism development of Elbasan region. *The Romanian Economic Journal, 15*(45).

Mennecke, B. E., & Crossland, M. D. (1998). Geographic Information Systems: Applications and Research Opportunities for Information Systems Researchers. *Proceedings of the 1996 Hawaii International Conference on System Sciences, HICSS-29*, 1060-3425.

Lee, B. S., & Buckley, D. J. (1992, June). Forestry Canada Applies GIS Technology to Forest Fire Management. *Earth Observation Magazine*.

Chang, K. (2007). *Introduction to Geographic Information System* (4th ed.). McGraw Hill.

Shahrak, F., Miri, G.R., & Esmailnejad, M. (2015). Eco-tourism zoning in Sistan and Baluchistan by using GIS. *Зборник радова ДГТХ, 44*(2), 160–171.

Jeffrey, S., & Estes, J. (1990). *Geographic Information Systems: An Introduction*. Englewood Cliffs, NJ: Prentice-Hall.

Wanyonyi, J. W., Imwati, A., & Boitt, M. (2016). GIS In Analysis of potential Sites For Ecotourism – A Case Study of Kwale County. *IOSR Journal of Environmental Science, Toxicology and Food Technology, 10*(10, 43-49.

Foote, K. E., & Lynch, M. (1995). *Geographic Information Systems as an Integrating Technology: Context, Concepts, and Definitions*. University of Texas.

Kim, D. (2002). GIS *for a Policy Decision Support in National Tourism Portal*. Retrieved April 10, 2010, from http://proceedings.esri.com/library/userconf/proc04/docs/pap1277.pdf

Kumari, S., Behera, M. D., & Tewari, H. R. (2010). Identification of potential ecotourism sites in West Distric: Sikkim using geospatial tools. *Journal of Tropical Ecology, 51*(1), 75–85.

Cekule, M. (n.d.). *GIS based interactive ecotourism and environmental information system in Internet*. Retrieved from www.sitna.ef.navara.es

Mazaher, M. (2010). Siting MSW landfill using weighted linear combination and analytical hierarchy process (AHP) methodology in GIS environment (case study: Karaj). *Waste Management (New York, N.Y.), 30*(5), 912–920. doi:10.1016/j.wasman.2010.01.015 PMID:20138748

Cetin, M., & Sevik, H. (2016). *Assessing Potential Areas of Ecotourism through a Case Study in Ilgaz Mountain National Park*. Business, Management and Economics. doi:10.5772/62573

Sousa, P. R., & De Brito Fernandez, S. (2008). *Geographic Information Systems in Tourism Management*. 4th IASME/WSEAS International Conference on Energy, Environment and Ecosystem and Sustainable Development (EEESD'08), Algarve, Portugal.

Rannveig, O., & Micael, C. (2009). A GIS approach to evaluating ecological sensitivity for tourism development in fragile environments: A case study from SE Iceland. *Scandinavian Journal of Hospitality and Tourism*, *9*(1), 22–38. doi:10.1080/15022250902761504

Ulmasova, S. (2011). *GIS and Remote Sensing Class*. NRS509.

Jovanovic, V. (2008). The application of GIS and its components in tourism. *Yugoslav Journal of Operations Research, 18*(2), 261-272.

Wei, W. (2012). Research on the Application of Geographic Information System in Tourism Management. *Procedia Environmental Sciences*, *12*, 1104–1109. doi:10.1016/j.proenv.2012.01.394

Mohammad, Z. H., & Ujang, U. (2012). Integrating multiple criteria evaluation and IS in Eco tourism: A review. *The International Archives of the Photogrammetry, Remote Sensing and Spatial Information Sciences*, *42*W1).

Chapter 4
Geographical Patterns in the Tourist City:
GIS for Spatiotemporal Analysis

Luis Encalada
University of Lisbon, Portugal

Carlos C. Ferreira
University of Lisbon, Portugal

Jorge Rocha
University of Lisbon, Portugal

Inês Boavida-Portugal
University of Groningen (RUG), The Netherlands

ABSTRACT

Understanding how tourists use space in urban environments has practical applications in destination management and planning. It reveals key issues for organizing facilities and essential services, creating new products, impact management, and other purposes related to urban tourism. The geotagged photos shared by Panoramio users, during their visit to the city of Lisbon between 2007 and 2014, allow the authors to incorporate into this chapter a quantitative and geographic reading on the tourist consumption of spaces within the city. In this chapter, they analyze the spatial distribution of tourists and its changes across time considering a period of 8 years. Furthermore, a regression analysis method was carried out in order to find the spatial relations between the observed pattern (geographical agglomeration of tourist's photos) and a set of 24 selected independent variables.

DOI: 10.4018/978-1-5225-5088-4.ch004

INTRODUCTION

Alongside the development of new information and communication technologies, less traditional data sources have been considered in the analysis of urban tourism, particularly for assessing the spatial behavior of tourists. Photo-sharing and instant messaging services, consumer review websites and other Web 2.0 platforms are examples of these new sources, characterized by the prominent participation of users, a high volume of user-generated content and its open access nature (Buhalis & Law, 2008; Herrera et al. 2015).

In the context of geographical information, the appearance of these new platforms came to expand the set of available sources, overpassing the condition in which the information was produced and distributed exclusively by official authorities - authoritative data. The social media and, in particular, social networks, are increasingly equipped with positioning and mapping tools.

Several studies have been led using data extracted from social networks platforms, as these services provide open and well-defined interfaces to interact with. They allow clients to extract the increasingly user-generated content, often with high spatiotemporal resolution, enriched with a set of other attributes such as geolocation, timestamps, user identification, individual profiles, etc.

Such crowdsourced data provides new insights and could be used to complement authoritative data. Considering the constraints about collecting data for urban tourism studies, some authors concluded that this information can be used as proxy to measure the attractiveness of places as well as the spatial distribution of tourists in the city (MacKay & Vogt, 2012; Tussyadiah, 2012). Content created by users (as opposed to top-down methodologies) has promoted individuals as generators of information (with high spatial and temporal resolution), expanding current alternatives for tracking their location. The tourists when they use their cell phones, their credit cards, or through access to social networks, they leave large amounts of digital tracks on their activities in a given target, resulting in a multidimensional data set known as "Big Data" (Buhalis & Amaranggana, 2014; Hawelka et al., 2014).

In this chapter, we analyze the spatial distribution of tourists in Lisbon based on data from 'Panoramio' social network. First, the data was retrieved from 'Panoramio', filtered and merged into a relational database. Then, the resulting features were mapped and time-space statistics were applied to find the tourists hot spots. Furthermore, a regression analysis method was carried out, in order to find the spatial relations between the observed pattern (geographical agglomeration of tourist́ s photos) and a set of 24 selected factors (explanatory variables).

SPATIO-TEMPORAL ANALYSIS OF TOURIST'S GEOTAGGED PHOTOS

Study Area

Located on the north side of the metropolitan area of Lisbon (LMA), the city of Lisbon has an area of 85 km2 and 547,773 inhabitants (INE, 2011).

According to Tourism statistics of the National Institute of Statistics, *i.e.* Statistics Portugal, LMA has 19.4% of the total beds capacity of the country, where it stands out the municipality of Lisbon with 43,505 beds. This value finds no parallel in the context of the LMA and is only surpassed at a national level by the municipality of Albufeira (Algarve) which features 47,685 beds (INE, 2014).

With regard to tourism demand, the LMA concentrates 24.95% of total overnight stays in the country. The region continues to be among the first choices of the major external markets, being the main destination for Spaniards, Brazilians, Americans and Italians. Concerning the internal market, is the second destination of choice for nationals (with 20.2% of total overnight stays). Lisbon is country city which boasts the largest number of overnight stays and guests, with more than 3.5 and 9 million, respectively (INE, 2014). The average stay of foreign guests in the city, is of 2.5 nights, below the average of LMA (2.6) and of Portugal mainland (3.1 nights).

According to the motivational inquiry conducted in 2014 by Tourism Observatory of Lisbon (TOL), both at the foreign and nationals level, 'City & shortbreak' is considered the largest motivation for visiting Lisbon. The city of Lisbon stands out as one of the favorite destinations of urban tourism in Europe, having been awarded with several international prizes.

Some areas of the Lisbon present better features for tourism activities. In this context, it is presented the main tourist areas (Figure 1) of the city, also called as micro-centralities (MC), following the proposed zoning in the Strategic Plan for Tourism in Lisbon Region (SPTLR) 2015-2019.

According to the inquiry of tourist's activities held in 2014 by TOL, for about two-thirds of those surveyed the city of Lisbon was the only travel destination (64.1% of these First-comers). Visiting monuments, attractions or museums, was the most practiced activity, mainly within the "First-comers" (81.3%), in comparison with the repeaters (66.3%).

Looking at the results of Table 1, one can note that there is a predominant interest for locations within the perimeter of the city, being the most visited, located in the MCs Belém and Baixa-Chiado. The importance of the historical-monumental nucleus is clear, as well, in which the Belém tower and Saint Jorge Castle, are the most visited attractions. In fact, the MCs shape the space experienced by the tourists. The

Figure 1. Tourism Micro-centralities in Lisbon (SPTLR, 2015)

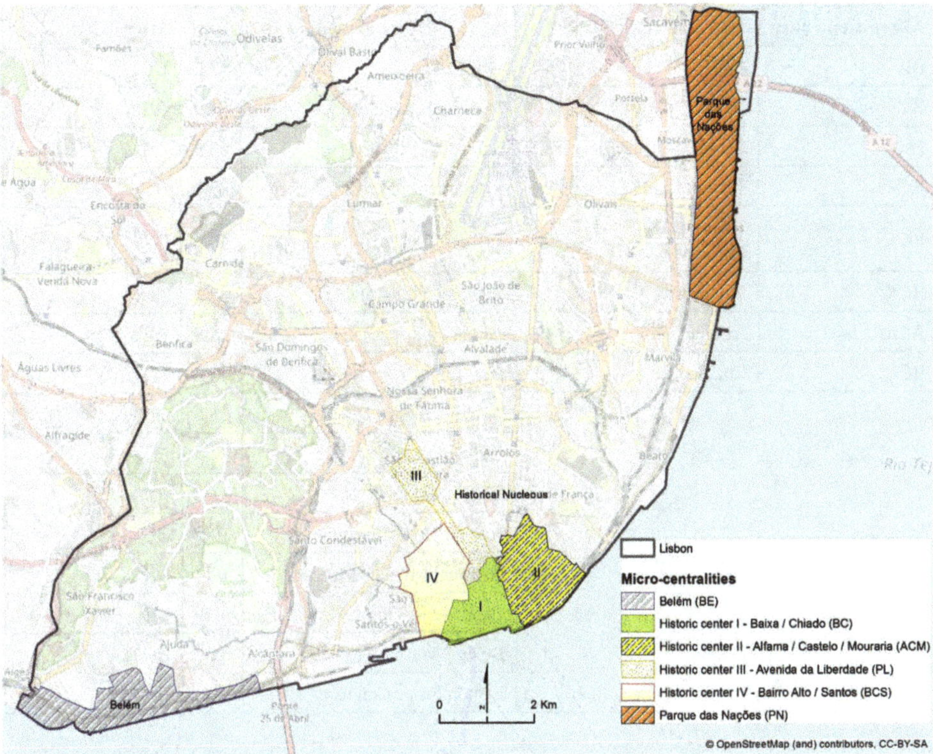

intensity of tourist activity overall the city extends from South to North, following the Riverside, between Belém and the Comércio Square, passing by the axis from Baixa to Avenida da Liberdade, and occasionally in the zone of Parque das Nações.

Data

Retrieving Process and Data Pre-Processing

In order to retrieve the photos from the Panoramio network the study area was segmented into smaller areas. For each of these sub-areas, we run a python API to extract the information. First, the study area was segmented into a mesh of 1000m x 1000m, resulting in 49,048 records. Then a 500m x 500m mesh was tested, resulting in 55,229 records. This process was repeated with increasable smaller meshes (in a total of 22 iterations) until the number of retrieved photos stabilized. The final database count with more than 70,000 records.

Table 1. Places of interest and attractions visited (TOL, 2014)

Micro-Centrality	Places of Interest	%	Places Visited	%	Micro-Centrality
BE	Belém	79.8	Belém Tower	87.1	BE
BC	Baixa	74.9	Jerónimos Monastery	71.9	BE
PN	Parque das Nações	72.6	Saint Jorge Castle	63.5	ACM
BC	Terreiro do Paço	72.1	Monument of the Discoveries	63.1	BE
PL	Avenida da Liberdade	70.3	Lisbon Oceanarium	40.6	PN
BCS	Bairro Alto	67.2	Lisbon Cathedral (Sé)	37.1	ACM
ACM	Alfama	64.8	Ajuda Palace	28.8	BE
BC	Chiado	57.0	Pavilion of Knowledge	23.8	PN
			Belém Cultural Centre	21.6	BE
			Gulbenkian Museum	14.7	PL
			Lisbon Casino	11.8	PN
			Ancient Art Museum	10.5	BCS
			National Museum of Cars	10.4	BE
			Chiado Museum	8.6	BC
			Design Museum (MUDE)	2.8	BC
			Lisbon Story Centre	2.2	BC

The method to differentiate the photos taken by visitors from the ones taken by locals is based on the work carried out by Girardin et al. (2008), Kádár (2014) and García-Palomares et al. (2015). The authors calculated the difference (in days) between the timestamp of the first and the last photo uploaded by each user. If the resulting number exceeded the average visit time for the studied place, the photos were considered as belonging to locals, otherwise the photos were catalogued as belonging to visitors.

The resulting dataset contained 17,604 photos taken by users considered as visitors. Spatial distribution of photographs is depicted in Figure 2. At first glance, some areas are clearly distinguishable from others, i.e. have a higher photos density.

Tourist's Photos Spatial Distribution

In the northeast of the city, the area of the 'Parque das Nações' (Figure 2) appears as an isolated hot spot, with a few photos on its neighborhood. Within 'Parque das Nações' the Oceanarium, the exteriors of the Vasco da Gama shopping centre and the Vasco da Gama Tower are highlighted points. In the inner city, emerges another

Figure 2. Photos taken by tourists and uploaded on 'Panoramio' between 2007 and 2014

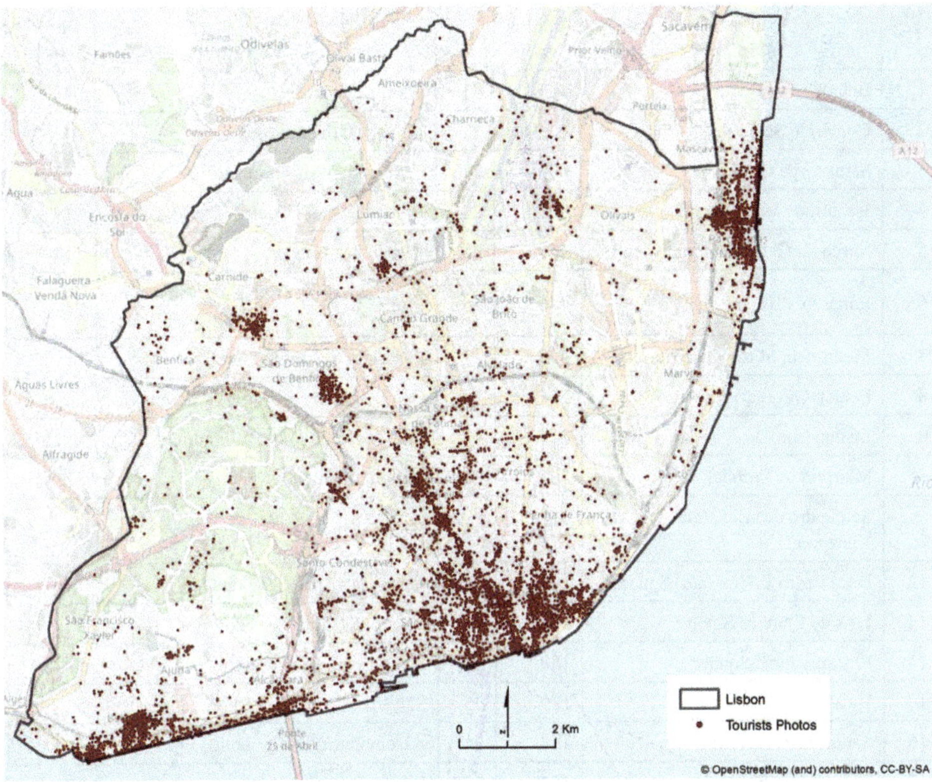

focus of photographs around the Airport of Lisbon, with photos near the building and on the runways.

The Belém Tower and the Comércio Square are, by far, the attractions that stand out (Table 2). The amount of pictures associated with these sites is very high, even when compared with other touristic sites within the city. However, if we count both values together, the two places, represent little more than 13% of total (tourists) photographs.

It is also evident the presence of photos out of the touristic areas (outside touristic MCs in Lisbon). Looking at the spatial distribution of these groups of photos, they tend to be more dispersed than the photos located within the MC areas. Of course, as these records are photos generated by tourists, in residential sectors, especially on the northeast side of the city, there are locations with total absence of records.

It turns out that, with the exception of the two football stadiums (Luz and José Alvalade) and the Marquês de Fronteira Palace, most of the sites are enclosed by the tourism micro-centralities.

Table 2. Number of photos taken on places of tourist interest (for functional reasons are only represented the locations with more than 45 photos).

ID	Place of interest	Photos (nº)	ID	Place of interest	Photos (nº)
1	Belém Tower	1,283	21	Glória Elevator	143
2	Comércio Square	1,124	22	Lisbon Cathedral (Sé)	141
3	Saint Jorge Castle	807	23	Gulbenkian Museum/garden	137
4	Jerónimos Monastery	801	24	Belém Cultural Centre	119
5	Vasco da Gama Shopping centre	479	25	Parque das Nações	114
6	Eduardo VII Park	446	26	River axe (Eixo Ribeirinho Alcântara - Belém)	92
7	Monument of the Discoveries	395	27	Portas do Sol Viewpoint	91
8	Lisbon Oncenarium	288	28	Campo Pequeno Bullfight Square	88
9	Lisbon Zoo	244	29	José Alvalade Football Stadium	85
10	Marquês de Pombal	207	30	National Pantheon	77
11	São Pedro de Alcântara Viewpoint	201	31	Fighter Museum	76
12	Dom Pedro IV (Rossio) Square	199	32	Santa Catarina Viewpoint	72
13	Luís de Camões Square	188	33	Vasco da Gama Tower	60
14	Restauradores Square	180	34	Santa Luzia Viewpoint	52
15	Bica Elevator	179	35	São Vicente de Fora Church	52
16	Santa Justa Elevator	168	36	Assembly of the Republic	51
17	Graça Church	167	37	Lisbon Botanic Garden	50
18	Rua (Street) Augusta	160	38	Santíssimo Coração de Jesus Convent	47
19	Luz Football Stadium	157	39	Marquês de Fronteira Palace	46
20	Império Square	145			

Temporal Analysis

From the temporal analysis of the data (Figure 4), one can see a downward trend in the number of photographs from 2,988 in 2007 to 862 in 2014 (Photos from year 2006 were excluded for temporal analysis). Furthermore, in 2014 the decrease was approximately 1,000 photos. Thus, the reduction of the volume of photos comes from the year 2013. This same trend is checked when analyzing the original dataset, with both the photos of locals and tourists. From 2007 to 2012 there is an increasing trend, but between 2013 and 2014, the trend reverses itself, passing from 12,949 to 6,005 photographs.

Regarding the number of users catalogued as visitors, with the exception of the amount registered in 2011, from 2007 to 2013, their number decreases continually. These results, of course, disagree with the numbers of tourism demand in Lisbon, and should not be interpreted as a reduction in the amount of current tourists, but only as a decrease in the proportion of users who shared their experience through this particular social network.

The distribution of photos along the time series is not regular (Figure 4), particularly in years 2007 and 2013, where the values deviate considerably from the annual average of 2,355 pictures (the average was calculated including the values from 2008 to 2013).

In many cases, the monthly evolution of the number of photos for each year is different. The values for the same month change significantly from one year to another (*e.g.* February, March, April, November, December). The last year presents some divergences regarding the monthly breakdown. In addition to the absence of records in the last four months in 2014, it presents the lower values of the data series in the months of April, May and June (Figure 4).

The user's monthly evolution (Figure 5), it is not similar to the number of photos. The number of users keep steadier, at least in the first seven months (presenting

Figure 3. Data taken from 'Panoramio', 2008-2014

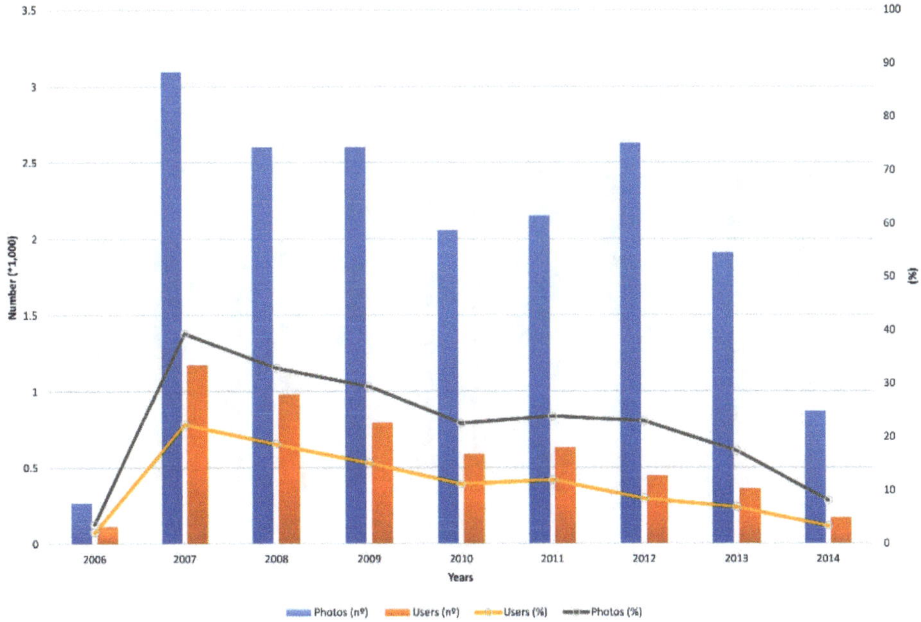

Figure 4. Monthly tourist's photos variation

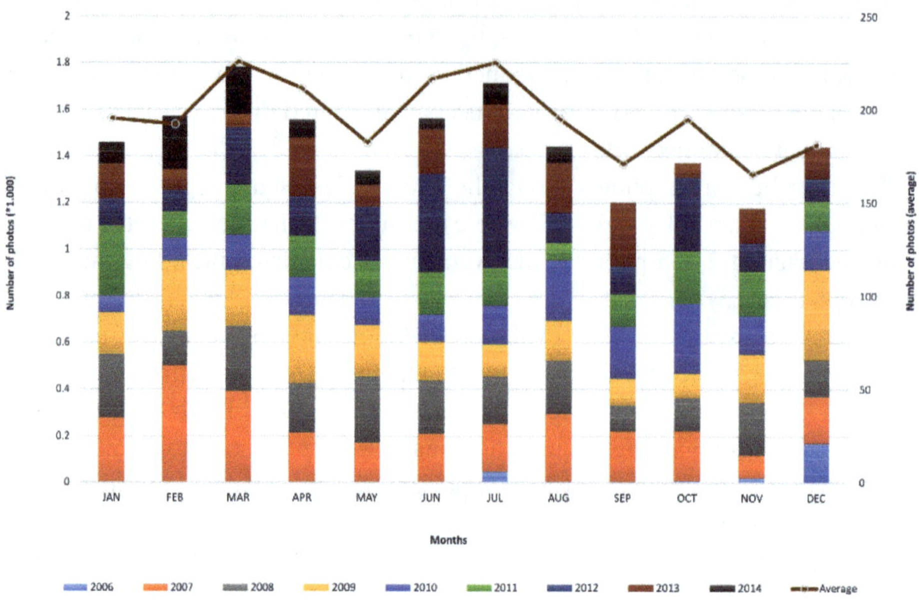

Figure 5. Monthly user's variation

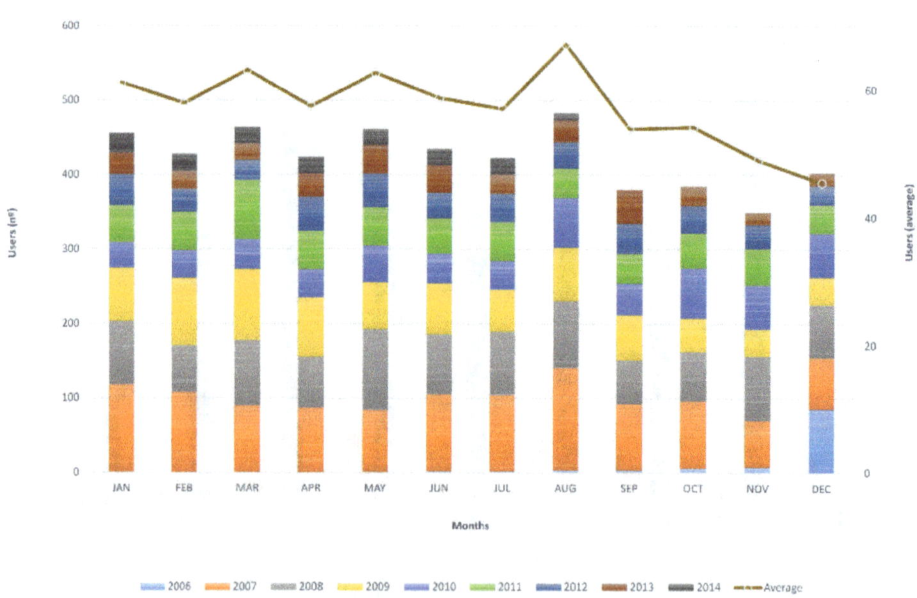

average values between 57 and 63 users). On average, there is a rise in August up to 67 users and in December the value falls to 45.

The trend in the number of tourists who upload photos does not differ from the rates of tourism seasonality projected by hotel occupancies. When compared, there are a similar growing rates in the same months as occupation increases. The correlation (R^2) between these two variables is 0.67 which can be considered satisfactory.

Spatial Analysis

Analysis by Micro-Centralities

Taking into account the micro-centralities proposed in the SPTLR, one can have an idea of tourist's preferences regarding some places of the city, viewed as areas of priority interest. In fact, 75% of the photos were taken within areas defined as micro-centralities. This amount is divided between the historic centre, the main focus of attraction with more than 40% of the total number of photos, Belém (about 20%) and Parque das Nações with 12% (Figure 6).

Despite the predominance of the historic centre, the distribution of the photographs within its limits is geographically asymmetric and can clearly be noted the predominance of two subareas over the others. In this context, MC I (Baixa-Chiado) and MC II (Alfama/Castelo/Mouraria) with, respectively, 35% and 32% of the total number of photos taken in the historical centre, differ from the others, MC III (Avenida da Liberdade) with 16% and MC IV (Bairro Alto/Santos) with 15%.

Figure 6. Distribution of photos by micro-centralities

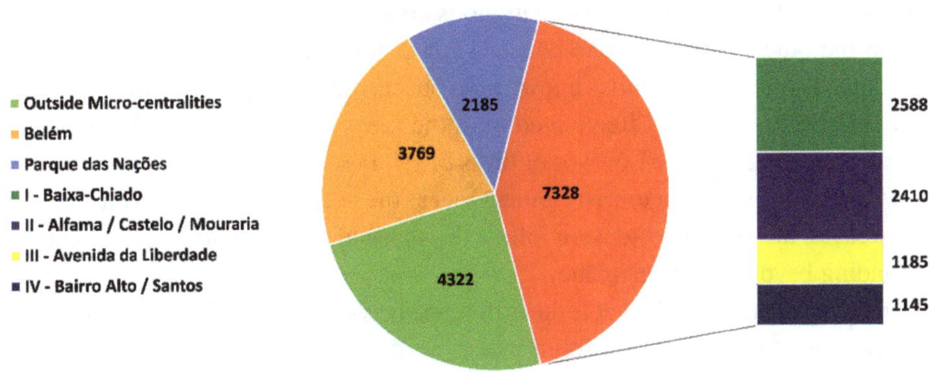

Clusters and Outliers Analysis

The method of cluster and outliers analysis is based on the difference between the values associated with the entities and spatial relationship between them. Calculating the Moran Local Index, one compares the value (number) of each entity or target features with the values of its neighbors, and the process is performed for each unit, continuously, until all the data is analyzed (Mitchell, 2005).

The neighborhood is implemented through spatial weights assigned to the neighbors in accordance with the nature of spatial relationship between entities (*e.g.* spatial weights or proximity matrixes).

Following Tobler's first Law of Geography, stating that everything is connected to everything but near things are probably more connected that distant ones, one choose the Inverse distance as the method to conceptualize the relationship between spatial observations. Thus, the influence of the neighboring features will decrease as the distance between them increases. Assuming that not all the observations within the study area are spatially related, it was defined as a threshold for the neighborhood radius of influence of 135m (this value was obtained throughout spatial autocorrelation measures).

As expected, the clusters are located in the tourist areas of the city, and the neighborhood of the most relevant attractions. In the previous sections when analyzing the absolute number of photos associated with the places of interest, were the well-known attractions of the city that stood out most. However, this approach did not include important aspects in the analysis, as the configuration of the local neighborhood or the statistical significance of the data.

Looking at Figure 7, the more salient fact has to do with almost all of the spatial clusters are of category HH (High-High), *i.e.* they are places where high values are surrounded by similar values (high). Therefore, there is a large turnout of visitors, both to these locations and in its contiguous areas.

On the other hand, only one cluster is of different category, LH (Low-High), and refers to a location that has a low value but surrounded by high values. So, this particular location is less frequented than the areas surrounding it. There are no clusters of type LL (Low-Low) or outliers of HL type (High-Low).

In fact, it turns out that tourist elements act as the main focus of spatial clusters. In many cases, the clusters extension follow the shape of the attractions (*e.g.* squares), expanding beyond their perimeter.

In order to understand what is the difference between spatial clusters with regard to their magnitude the Local Moran Index was calculated (Figure 8).

The higher values of the index are associated with locations on or in the neighborhood of the attractions (*e.g.* the monuments of the Belém Tower, monument

Figure 7. Clusters and outliers of tourists photos in Lisbon

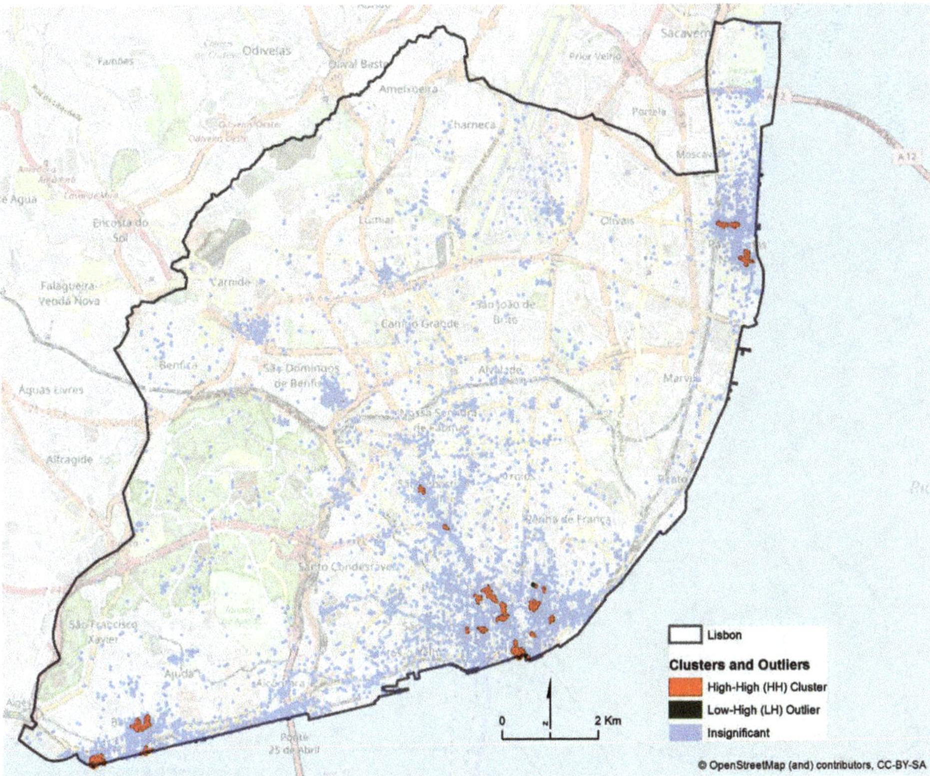

of the discoveries and Marquês de Pombal). In the case of the Jerónimos Monastery, and the Lisbon Oceanarium, the more differentiated values correspond to places located on the front, in the area near the entrances.

With regard to the clusters on the squares (*e.g.* 'Restauradores', 'Dom Pedro IV', 'Comércio' and 'Luís de Camões' squares), the most salient values of the index tend to be closer to the center.

Multiple Linear Regression

The first fact to consider in the implementation of a linear regression model is the selection of the explanatory variables that are going to be used. One additional independent variable can improve the prediction of the dependent variable so they must be chosen carefully. In this case, 24 variables were selected as independent variables (Table 3), *i.e.*, as explanatory factors of geographical agglomeration of photos.

Figure 8. Local Moran Index from cluster and outliers analysis

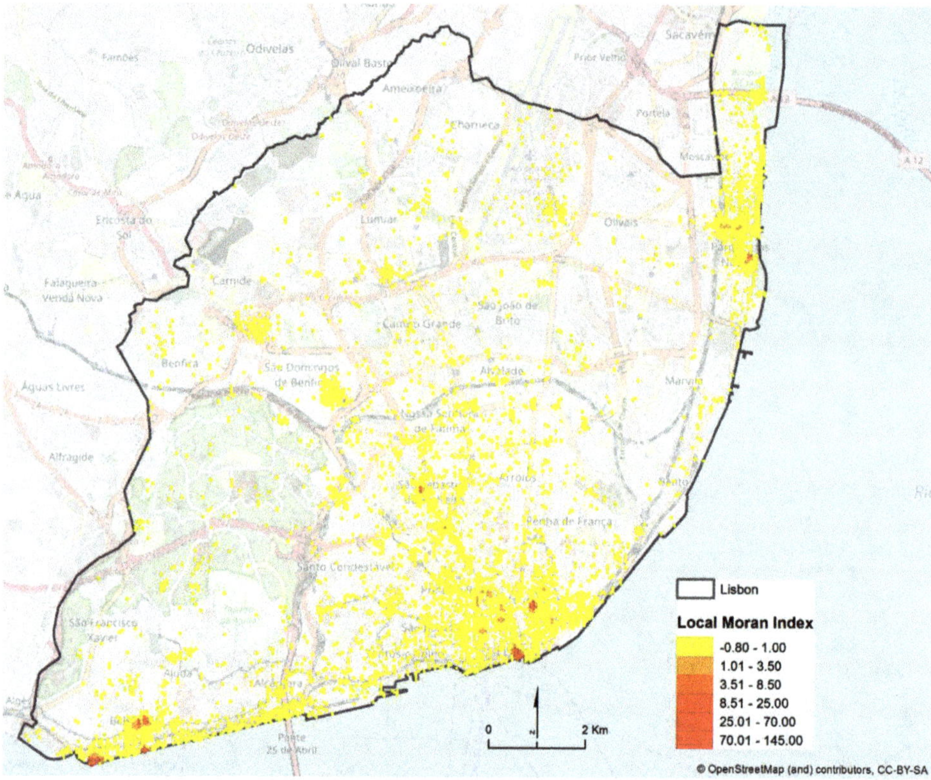

Table 3. Pre-selected explanatory variables

Code	Variable	Code	Variable
V1	Civil Architecture	V13	Hospitality
V2	Industrial Architecture	V14	Churches
V3	Military Architecture	V15	Markets
V4	Noble Architecture	V16	Viewpoints
V5	Award-Winning Architecture	V17	Monuments of Public Interest
V6	Religious Architecture	V18	National Monuments
V7	Cemeteries	V19	Museums
V8	Blocks of public interest	V20	Playgrounds
V9	Recreational docks and Marinas	V21	Picnic Parks
V10	Elevators and Lifts	V22	Recreational Parks
V11	Fairs	V23	Urban Parks
V12	Geo-monuments	V24	Theatres

Spatially, all these variables were represented through point layers. All variables were transformed into reason scales, by applying the Inverse Euclidean distance to the parsed element. As the unit of measurement is always the same (meters) it was not necessary to standardize the values. In order to illustrate the variables transformation, the map with the Inverse Euclidian distance to the features of the variable ' viewpoints ' (V16) is presented in Figure 9.

The predictive performance of a linear regression model is not only related to the correlation of the dependent variable with a specific (independent) variable, but also to the correlation of the former with all other independent variables of the model. Thus, one must assess the possible dependencies among all the variables.

In cases where these dependencies are strong, one can state that there is multicollinearity. The multicollinearity can be understood as the existence of two or more independent variables in a multiple regression analysis that are highly correlated. The selection and use of independent variables in the model must take into account its possible multicollinearity

Figure 9. Inverse Euclidean distance for the Viewpoint variable (For illustration the distance was classify in 5 intervals)

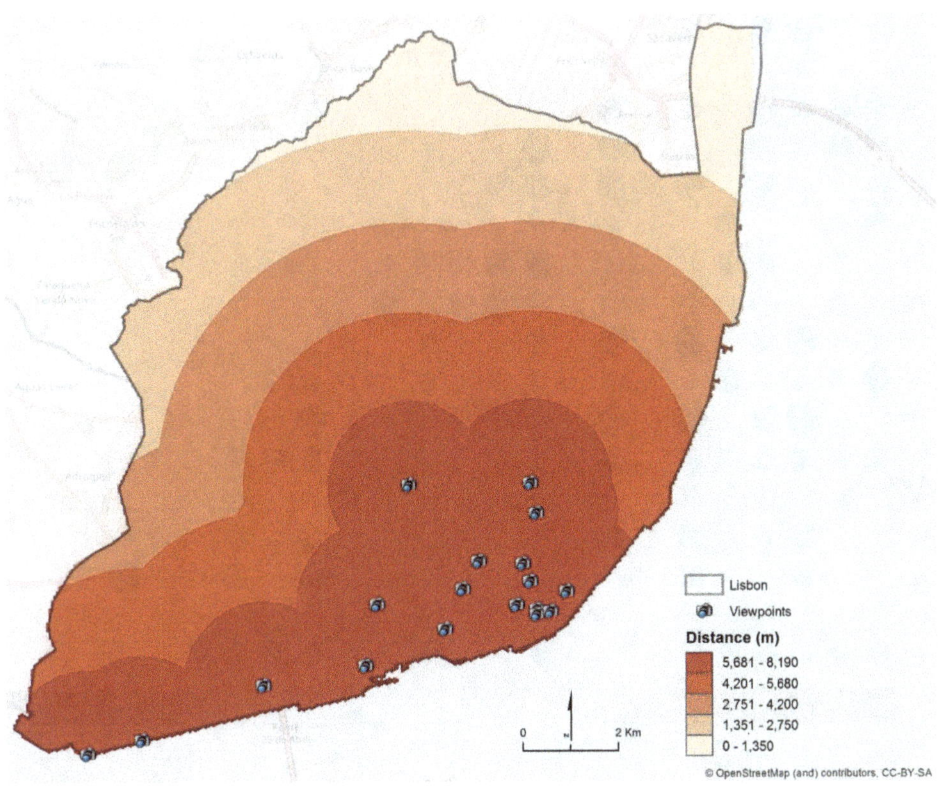

A classic case of multicollinearity occurs when none of the explanatory variables in the regression is statistically significant. As a result, it is said that two or more explanatory variables are perfectly collinear if one or more variables can be expressed as a linear combination of the other variable(s).

As collinearity means that the independent variables are correlated, this can be detected, among other ways, through the array of correlation between them. So, the Pearson correlation coefficients between independent variables was calculated (Figure 10). Only variables with correlations $r > -0.85$ or $r < 0.85$ should be included in a regression model without the risk of high multicollinearity.

Figure 10. Pearson correlation coefficient between pre-selected variables

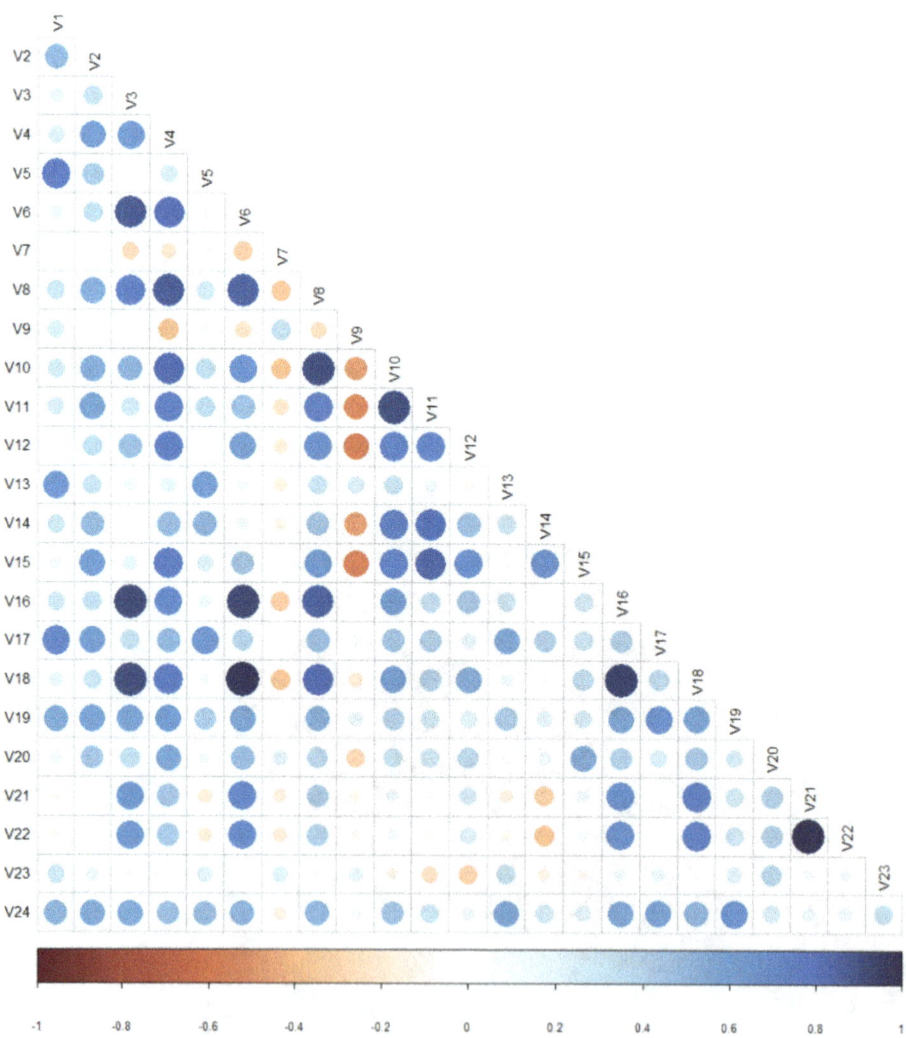

Analyzing the regression graph, it can be observed that strong negative relations predominate, whereas the positive relations are only moderate. The first case which requires special attention is the relationship between the variable V18 (National Monuments) and V3 (Military Architecture), V5 (Award-winning Architecture) and V16 (Viewpoints); the V16 with V3 (Military Architecture) and the V6 (Religious architecture); the V11 (Fairs) with the V10 (Elevators and Lifts); the V10 with the V8 (Blocks of Public Interest); the V8 with the V4 (Noble Architecture) and the V6; and the V6 with the V3 (Military Architecture).

The presence of multicollinearity can be detected by using the Variance Inflation Factor (VIF), whose high values indicate the existence of multicollinearity.

$$VIF = \frac{1}{1 - R_j^2}$$

Where R_j comes from the regression of x_j with all the other variables, *i.e.* the determinant coefficient of the regression equation.

The VIF is a measure of the degree to which each independent (temporarily transformed in dependent) variable is explained by all the other independent variables. The greater the variance inflation factor, the more severe will be the multicollinearity. Indeed, the VIF measures the redundancy between the explanatory variables. As a general rule, the explanatory variables with VIF values exceeding 7.5 should be removed (one by one) from the regression model. The Table 4 accounts for the VIF values for each of the 24 pre-selected variables. Of these, 12 were excluded for presenting VIF values above the defined threshold.

Complementarily, an exploratory analysis was performed in which the dependent variable was modeled against sets of 1 to 5 independent variables. Overall, 35,526 combinations were tested of which 6,866 presented an R^2 greater than 0.9. Within this narrower group, the 38 best performing models used always 5 variables, with predictive performances between 0.95 and 0.96. The Table 4 also identifies the variables that were most frequently selected in the testing models.

The cross-tabulation of VIF and weight fields, retrieved a final set of 7 independent variables to be used in the model (Table 5).

The R^2 and adjusted R^2 measures were used as performance indicators of the model. As final result the regression achieved values of $R^2 = 0.6828$ and adjusted $R^2 = 0.6827$, which means the model represents about 68% of the spatial variation of tourist's photos.

From MLR analysis, it is possible to identify the most important variables in explaining the spatial distribution of photographs and their importance (weight). The variables were the proximity to: Public Monuments (0.29); Recreational Docks

Table 4. VIF and frequency values

Code	Variable	VIF	Frequency
V1	Civil Architecture	5.39	0
V2	Industrial Architecture	5.87	0
V3	Military Architecture	20.85	0
V4	Noble Architecture	16.83	1
V5	Award-Winning Architecture	4.50	1
V6	Religious Architecture	45.08	0
V7	Cemeteries	3.50	0
V8	Blocks of public interest	104.93	0
V9	Recreational docks and Marinas	6.70	6
V10	Elevators and Lifts	114.61	29
V11	Fairs	22.37	0
V12	Geo-monuments	5.70	21
V13	Hospitality	6.16	33
V14	Churches	16.87	32
V15	Markets	10.31	8
V16	Viewpoints	58.24	2
V17	Monuments of Public Interest	3.96	1
V18	National Monuments	58.77	0
V19	Museums	6.25	0
V20	Playgrounds	4.14	3
V21	Picnic Parks	359.06	5
V22	Recreational Parks	442.87	5
V23	Urban Parks	2.99	0
V24	Theatres	4.27	1

Table 5. MLR final variables selection

Code	Variable	Weight
V13	Hospitality	0.10
V12	Geo-monuments	0.12
V9	Recreational docks and Marinas	0.21
V20	Playgrounds	0.03
V5	Award-Winning Architecture	0.15
V17	Monuments of Public Interest	0.29
V24	Theatres	0.10

and Marinas (0.21); Award-winning Architecture (0.15); Geo-monuments (0.12); Hospitality (0.10); and Theatres (0.10). However, more than the analysis of the factors and prediction model, what can be more interesting in terms of tourist planning is an analysis of the MLR model residuals.

Indeed, in MLR the assumptions of the adjusted model need to be validated for the results to be reliable. The residuals analysis is precisely a technique used to investigate the suitability of a regression model. The residual is given by the difference between the observed variable response and the variable estimated reply. The basic idea of residuals analysis is that, if the model fits, the residuals should reflect the properties of the model error, *i.e.*, positive residuals indicate results above the predicted by the model and negative residues indicate the opposite.

In practice, what residual mean is that, given the observed conditions, *i.e.*, explanatory variables, there are areas where the number of pictures is higher than expected (positive residuals) and others where there are fewer pictures than it would be expected (negative residuals). It is precisely those deficit areas that may require assistance in the framework of planning so that all their potential can be used.

In the Figure 11, Belém stands out with values above what its potential would justify. Unexpectedly, the downtown area (Baixa-Chiado), which although in all previous analyses appear as a major hot spot, is still below what it potentially could reach. The Parque das Nações area follows the same trend but not with a disparity as big. However, they are clearly two areas, two micro-centralities identified by the Office of tourism, which may be the subject of planning actions, with the aim of attracting more visitors.

CONCLUSION

Photographs taken by tourists make it possible to identify and analyze visual tourist attraction areas in the city. Most studies based on information from social networks have been limited to create products, *e.g.* maps and animations. The results show that the observed distribution patterns of the photos uploaded by tourists and residents share some similarities. However, the photographs of the tourists revealed higher spatial concentration than those uploaded by residents. The comparison of spatial clusters of photos with the main points of attraction indicate that the results are consistent, since the spatial clusters match the well-known tourist attractions.

The dataset used for analysis provides a robust sample that replicates the space experienced by Lisbon city tourists. The results of the statistical evaluation of the general (global) trend highlighted a predisposing for clustering, leaving aside any possibility that is genesis is a product of randomness.

Figure 11. Residuals of the Multiple Linear Regression model

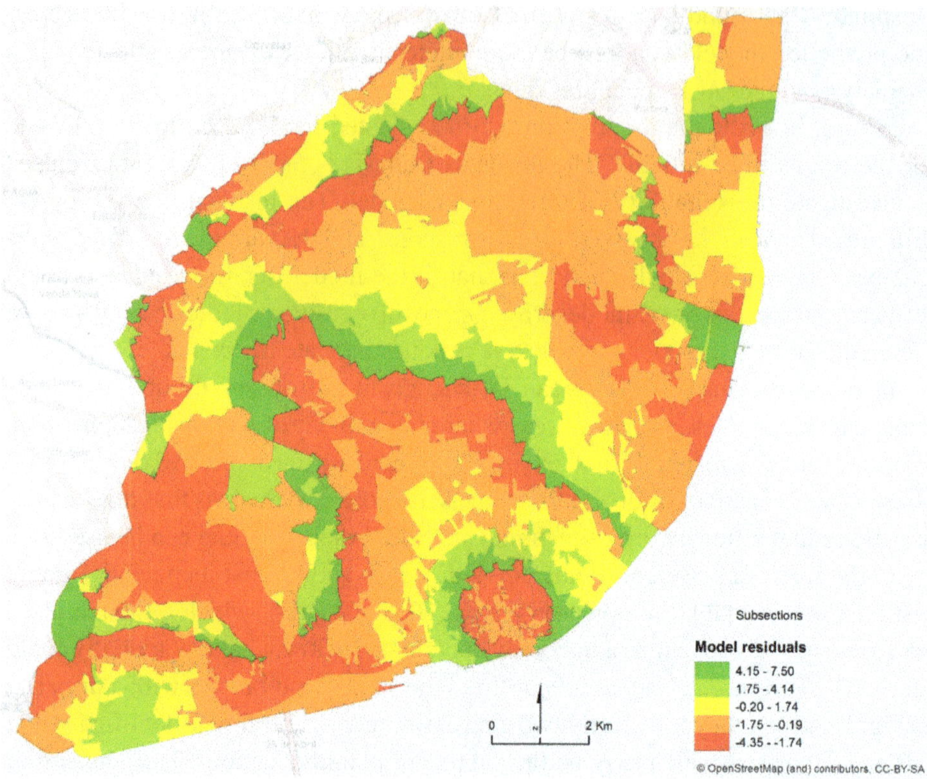

Touristic areas can be well identified by analyzing the concentration of tourist's photographs. Yet, some areas may not be regarded as tourist attractions despite presenting high potential for tourism activities. Thus, it is possible to disclose which areas are more attractive to visitors. Indeed, knowing which areas are under touristic pressure or not, contributes effectively to tourism management.

REFERENCES

Buhalis, D., & Amaranggana, A. (2014). Smart tourism destinations. *Proceedings of the international conference: Information and communication technologies in tourism 2014*, 553-564.

Buhalis, D., & Law, R. (2008). Progress in information technology and tourism management: 20 years on and 10 years after the internet - The state of eTourism research. *Tourism Management*, *29*(4), 609–623. doi:10.1016/j.tourman.2008.01.005

Entidade Regional de Turismo da Região de Lisboa. (2015). *Plano Estratégico para o Turismo na Região de Lisboa 2015-2019*. Available online: http://www.ertlisboa.pt/pt/institucional/plano-estrategico-para-o-turismo-na-regiao-de-lisboa-2015-2019/

García-Palomares, J. C., Gutiérrez, J., & Mínguez, C. (2015). Identification of tourist hot spots based on social networks: A comparative analysis of European metropolises using photo-sharing services and GIS. *Applied Geography (Sevenoaks, England)*, *63*, 408–417. doi:10.1016/j.apgeog.2015.08.002

Girardin, F., Fiore, F. D., Ratti, C., & Blat, J. (2008). Leveraging explicitly disclosed location information to understand tourist dynamics: A case study. *Journal of Location Based Services*, *2*(1), 41–56. doi:10.1080/17489720802261138

Hawelka, B., Sitko, I., Beinat, E., Sobolevsky, S., Kazakopoulos, P., & Ratti, C. (2014). Geo-located Twitter as proxy for global mobility patterns. *Cartography and Geographic Information Science*, *41*(3), 260–271. doi:10.1080/15230406.2014.890072 PMID:27019645

Herrera, A., San Martín, H., & Hernández, J. (2015). Perceived influence on behavior of user-generated content on social network sites: An empirical application in the hotel sector. *Revista Española de Investigación de Marketing ESIC*, *19*(1), 12–23. doi:10.1016/j.reimke.2014.09.001

Instituto Nacional de Estatística. (2011). *Censos 2011 Resultados Definitivos - Região Lisboa*. INE. Available in: http://censos.ine.pt/xportal/xmain?xpid=CENSOS&xpgid=ine_censos_publicacoes

Instituto Nacional de Estatística. (2014). *Estatísticas do Turismo 2014*. INE. Available in: https://www.ine.pt/xportal/xmain?xpid=INE&xpgid=ine_publicacoes&PUBLICACOESpub_boui=139601&PUBLICACOESmodo=2&xlang=pt

Kádár, B. (2014). Measuring tourist activities in cities using geotagged photography. *Tourism Geographies*, *16*(1), 88–104. doi:10.1080/14616688.2013.868029

Mackay, K., & Vogt, C. (2012). Information technology in everyday and vacation contexts. *Annals of Tourism Research*, *39*(3), 1380–1401. doi:10.1016/j.annals.2012.02.001

Mitchell, A. (2005). The ESRI guide to GIS analysis, Volume 2: Spatial measurements and statistics. ESRI Press. ISBN: 9781589481169

Observatório de Turismo de Lisboa (TOL). (2014). *Survey of tourism motivation – Lisbon City 2014*. OTL. Available in: http://www.visitlisboa.com/Observatorio.aspx

Observatório Turismo de Lisboa (TOL). (2014). *Visitor's activities and information survey in Lisbon Region 2014*. Available in: http://www.visitlisboa.com/Observatorio.aspx

Plano Regional de Ordenamento do Território da área Metropolitana de Lisboa. (2009). *Diagnóstico Sectorial Sistema Urbano, Habitação e Equipamentos Constitucional. Comissão de Coordenação e Desenvolvimento Regional de Lisboa e Vale do Tejo (CCDR LVT)*. Accessed in June, 2017. Available in: http://www.ccdr-lvt.pt/pt/plano-regional-de-ordenamento-do-territorio-da-area-metropolitana-de-lisboa/54.htm

Tussyadiah, L. (2012). An assessment of contagion on social networking sites. *Proceedings of the international conference: Information and communication technologies in tourism 2012*, 25-35. doi:10.1007/978-3-7091-1142-0_3

KEY TERMS AND DEFINITIONS

Big Data: Data sets that are so large or complex that traditional data processing techniques are inadequate to deal with them. The term often refers to data capturing, storage, analysis, search, sharing, transfer, visualizing, querying, updating, etc. Recently, the term leans towards the use of predictive, user behaviour or certain other advanced data analytics methods that extract value from data.

Data Mining: Computing process for discovering patterns in large data sets including methods of machine learning, statistics, and database systems. It is a process where intelligent methods to extract data patterns are applied. It is an interdisciplinary subfield of computer science.

Geotagged Photos: A photo associated with a geographical location by geotagging. Usually done by assigning at least a latitude and longitude to the image. Other information fields may be included.

Open Data: Freely available data to everyone to use and republish as they wish, without restrictions from copyright, patents, or other control mechanisms. The objectives are similar to those of other "open" movements such as open source, open content, and open access.

Spatial Autocorrelation: Spatial autocorrelation is a method of exploratory spatial data analysis. Is the correlation between the values of a single variable that is rigorously due to the proximity of these values in geographical space by introducing a deviation from the assumption of independent observations of classical statistics.

Spatial Statistics: Techniques that study entities using their topological, geometric, or geographic properties additionally to all the other features that characterize traditional statistics.

Spatiotemporal Analysis: Emerging research area due to the development and application of new computational techniques for the analysis of large spatiotemporal databases. Spatiotemporal models stand up from data collected across time and space and have at least one spatial and one temporal property.

Timestamps: Sequence of information identifying when a certain event occurred, usually giving date and time of occurrence. Nowadays the practice of the term has expanded to refer to digital date and time information attached to digital data.

User-Generated Content: Also known as user-created content. Any kind of content created by users of a system or service and made available publicly. Most frequently appears as add-ons to online platforms, such as social media, and may include blogs, wikis, videos, comments, or ecommerce.

Chapter 5
Reigniting GIS's Application in Ecotourism:
A Case Study of Sundarbans in Bengal

Sovik Mukherjee
Shri Shikshayatan College, India

ABSTRACT

The chapter brings out a brief note on the tourist attractions, hotels and lodges, NGOs/travel agencies operating in that region, railway/bus stations, land use profile, etc. in the Sundarban area of West Bengal in conjunction with exploring the potential of ecotourism using GIS and some secondary source data. Moving onto the analysis part, by making use of geo-spatial data, the attributes of ecotourism potential in the Sundarbans has been explored. The author makes use of the Euclidean distance mechanism and principal component analysis to rank the ecotourism sites in Sunderbans (i.e., based on the construction of ecotourism potential index [EPI]). The novelty of the chapter lies in comparing the ranks obtained by constructing the EPI following the principal component analysis and the Euclidean distance function. It needs to be mentioned here that these tourist spots have been selected based on the information collected on the inflow of both domestic and foreign tourists to these spots. The chapter concludes by discussing the future scope of research in this regard.

All you need is the plan, the road map, and the courage to press on to your destination.
— Earl Nightingale

DOI: 10.4018/978-1-5225-5088-4.ch005

INTRODUCTION

Geographic Information System (GIS) uses a convex combination of various disciplines which include geography, mathematics, statistics, computer science, management and mapping science. On the basis of collected geospatial data, the technology offers a dynamic analysis for not only research but serves as a decision making instrument in the context of urban planning, disaster prevention and mitigation, tourism and other commercial areas, etc. An advanced form of GIS in this context is — Web based Geographic Information System (WGIS). It is fast becoming an effective tool for tourism and has developed itself into a platform that encourages ecotourism by bringing in the notion of area assessment, service management, resource use management, visitor flow management, and assessing impacts of the development of tourism. Therefore, the prospect of GIS applications in ecotourism is noteworthy. The study brings out a brief note on the tourist attractions, hotels and lodges, NGOs/travel agencies operating, railway/bus stations to land use profile, etc. in the Sundarbans in conjunction with exploring the potential of ecotourism using GIS based data. It should be also noted that this study will frequently make use of satellite data on digital image classification.

To introduce the readers to the landscape of Sundarbans, the study area of the Indian Sundarbans is of 130 kms in length out of the total 180 kms of total coastal length of West Bengal.

The location of Sundarbans in the Indian part is between 21° to 22°-65/ N and 87° to 88°-93/E. Indian Sundarbans covers the areas of south-eastern part of both North and South 24 Parganas in West Bengal. Southeast extreme region of three districts namely Satkhira, Khulna and Bagerhat constitute the Bangladesh part of Sundarbans. The entire Indian Sundarbans covers an area of 9630 sq.km out of a total area of 25500 sq.km and the remaining part lies in the area covered by the Sundarbans in Bangladesh. The main objective of this paper is to select a few tourist spots in Sundarbans and draw out the potential of ecotourism based on a set of three parameters, i.e. the number of foreign and domestic tourists visiting the spots, number of Local/ NGOs/Government agencies participating and infrastructure availability.

Coming to the profile of the tourist spots that the author has selected in this regard. One can refer to the location map of Sundarbans in figure 1 to pinpoint the exact location of these places within the periphery of Sundarbans. The first spot selected is the Sajnekhali Watch Tower at Sajnekhali which is about 130 kms from West Bengal, located at the confluence of River Matla and Gumdi. It is one of the most renowned places to visit in Sundarbans and the head office of the forest department is also located here. The splendid sight for exploring the wilderness of the avifauna species like Pelican, Kingfishers, Plovers, Grey and Purple Heron, White Ibis, White Bellied Sea Eagle, Whimprel, Curfews, and Sandpipers. However,

Figure 1. The Location Map of Sundarbans, India

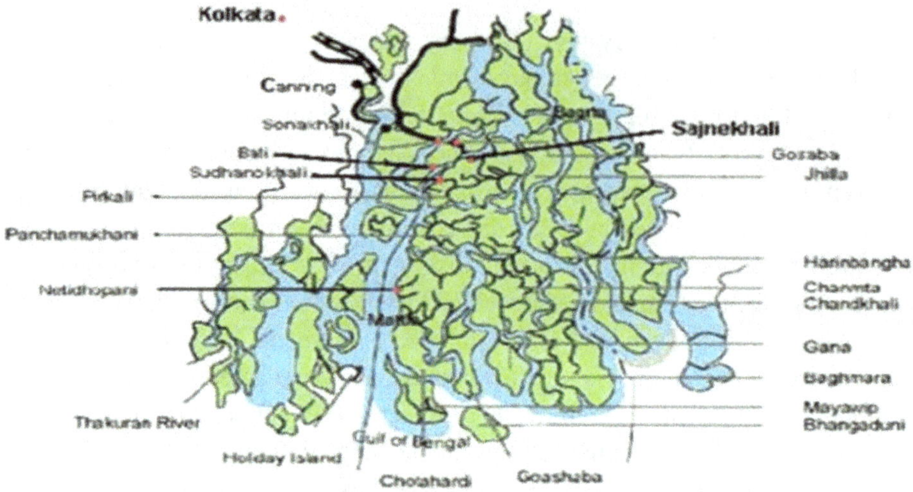

the accommodation capacity of the watch tower is 20 people at a time. The major attractions in this region other than the wildlife beauty are Bono Bibi Temple, Mangrove Interpretation Center and the tourist lodge of the West Bengal Tourism Department. Next is the Sudhanyakhali Watch Tower. This place is a treat for wildlife lovers where one can witness the refreshing sight of a tiger, wild boars or crocodiles coming to drink sweet water from a pond behind this watch tower. The third in line is also a watch tower in the series of watch towers and perhaps the most famous among these known as the Netidhopani Watch Tower. There is a famous legend as per Hindu mythology associated with this. According to the Hindu mythology legend, Behula while accompanying her dead husband, Lakhsindar on his last journey on the boat, was passing the bank of what is now called Netidhopani, witnessed an interesting thing. It is said that a woman was washing clothes and a child was disturbing her continuously. Being annoyed, she sprinkled some water on the boy and within a fraction of a second the boy became lifeless. The woman finished her laundry and then again sprinkled some water on the boy, chanting some mantras, to bring him back to life. Behula decided in a flash that this was the person who could bring her husband back to life. She rowed the boat to the bank and asked the lady (Netidhopani) to teach her the verses. The lady asked Behula to be her maid and serve her with diligence. Only then would she grant her wish. Many months passed-by as Behula served her unquestionably. Finally, one day she put her foot down and asked Netidhopani to tell her the secret verses by which she could bring Lakhinder back to life. At this, Netidhopani told that her verses do not have the capability to be effective so as Lakhinder died due to the curse of a goddess. However, she offered to take Behula to heaven instead. Behula agreed and came to heaven. There, she prayed to Lord Shiva for Lakhinder's life and then Lord Shiva told her the way that her father-in-law would have to pray to goddess Manasa only then her husband would get back her life. Consequently, Netidhopani plays a crucial role in this legend. This watchtower stands amidst the ruins of a 400 year old Shiva Temple. Also, the road on the extreme right of this tower is believed to have been built by King Pratapaditya [(1561–1611 CE) was a zamindar, the Hindu king of Jessore who fought against the Mughals and is famous for the Battle of Salka and Kagargahat] to guard the coastal area. The Bhagatpur Crocodile Project is an abode of the largest estuarine crocodiles. One of the most well liked spots in the Sundarbans, famous for its hatchery and bird sanctuary. From watch towers there is the Halliday Island, positioned on the southern side of Sundarban Tiger Reserve. Halliday Wildlife Sanctuary is also located within the Sundarban Tiger Project Reserve. The sanctuary has the honor of being the last retreat of the rare barking deer. The last in the list of top tourist attractions in this region is Kanak. The crystal clear waters and shores in Kanak are counted among the most preferred basking spot for Olive Ridley Turtles. It is said that these creatures travel in an upright position for almost 100 kilometers from the mouth of the sea to the Sundarban National Park. It should be mentioned that these tourist spots have been selected based on the the inflow of both domestic and foreign tourists to these places.

The rest of the paper has been organized as follows. To start with, a brief evaluation of the select literature on the concept of GIS and the application of GIS data in analyzing several problems across the globe. In Section 3, the author presents a profile of the tourist spots under deliberation. Section 4 illustrates the methodology employed and puts forward the results and the discussions thereof. It comes to a close by highlighting the future research possibilities in this regard.

A BRIEF REVIEW OF THE EXISTING PERSPECTIVES

The application of WGIS in tourism research includes ecotourism planning, resource assessment and management, recreational amenities, facility monitoring and suitable geographic location identification which have been discussed at length over the last few decades. Mainly, the value addition lies in its ability to manipulate both types of data, spatial as well as attributes (Bahaire & Elliott-White, 1999). In this context, the study brings out a comprehensive analysis starting with tourist places, hotels, NGOs/travel agencies operating in the area, railway/bus stations, land use profile, accessibility of the locations, etc. in the Sundarbans of West Bengal in conjunction with exploring the potential of ecotourism by making use of GIS data. Nevertheless, the application of WGIS in ecotourism remains mostly unexplored on account of a budget constraint and improper databases in this regard. For instance, there is very little site-specific information about sources of origin of the visitors, their travel plan, why did they choose such a place or motivation for travel, expenditure pattern of tourists, spatial patterns of recreation and tourism use, levels of use and impacts, suitability of sites for the purpose of tourism, etc. — all can be thought of as fields of application of GIS. Consequently, the application of GIS in ecotourism development becomes limited (Giles, 2003). Coming to some of the cross country studies using GIS data, Dogru et al. (2007) pioneered the development of a map based browser for providing information about earthquakes in Turkey. Given user queries, the application displays maps drawn by the program. Also, in the context of Turkey, Sertel et al. (2008) applied WGIS techniques for supporting the planning activities for tourism in Manavgat, located on the southern coast of Turkey. Jovanović (2016) and Thecla I. et al. (2014) have done a study related to the application of GIS and it's components of tourism for Zlatar areas of Serbia and Anambra State of the Nigerian province respectively. In essence both the studies looked at how the economic benefits for tourism industry can be reaped by the state concerned. For Langkawi islands in Malaysia, Marson et al. (2015) developed a spatial database decision support system (SDSS) application integrated with GIS functions using user friendly graphic user interfaces (GUI) that can help the tourists to choose and plan their activities, according to their convenience. Kommana (2013) used the open source software mainly, Geoserver, to design a web map application and making it available on the internet that can be used by any institute or NGO across the globe. The user can overlay the Geoserver WMS layers, by choosing background e.g. Google Earth, Open Street Map, etc. and adding the zoom in and out, pan navigation, switch layer, layer overview styles. Coming to some of the studies related to India, Mustafa et al. (2011) makes an analysis on the land suitability for different crops using remote sensing and GIS data which integrates spatial database, statistical data with textual information for developing a data model that can provide actual and full information for analysis and planning for the Agra region. Tyagi (2014) developed a Web based customized Tourism Information System (TIS) for Eastern U.P., built on the GIS platform. The application provides a dynamic mapping by showing the location of tourist spots as well as location of restaurants, travel agencies, hospitals, passport office, information bureau etc. In the same spirit, Hussain (2006) developed an interactive web based GIS application prototype for Hyderabad city to promote the city's tourist activities and provide the users with an innovative way to access the spatial content of the city. Rawat and Kumar (2015) created a WGIS prototype for geospatial information of Bhopal city in Madhya Pradesh, India. The design of the web page was done with HTML, AJAX and Java Script and Open Layers. Krishan et al. (2016) prepared an interactive tourist map for Chandigarh city by making use of web GIS techniques. Using GIS data, Mujabar and Chandrasekar (2013) carries out a shoreline change analysis for the coastline between Kanyakumari and Tuticorin in the Indian context. Agrawal and Gupta (2014) focus on a WGIS framework on WAMP, ALOV and MySQL for Sivsagar district in Assam.

Moving to the use of GIS application in tourism planning, Culbertson et al. (1994) discusses the use of GIS in regional tourism planning in Banff, Canada and Colorado, USA. Berry (1991) provides an early illustration of utility of this kind of map analysis in the context of US Virgin Islands. Similarly, Williams et al. (1996) demonstrated how GIS can be used to identify areas suitable for ecotourism in British Columbia. Bahaire and Elliott-White (1999) included in their study has provided a brief description of the various applications of GIS in tourism planning in UK. The study by Farsari and Prastacos (2004) highlighted particular features of applications of GIS in the tourism suitability analysis, landscape resource inventory and designation of tourist areas in terms of use levels in the pre- and post-tourism visual impact analysis in the US. Boyd and Butler (1996) in their study exhibit the application of GIS in identifying areas suitable for ecotourism in Northern parts of Ontario, Canada; in particular natural areas, containing rare or endangered species or habitats in remote areas. Essentially, their work analyzes satellite images to produce an account of a variety of characteristics associated with natural landscapes. By analyzing satellite images it is possible to identify complex areas inside the deep forest or rugged mountain terrains even in the deep ocean where a general survey method is next to impossible. Here lies the importance of GIS in tourism development in the context of natural environment. In ecotourism planning the first issue that emerges is the environment and its conservation. An ecotourism destination must in no way be developed without planning in terms of environmental concern. In reference with the Gray rocks reservoir of USA, Gribb (1991) took up a planning endeavour to come up with a recreational development oriented plan which would look after the conservation of the reservoir. In the United Kingdom (UK) for instance, fifty-three percent of the major retailers had adopted GIS by 1998 and Chen (2007) shows comprehensively how brick and mortar businesses can also use GIS applications to compete with internet retailers. Though studies with GIS data have happened across the globe in different aspects, studies pertaining to Sundarbans in India are something very rare and moreover, the literature does not talk about the development of an Ecotourism Potential Index (EPI) for a place concerned with respect to certain parameters as already stated. This paper takes a modest step in this regard.

A BRIEF NOTE ON ACCOMMODATION FACILITIES AND LAND USE PATTERN OF THE CONCERNED TOURIST SPOTS

Starting with hotels, lodges and guest houses as reported in figure 2, most of the village resorts, tourist lodges, hotels and government guest houses are concentrated near the Sajnekhali area. Interestingly, Halliday Island, Kanak and even Netidhopani do not have any accommodation facility. But for tourists, other than the tiger reserve, one of the major attractions is Netidhopani as compared to the other spots that the author considers here. The rates per night have also been provided for the private resorts while Sajnekhali tourist lodge belongs to West Bengal Tourism Department and hence the rates have not been provided.

The comparison of the land use in 2000 with that for 2015 shows the change that has taken place during the past 5 years. Bare land has been increased by 2 per cent during this period. However, the major trees of this forest are Sundri have dramatically decreased 34.19 per cent. Only Gewa trees have increased by 21.75 per cent. This will be very clear if one refers to the land use pattern diagrams in figure 3 and figure 4. The local people are fully dependent on this forest and the major activities of localites are essentially — fishing, honey gathering, wood cutting and processing for timber production. The pattern of change in land use in figure 3 can be attributed to issues like climate change, flooding, effects of disastrous cyclones which have affected mangrove forests severely. The role of NGOs become crucial in this regard and to name some of the NGOs operating in this region, like,—

Figure 2. Accommodation facilities in Sundarbans, India

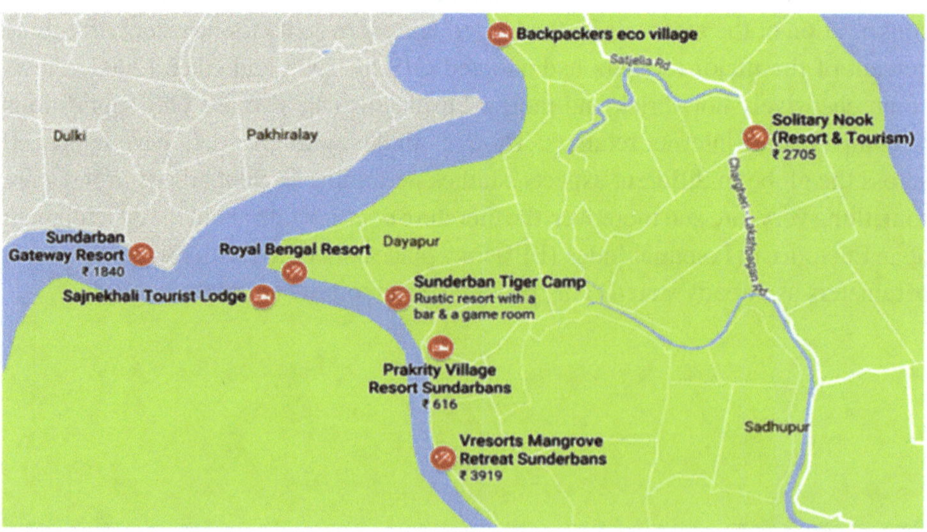

Figure 3. The land use patterns in 2015

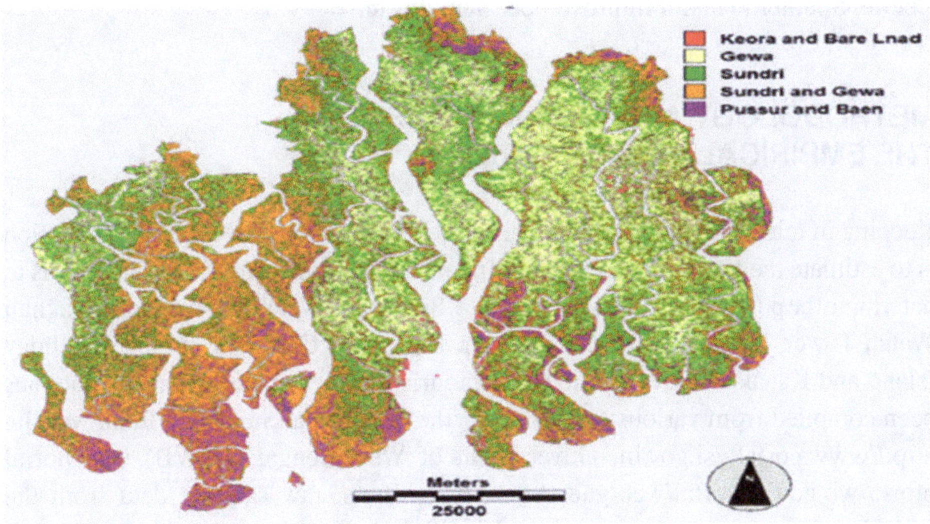

Figure 4. The land use patterns in 2000

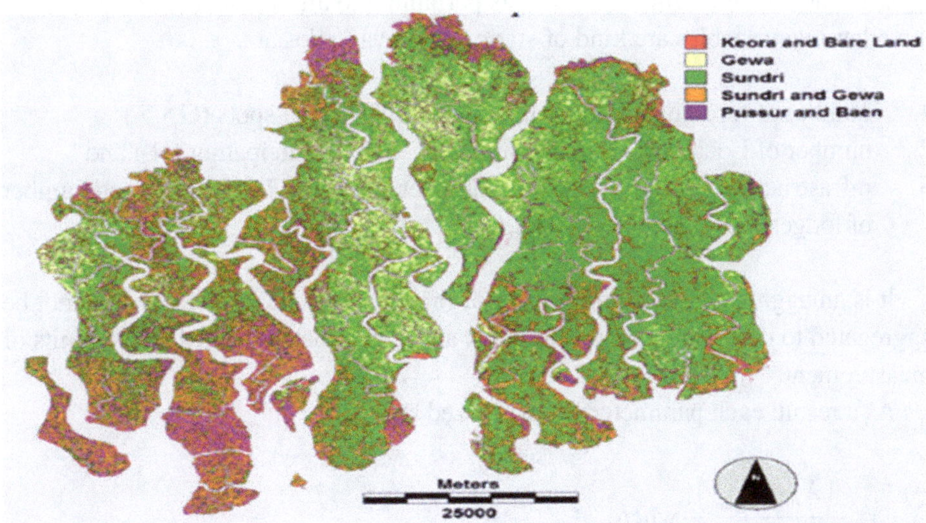

Sundarban Yuba-O-Samaj Kalyan Samity (Boys), Sundarban Social Development Centre, Southern Health Improvement Samity, etc.

METHODOLOGY AND FINDINGS THEREOF: THE EMPIRICAL FRAMEWORK

Keeping in mind the importance of this issue, the research objective of this section is to estimate the Ecotourism Potential Index (EPI) for the year 2015. The idea is to determine the position of the tourist spots, i.e. Sajnekhali Watch Tower, Sudhanyakhali Watch Tower, Netidhopani Watch Tower, Bhagatpur Crocodile project, Halliday Island and Kanak with respect to the ecotourism achievement level. The data has been compiled from various sources, *viz.,* the Geological Survey of India website http://www.portal.gsi.gov.in, Government of West Bengal (GoWB) web portal https://wb.gov.in/portal/web/guest/sundarban-affairs, raw satellite data from the ESRI Landsat app, google maps, google earth, also from *https://earthobservatory.nasa.*gov and raw data from different administrative offices, forest officers, hotels and travel agencies in that particular area of investigation. The present paper has collated the data for number of tourists pertaining to all the months of 2015 while the other two variables are kind of static. These variables are,

1. number of foreign and domestic tourists visiting the spots (G_1),
2. number of Local/ NGOs/Government agencies participating (G_2) and
3. infrastructure availability (G_3) i.e. number of railway/bus stations and number of lodges, guest house or hotels.

It is an aggregate index comprising of three parameters as stated, so cannot be aggregated to derive the composite index as the parameters have different units of measurement.

As a result, each parameter is normalized by –

$$X_{iN} = \frac{\left(X_i - X_{\min}\right)}{\left(X_{\max} - X_{\min}\right)} \text{ where,}$$

X_{iN} is the normalized value, X_{min} is the minimum value observed across the 6 spots considered for some specific parameter, X_{max} is the maximum value observed across the 6 spots for a particular parameter and X_i is the value a particular parameter for tourist spot *i*. Normalization yields a value for every parameter for every tourist spot which lies between 0 and 1. The value '0' depicts the worst case and '1' depicts the best case scenario. Following the concept of Human Poverty Index concept, as propounded by Sen and Anand (1997), here, the author considers the following achievement index function for a particular tourist spot as,

$$DI_j(\alpha) = \left[\frac{\sum w_i G^{\alpha}_{\ i}}{\sum w_i}\right]^{\frac{1}{\alpha}} \tag{1}$$

where, G_i = proportion of achievement in the ith parameter in the jth tourist spot, α here stands for the order of the average, w_i be the weights attached to G_i parameter, i = 1, 2, 3 in the jth state. Equation (1) does follow some important properties of a good index but it needs to be mentioned that $DI(\alpha)$ suffers from the problem of overlapping dimensions and multicollinearity. It might so happen that the fall in one parameter may also lead to a fall in the other parameter. To rule out such possibilities, employing Principal Component Analysis (PCA) is the best procedure. The weights attached to each of the deprivation parameters are not arbitrarily given; it is determined endogenously from the data matrix through PCA (Bhattacharya & Haldar, 2014). The weights are given by the eigenvalues corresponding to the principal components extracted.

That is, $w_i = \dfrac{\lambda_i}{\sum \lambda_i}$

The author makes an assumption that the value of $\alpha = 2$. There is a justification for considering $\alpha = 2$ as it corresponds to Euclidean distance, i.e., the common understanding of distance. In addition to this, it is justified on the ground that a higher weight is attached to the dimension in which the achievement is more (Reddy and Nathan, 2011).

Now, the question is how many PCAs to extract? In this article, the author has extracted two PCs based on the Scree Test criteria. The Scree-Plot criterion, as proposed by Cattell (1966), plots the successive eigenvalues, which drop off sharply and then tend to level off and hence become flat (Dunteman, 1989). The elbow rule suggests retaining all the eigenvalues in the sharp descent before the first one on the line where they start to level off. This implies that the eigenvalues on the line before they start to level off explain a major proportion of the variability and the others explain only a negligible portion and are likely to be less significant. The results here show that the first principal component explains 64 per cent of the total variance and the second principal component explains 34 per cent of the total variance and together they have explained almost 98 per cent of the total variance. In addition to the Scree Test criteria, the author has employed the Kaiser (1958) method of retaining those components having eigenvalues greater than one. So, table 3 shows results consistent with the Scree Plot criteria fact there should be two principal components in this analysis. Hence, it is possible to represent the achievement of the concerned tourist spots with respect to different parameters in a reduced space dimension with limited loss of information.

The Ecotourism Potential Index for the jth tourist spot has the form of –

$$EPI_j(\alpha) = \left[\left(\frac{w_1}{w_1 + w_2}\right)\left\{\left(PC_1\right)_j\right\}^{\alpha} + \left(\frac{w_2}{w_1 + w_2}\right)\left\{\left(PC_2\right)_j\right\}^{\alpha}\right]^{\frac{1}{\alpha}} \tag{2}$$

It can also happen that these three parameters are not really correlated among themselves. So, in that case one can do away with principal components. So, an alternative is the Euclidean distance function. The author here attaches equal weights to all the parameters, namely, *number of foreign and domestic tourists visiting these tourist spots, number of Local/NGOs/Government agencies participating and infrastructure availability in the concerned spot*. Given the fact that all the parameters are of equal importance in judging the quality of the ecotourism potential, the three dimensional index may be represented in a three-dimensional space with the value '0' being the minimum value and '1' as the required value. The EPI uses inverse of the weighted Euclidean distance from the ideal point of (1,1,1). As a result, the EPI calculation for spot j is given by:

$$EPI_j = 1 - \sqrt{\frac{\left[\left(1 - ER_j\right)^2 + \left(1 - LNG_j\right)^2 + \left(1 - I_j\right)^2\right]}{3}} \tag{3}$$

The numerator of the term within the square root gives the Euclidean distance of spot j from the ideal point (1,1,1). The inverse distance has been calculated to

Table 1. Kaiser Method Results

Principal Component	Eigenvalue
1	1.946
2	1.020
3	0.041

show that higher is the value of EPI, better will be the ecotourism potential and higher will be the rank given to that particular spot. Following Nathan et al. (2008), this particular index satisfies the properties of NAMPUS that is **N**ormalization, **A**nonymity, **M**onotonicity, **P**roximity, **U**niformity and **S**ignaling.

Coming to the comparison of the results using the two methods,

The results are indeed interesting as both the methods have assigned a similar rank to these tourist spots. First, it should be noted that Bhagatpur Crocodile project and Kanak there are not many hotels or guest houses to accommodate tourists as compared to the other spots. Tourists kind of stay somewhere else and come to visit either Kanak or Bhagatpur. This is consistent with what we see in Sundarbans. Given the tourist attractions and facilities available at Netidhopani as already stated, it has bagged the number one rank in terms of ecotourism potential.

The one land that all men desire to see and having seen once, by even a glimpse, would not give the glimpse for all the shows of all the rest of the globe combined.
— Mark Twain on India

CONCLUDING REMARKS AND FUTURE RESEARCH POSSIBILITIES

Summing up, this article has broadly reviewed the potential of ecotourism in Sundarbans in India. Ecotourism is the unique selling proposition (USP) of the Sundarban region and the novelty of the paper lies in trying to figure out quantitatively its potential value. Moreover, exploring the relative position of these tourist spots (taken into consideration in the Sundarban region) in terms of their ecotourism potential has added another dimension to this issue. Any researcher can make an attempt to compare the value of this Ecotourism Potential Index (EPI) over a period of time to see whether the values actually fall or rise i.e. whether there is any change

Table 2. Scores and Ranks

State	EPI Score (PCA)	Rank	EPI Score (Euclidean distance)	Rank
Sajnekhali	20.55	2	0.512	2
Sudhanyakhali	13.60	4	0.489	4
Netidhopani	23.82	1	0.534	1
Bhagatpur Crocodile Project	13.49	5	0.427	5
Halliday Island	15.01	3	0.509	3
Kanak	13.01	6	0.387	6

in the ranks over time. Ten years down the line, researchers can comment on the significance of policies propagated by the government associated to ecotourism. Only time will tell, whether Sundarbans can attain its full potential and as already discussed how GIS can aid in this process.

There are some specific limitations, which should be addressed as a means of improvement for further study. It was not possible for the author to locate every possible detail on the maps and only certain selected components have been looked at. This study was conducted based on raw satellite data in conjunction with compilation of data from different administrative offices, hotels and travel agencies in that particular area of investigation. But going for a questionnaire based study among the localites has not been carried out. The localities of the region are an important pillar because their livelihood entirely depends on the forest. This remains as one of the future agenda.

The policymakers can develop a GIS based web portal from where one can extract the necessary information about the tourist spot concerned. In the author's opinion, this can definitely be of great help in developing the ecotourism potential in the Sundarbans. This preliminary research work is only the foundation and a holistic judgement can be given only when the opinion of the localites is taken into consideration in this regard. Nevertheless, it will bring up opportunities for further research and investigation, and help policy makers to evaluate the existing options for further improvement of ecotourism facilities in the Sundarban region.

REFERENCES

Agrawal, S., & Gupta, R. D. (2014). Development and comparison of open source based Web GIS Frameworks on WAMP and Apache Tomcat Web Servers. *The International Archives of the Photogrammetry, Remote Sensing and Spatial Information Sciences*, *40*(4), 1.

Akukwe Thecla, I. (n.d.). Determinants of Flooding in Port Harcourt Metropolis. *Journal of the Humanities and Social Sciences*, *19*(11), 62–72.

Anand, S., & Sen, A. (1997). Concepts or Human Development and Poverty: A Multidimensional Perspective. *United Nations Development Programme, Poverty and human development: Human development papers*, 1-20.

Arnold, J. G., Srinivasan, R., Muttiah, R. S., & Williams, J. R. (1998). Large area hydrologic modeling and assessment part I: Model development. *JAWRA Journal of the American Water Resources Association*, *34*(1), 73–89. doi:10.1111/j.1752-1688.1998.tb05961.x

Bahaire, T., & Elliott-White, M. (1999). The application of geographical information systems (GIS) in sustainable tourism planning: A review. *Journal of Sustainable Tourism*, *7*(2), 159–174. doi:10.1080/09669589908667333

Berry, J. K. (1991). GIS in island resource planning: A case study in map analysis. *Geographical Information Systems*, *2*, 285–295.

Bhattacharya, G., & Haldar, S. K. (2014). Trend, Differential and Determinants of Deprivation of Reproductive and Child Health in the Districts of West Bengal, India. *Journal of Health Management*, *16*(1), 93–112. doi:10.1177/0972063413518686

Boyd, S., & Butler, R. (1996). Seeing the forest through the trees: Using GIS to identify potential ecotourism sites in Northern Ontario. In L. C. H. Harrison & W. J. Husbands (Eds.), *Practicing Responsible Tourism: International case studies in tourism planning, policy and development* (pp. 380–403). New York: John Wiley and Sons.

Cattell, R. B. (1966). The scree test for the number of factors. *Multivariate Behavioral Research*, *1*(2), 245–276. doi:10.1207/s15327906mbr0102_10 PMID:26828106

Chen, R. J. (2007). Geographic information systems (GIS) applications in retail tourism and teaching curriculum. *Journal of Retailing and Consumer Services*, *14*(4), 289–295. doi:10.1016/j.jretconser.2006.07.004

Dogru, A. O., Ulugtekin, N. N., & Alkoy, S. (2007, August). GIS applications on epidemiology with cartographic perspective in Turkey. *23rd International Cartographic Conference*.

Dunteman, G. H. (1989). *Principal components analysis (No. 69)*. Sage. doi:10.4135/9781412985475

Farsari, Y., Butler, R., & Prastacos, P. (2007). Sustainable tourism policy for Mediterranean destinations: Issues and interrelationships. *International Journal of Tourism Policy*, *1*(1), 58–78. doi:10.1504/IJTP.2007.013898

Giles, W. (2003). *GIS applications in tourism planning*. GIS 340-GIS Seminar. Retrieved from https://www.scribd.com/document/72126092/GIS-Application-in-Tourism-Planning

Gribb, W. (1991). Integrating visitor surveys, environmental factors, and GIS into recreation planning: Site development at Grayrocks reservoir, Wyoming. GIS/LIS'91 Proceedings, 1, 177-86.

Hussain, M., Arsalan, M. H., & Mehdi, M. R. (2006). Role of GIS in public health management in Pakistan. In *Advances in Space Technologies, 2006 International Conference on* (pp. 102-107). IEEE. doi:10.1109/ICAST.2006.313807

Jovanović, V. (2016). The application of GIS and its components in tourism. Yugoslav Journal of Operations Research, 18(2).

Kaiser, H. F. (1958). The varimax criterion for analytic rotation in factor analysis. *Psychometrika*, *23*(3), 187–200.

Kommana, K. (2013). *Implementation of a Geoserver Applicatoin For GIS Data Distribution and Manipulation*. Department of Physical Geography and Quaternary Geology, Stockholm University. Retrieved from http://www.diva-portal.org/smash/get/diva2:640096/fulltext01.pdf

Krishan, G., Sharma, L. M., Yadav, B. K., & Ghosh, N. C. (2016). Analysis of Water Level Fluctuations and TDS Variations in the Groundwater at Mewat (Nuh) District, Haryana (India). *Current World Environment*, *11*(2), 388–398. doi:10.12944/CWE.11.2.06

Mujabar, P. S., & Chandrasekar, N. (2013). Shoreline change analysis along the coast between Kanyakumari and Tuticorin of India using remote sensing and GIS. *Arabian Journal of Geosciences*, *6*(3), 647–664. doi:10.1007/s12517-011-0394-4

Mustafa, A. A., Singh, M., Sahoo, R. N., Ahmed, N., Khanna, M., Sarangi, A., & Mishra, A. K. (2011). Land suitability analysis for different crops: A multi criteria decision making approach using remote sensing and GIS. *Researcher*, *3*(12), 1–24.

Nathan, H. S. K., Mishra, S., & Reddy, B. S. (2008). *An alternative approach to measure HDI*. Indira Gandhi Institute of Development Research (IGIDR), Working Paper, WP-2008-001.

Rawat, J. S., & Kumar, M. (2015). Monitoring land use/cover change using remote sensing and GIS techniques: A case study of Hawalbagh block, district Almora, Uttarakhand, India. *The Egyptian Journal of Remote Sensing and Space Sciences*, *18*(1), 77–84. doi:10.1016/j.ejrs.2015.02.002

Reddy, B. S., & Nathan, H. S. K. (2011). *Emerging energy insecurity — the Indian dimension*. Academic Press.

Sertel, E., Findik, N., Kaya, S., Seker, D. Z., & Samsunlu, A. (2008). Assessment of landscape changes in the Kizilirmak Delta, Turkey, using remotely sensed data and GIS. *Environmental Engineering Science*, *25*(3), 353–362. doi:10.1089/ees.2006.0149

Tyagi, N. (2014). Web GIS application for customized tourist information system for Eastern UP, India. *Journal of Genomics*, *8*(1), 1–6.

ADDITIONAL READING

Boers, B., & Cottrell, S. (2007). Sustainable tourism infrastructure planning: A GIS-supported approach. *Tourism Geographies*, *9*(1), 1–21. doi:10.1080/14616680601092824

Chang, G., & Caneday, L. (2011). Web-based GIS in tourism information search: Perceptions, tasks, and trip attributes. *Tourism Management*, *32*(6), 1435–1437. doi:10.1016/j.tourman.2011.01.006

da Cruz Albuquerque, H. C., Cardoso Pedrosa Martins, F. M., & Tomás Galiza Cardoso, L. M. (2014). *Geographical information systems, tourism* (pp. 1–3). Springer International Publishing.

Dondo, C. H., Bhunu, S. T., & Rivett, U. (2002). GIS in tourism-A Zimbabwean perspective. *International archives of photogrammetry and remote sensing, 34*(6/W6), 197-200.

Fajuyigbe, O., Balogun, V. F., & Obembe, O. M. (2007). Web-based geographical information system (GIS) for tourism in Oyo State, Nigeria. *Information Technology Journal*, *6*(5), 613–622. doi:10.3923/itj.2007.613.622

Feng, R., & Morrison, A. M. (2002). GIS applications in tourism and hospitality marketing: A case in Brown County, Indiana. *Anatolia*, *13*(2), 127–143. doi:10.1080/13032917.2002.9687129

Kumar, T., Gautam, A. K., & Kumar, T. (2014). Appraising the accuracy of GIS-based Multi- criteria decision making technique for delineation of Groundwater potential zones. *Water Resources Management*, *28*(13), 4449–4466. doi:10.1007/s11269-014-0663-6

Kumari, S., Behera, M. D., & Tewari, H. R. (2010). Identification of potential ecotourism sites in West District, Sikkim using geospatial tools. *Tropical Ecology*, *51*(1), 75–85.

Law, R., Leung, R., & Buhalis, D. (2009). Information technology applications in hospitality and tourism: A review of publications from 2005 to 2007. *Journal of Travel & Tourism Marketing*, *26*(5-6), 599–623. doi:10.1080/10548400903163160

Mahmoud, S. H., Alazba, A. A., & T, A. M. (2014). Identification of potential sites for groundwater recharge using a GIS-based decision support system in Jazan region-Saudi Arabia. *Water Resources Management*, *28*(10), 3319–3340. doi:10.1007/s11269-014-0681-4

Sadoun, B., & Al-Bayari, O. (2009, May). A GIS system for tourism management. In *Computer Systems and Applications, 2009. AICCSA 2009. IEEE/ACS International Conference on* (pp. 226-232). IEEE. doi:10.1109/AICCSA.2009.5069329

Wei, W. (2012). Research on the application of Geographic Information System in tourism management. *Procedia Environmental Sciences*, *12*, 1104–1109. doi:10.1016/j.proenv.2012.01.394

KEY WORDS AND DEFINITIONS

Biodiversity Conservation: The efforts taken by various nodal agencies for not only deriving sustainable benefits for the present and future generations but also ensuring that the protection, upliftment, and scientific management of the variety and variability of species, their genetics, and at the ecosystem level life on earth gets maintained at its threshold level.

Climate Change: Climate change can be defined as the change in the statistical distribution of weather patterns or the rise in the earth's temperature over an extended period of time ranging from decades to millions and millions of years. The factors responsible are volcanic eruptions, tectonic movements, from the solar radiations received by the earth or may be from the human activities, etc.

Ecotourism: Ecotourism is a kind of sustainable tourism involving environmentally responsible journey to comparatively undisturbed, immaculate natural areas. The purpose for such a travel is to not only educate the traveler but also provide resources for the ecological conservation, socio-economic development, and consequently an upliftment in the living standards of the localites thereof.

Ecotourism Potential Index (EPI): The EPI uses inverse of the weighted Euclidean distance from the ideal point of (1,1,1,). As a result, the EPI calculation for spot *i* is given by: $EPI_i = 1 - \sqrt{\frac{\left[\left(1 - ER_i\right)^2 + \left(1 - LNG_i\right)^2 + \left(1 - I_i\right)^2\right]}{3}}$. The numerator of the term within the square root gives the Euclidean distance of state *i* from the ideal point (1,1,1). The inverse distance has been calculated to show that higher is the value of EPI, better will be the ecotourism potential and higher will be the rank given to that particular spot. This particular index satisfies the properties of — NAMPUS that is Normalization, Anonymity, Monotonicity, Proximity, Uniformity and Signaling. OR It can be,

$$EPI_j(\alpha) = \left[\left(\frac{w_1}{w_1 + w_2}\right)\left\{\left(PC_1\right)_j\right\}^{\alpha} + \left(\frac{w_2}{w_1 + w_2}\right)\left\{\left(PC_2\right)_j\right\}^{\alpha}\right]^{\frac{1}{\alpha}}$$

using Principal Component analysis (Refer to the text for details).

Euclidean Distance: The Euclidean distance in Euclidean-n space is the distance (dist) between two vectors **a**, a = (a_1, a_2, … a_n) and **b**, b = (b_1, b_2, … b_n) is given by:

$$dist(a,b) = dist(b,a) = \sqrt{\left(a_1 - b_1\right)^2 + \left(a_2 - b_2\right)^2 + \ldots\ldots\left(a_n - b_n\right)^2}.$$

Landuse Change: It refers to the course of action by which human-induced activities transform the landscape resulting in a change in the composition of greenhouse gas emissions from the land.

Principal Component Analysis: Principal component analysis (PCA) is a statistical procedure that converts a set of possibly correlated variables into an array of linearly uncorrelated variables, called principal components by making use of an orthogonal transformation. This transformation is defined in such a way that the first principal component has the largest possible variance (i.e., it explains a major proportion of the variability in the data) and each succeeding component in turn has the next highest variance possible but it is orthogonal to the preceding components. Principal components are sensitive to the relative scaling of the original variables so normalizing the data and making it lie between 0 and 1 will yield robust results.

Tourism Marketing: It can be defined as the business of attracting visitors to a specific location by taking strategies thereof. In a much broader sense, the term tourism marketing is the business of attracting visitors to a specific location.

Chapter 6
TCloud:
Cloud SDI Model for Tourism Information Infrastructure Management

Rabindra K. Barik
KIIT University, India

Rakesh Kumar Lenka
IIIT Bhubaneswar, India

Harishchandra Dubey
University of Texas at Dallas, USA

Kunal Mankodiya
University of Rhode Island, USA

ABSTRACT

This chapter proposes and develops a cloud-computing-based SDI model named as TCloud for sharing, analysis, and processing of spatial data particularly in the Temple City of India, Bhubaneswar. The main purpose of TCloud is to integrate all the spatial information such as tourism sites which include various temples, mosques, churches, monuments, lakes, mountains, rivers, forests, etc. TCloud can help the decision maker or planner or common users to get enough information for their further research and studies. It has used open source GIS quantum GIS for the development of spatial database whereas QGIS plugin has been linked with quantum GIS for invoking cloud computing environment. It has also discussed the various spatial overlay analysis in TCloud environment.

DOI: 10.4018/978-1-5225-5088-4.ch006

INTRODUCTION

Spatial Data Infrastructure (SDI) has facilitated the sharing and exchanging of spatial data holding by different stakeholders. It has initiated to create an environment that enable a wide diversity of common users to access, retrieve and disseminate spatial as well as non-spatial data in secure way (Georis-Creuseveau et al., 2016; Smith et al., 2004). The technological growing need to organise the data across various disciplines and organisations and also the need to create multi-participant and decision-supported environments that have resulted in the concept of SDI. It enable users to save resources, time and effort when trying to acquire new datasets by avoiding duplication of expenses associated with generation and maintenance of spatial data (Mansourian et al., 2005; Giuliani et al., 2016; Coleman et al., 2016).

SDI can be used for decision making, storage of various kinds of spatial data, bringing data and maps to a common scale as per the user needs. It has the ability to superimposing, querying, analyzing, designing and presenting final maps/ reports to the administrators and planners (Yue et al., 2016; Senaratne et al., 2017; Koswatte et al., 2015). The utility of SDI has become widely popular and are being used for a wide range of applications (Pun-Cheng et al., 2016; Mwange et al., 2016). There are five important components of SDI which has been encapsulated. The component of SDI has shown in Figure 1. In the dynamic nature of SDI, technologies, polices and accessing network are the core dynamic components where people and data are the static components (Georis-Creuseveau et al., 2016).

SDI can play a vital role in various applications such as environmental monitoring, natural resource management, healthcare, land use planning and urban planning. SDI integrates common database operations such as query formation, statistical

Figure 1. Five important components of SDI (Georis-Creuseveau et al., 2016; Laura et al., 2017)

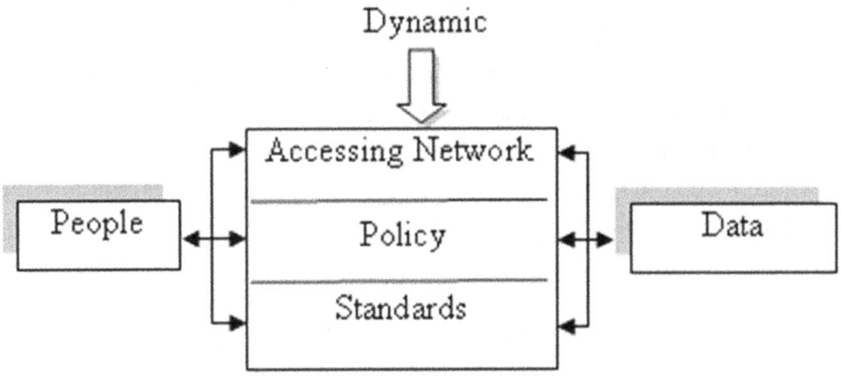

computations and overlay analysis with unique visualization and geographical functionalities. These characteristics are distinguish SDI from other information systems and make it valuable to a wide range of public and private enterprises for explaining events, predicting outcomes and designing strategies (Barik and Samaddar, 2014(a); Barik and Samaddar, 2014(b)).

SDI with specific model i.e. SDI Model has also played an important role in mineral resources, water resources and coastal management in which it has potential to share and analysis all the hydrologiacal data, coastal data and mineral data related spatial data in a common platform. SDI Model has the possiblity to improve the spatial dimension of coastal zone aswell as water resources management. It can allow to access interoperability of data from river, coastal, marine and terrestrial environments (Strain et al., 2004; Hamylton and Prosper, 2012; Idrees, 2015). For seamless development of SDI Model, it has integrated with cloud computing. SDI Model has merged with cloud computing to perform a value added services that give rise to spatial cloud computing. The spatial data have rich information about temporal as well as spatial distributions. In traditional setup, we send the data to the cloud where these are going for further processing and analysis (Yang et al., 2017;Yang et al., 2011).

For the development and implementation of SDI Model, there are many open source software available as compete with the proprietary software (Raghunathan et al., 2005; Barik et al., 2009). Developers have created many Open Source Libraries and Plugin to handle huge spatial data. Open Geospatial Consortium (OGC) aims to strengthen utility of development of community-led projects and Open Source GIS Standards (Guan et al., 2012;Nasr 2007). For the development of seemless SDI Model, the spatial web services, spatial database creation and spatial modeling services are required. So for implementing these services, it can use the open source GIS software for full phase operations (Brovelli et al., 2016).

In the present scenario, SDI Model can be implemented in cloud computing environments which formulated as Cloud SDI Model. It provides high processing capabilties and infrastructes that can increase throughput and reduces latency near the edge of the spatial clients. Cloud SDI Model also reduces the storage needed for spatial data in the cloud. In addition, reduction in the required transmission power results in overall improvement in efficiency. In this book chapter, we let the spatial data be processed at the edge using Cloud SDI environment. The present book chapter made the following contributions to the Cloud SD Model in Tourism Information Infrastructure Management:

- Cloud SDI Model i.e. TCloud framework is projected to reduced latency and improve throughput for stroring, retrieving and analysis of various components associated with the tourism spatial data

- It has been performed a case study of all the related tourism information like temples, church, atms, bus stops and railway network associated with the temple city, Bhubaneswar, India.
- Investigation of optimum adoption of Open source GIS software for tourism spatial database creation
- Spatial overlay analysis in thick, thin and mobile clients environments was performed in TCloud environment

So by integrating cloud computing and SDI Model, the need of the hour is to provide easy to perceive and use of temple city, Bhubaneswar for tourism information network in India. This should also fulfil the much awaited need those are wish to pursue the analysis on tourism information management. The transparent working of this network should also put pressure on the decision makers for proper management of temple city, Bhubaneswar, India. This has necessitated the design, development and implementation of TCloud framework.

RELATED WORK

SDI provides an environment within which organizations interact with technologies to foster activities for using, managing and producing geographic data. It is the technology, policies, standards, human resources, and related activities necessary to acquire, process, distribute, use, maintain and preserve spatial data. The core components of SDI can be viewed as policy, networking, standards, people and data. These can be grouped into the different categories which are based on the various spatial web services interactions within the SDI Model (Internet-1, 2016; Puri et al., 2007).

A spatial web service is one of the web services that can access the data from GIS database. It also provides the geographic information through a common browser interface to the end users. Thus, spatial web capabilities indicate to a web based GIS which can be modeled using the client-server architecture. A thin client model is where most of the processing work is done on demand in the server and the client does not perform any task other than to display the data on screen. The thin client-server architecture has been used for spatial web capabilities (Harper, 2006). The thin client system comprises of two parts, the client side and the server side. Figure 2 shows the thin client GIS architecture model.

The Service Oriented Architecture tries to construct a flexible, dynamic, re-configurable service system and distributed over Internet that can meet service requirements and information for the development of SDI. The prime component in the SOA based SDI is spatial web service which is stateless, self contained, well

Figure 2. Thin client GIS Architecture (Barik et al., 2011)

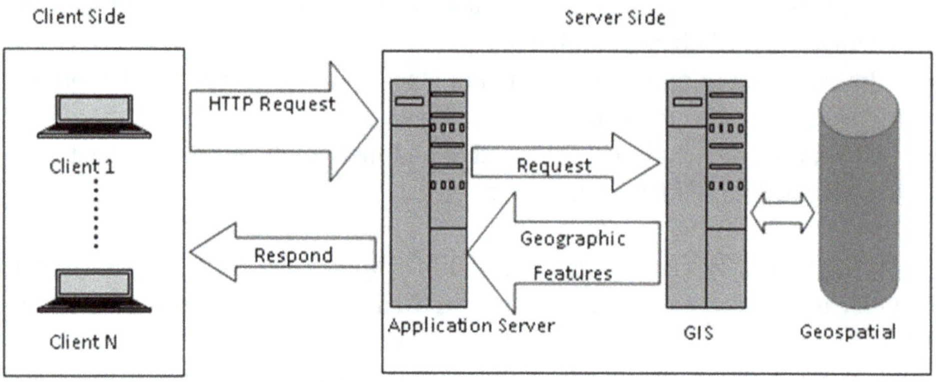

defined set of actions and does not depend upon the state of other services associated with the system (Xiaolin, 2005). The basic operations in SOA based SDI Model include publish, find and bind. To be able to integrate any services into SOA based architecture, it should provide at least one of the SOA's major operations. Figure 3 shows the three major functionalities in SOA based SDI Model (Barik et al., 2016).

In SOA based SDI Model, there are three types of key actors i.e. service requester or GIS user community, GIS service provider and GIS catalog service. Catalog

Figure 3. Three functionalities and architecture of SOA based SDI Model

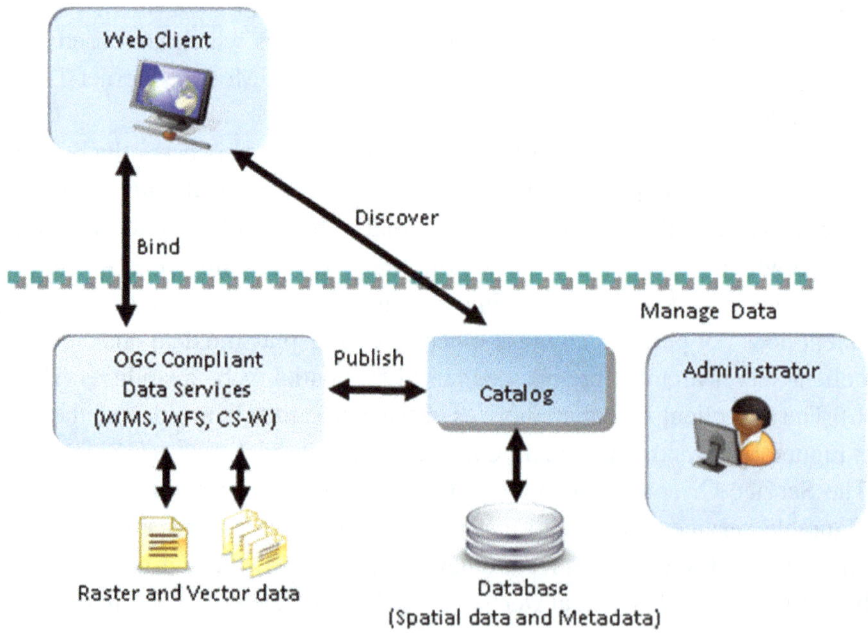

service can be called registry service or broker but their main functionality is almost same in most of the applications. Catalog service helps the requestors to discover or find the right services. When a service provider sets up a service over the internet and wants the users to use their service, it needs to publish their services. When a service requests by the requestor, service brokers and the requestors need to collaborate for finding desire services. The service provider and requestor bind as to the format of the request, after the required service is found. Then, the requestor access and invoke the services of the providers (Vaccari et al., 2009). In development and implementation of SDI Model, the major focus has been on SOA based spatial web service by using spatial data. According to services, it can be grouped data service, processing service and catalog service. Data service is tightly coupled with specific data sets and offers WFS, WMS and WCS. Processing Service provides operations for transforming or processing the data in a manner determined by the user-specific parameters. CS-W allows users and applications to classify, register, describe, search, maintain, and access information about Web Services (Li et al., 2008; Supak et al., 2014).

Cloud computing environment has been provided an enormous computable assets and store for implementation of various spatial analyses. The cloud computing Model (Buyya et al., 2009) has been provided a transition from simple desktop applications to computable web servers. Cloud computing and other web processing architectures have been delivered an open environment in web to share different assets (Chen et al., 2012; Huang et al., 2013).

Cloud computing has delivered four different kinds of services: Infrastructure as a service (IaaS), Database as a Service (DaaS), Software as a Service (SaaS) and Platform as a Service (PaaS) (Yang et al., 2010). Figure 4 shows the graphical representation of the different service Models and their components associated with each component. These different services are available to the users in a Pay per-use-on-demand Model.

SDI defines the cumulative of technology, standards, strategies, policies and manpower required to attain, allot, sustain, process, use and reserve spatial data. The basic constituents of SDI have been observed as data, networking, public, policy and standards. Further, SDI Model can be implemented through Service Oriented Architecture (SOA) or Cloud Computing technologies for better and efficient use. The SOA tries to construct dynamic, distributed and flexible facility system over the web that can meet information and service requirements for development of SDI. The components in the SOA based SDI are spatial web services i.e. a well-structured set of activities which are stateless, self-confined, and independent upon the state of other services (Leidig et al., 2015;Patra and Barik, 2015). After SOA based SDI Model, Cloud SDI Model comes into picture. Cloud SDI Model has been delivered a platform in which organizations interrelate with technologies, tools and expertise

Figure 4. Cloud computing service Models (Patra and Barik, 2015)

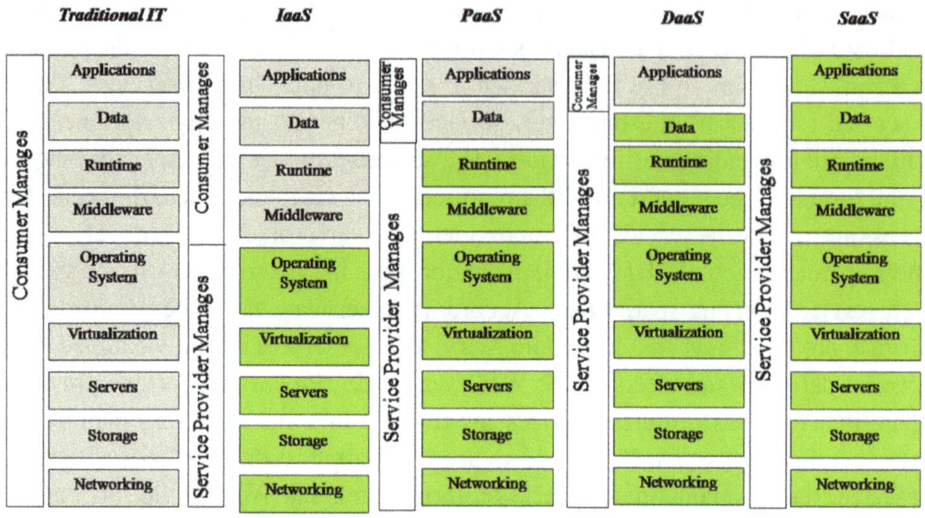

to nurture deeds for producing, handling and using geographical statistics and data on cloud computing environment.

Cloud SDI Model deploys a unique-instance, multitenant design and permitting more than one client to contribute assets without disrupting each other. This integrated hosted service method helps installing patches and application advancements for user's transparency. Spatial Cloud another characteristic is embrace of web services and SOA, a wholly established architectural methodology in the engineering. Many Cloud platforms uncover the applications statistics and functionalities via web service. This permits a client to query/update different types of cloud services and applications data programmatically, along with the provision of a standard mechanism to assimilate different cloud applications in the software cloud with enterprise SOA infrastructure (Evangelidis et al., 2014;Schaffer et al., 2010). Figure 5 illustrates the system architecture for Cloud SDI Model.

According to the system architecture, the client tier layer has been divided into thick, thin and mobile clients for capabilities of visualising the spatial information. Mobile clients are those users those are operating through the mobile devices where as the users those are working on the simple web browsers is defined as thin clients. In thin clients, user does not require additional plugin/software for the required operation. Thick clients are those users, which are processing or visualising the spatial data in desktop where it requires additional software for full phase operation. The Application Tier comprises the main spatial services executed by servers. This application tier layer enables intermediate amongst the different clients and

Figure 5. System architecture for Cloud SDI Model

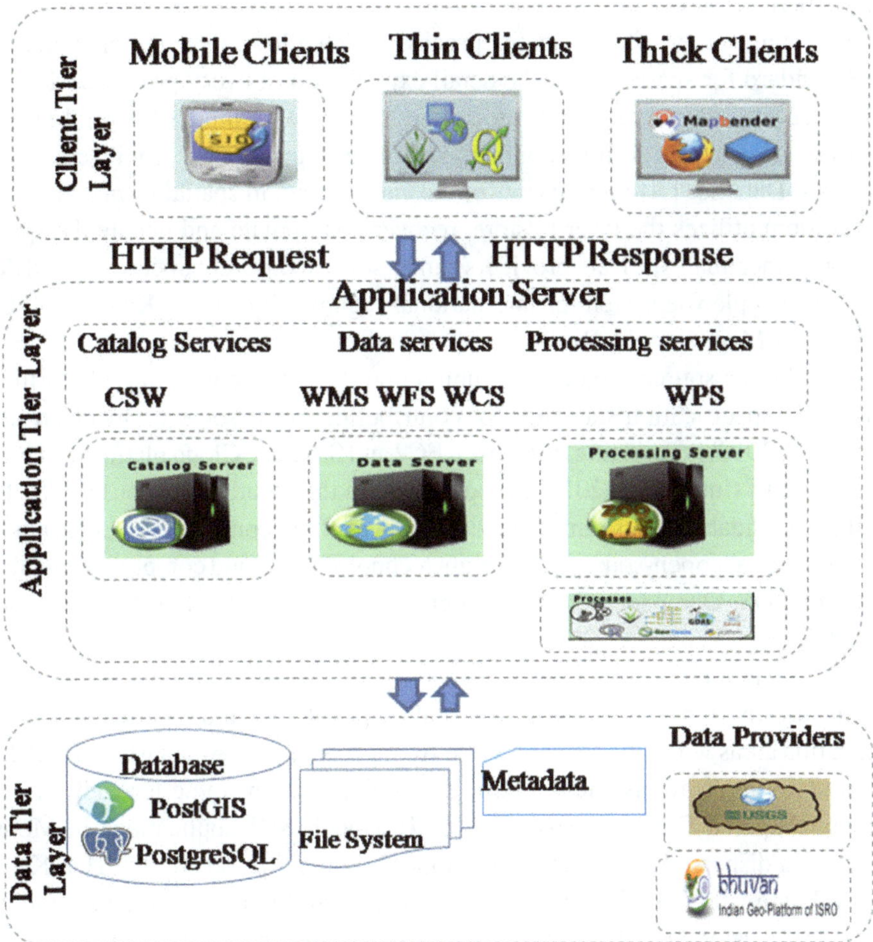

providers. In top of the application tier, dedicated server for application has been operated for different services i.e. Web Map Service (WMS), Web Coverage Service (WCS), Web Features Service (WFS), Catalog service for Web (CS-W) and Web Processing Service (WPS). The dedicated application server has the responsible for requests to and response from client to application server. In addition, application services include three types of server application i.e. catalog servers, data servers and processing servers.

Catalog severs are used to search the metadata information regarding the stored spatial data. Catalog server is one of the important system components for controlling spatial information in cloud environment. In the catalog service, a standard publish-find-bind service framework are implemented which has been defined by OGC web service architecture. Data server deals with the WMS, WCS and WFS (Yue et al., 2013).

Processing server offers a spatial process which allows different clients to use WPS standard for spatial data. The detail explanation of every processes done by client request, forward the desire processing service with input of several factors, specifies and provides definite region in leaping box and feedbacks with composite standards. Data tier Layer comprises of the various data in spatial form and related info. System utilizes the layer to store, recover, manipulate and update the spatial data for further analysis. Data providers can be store in different open source DBMS packages, simple file system or international organizations (e.g., Bhuvan, USGS, Open Street Map, Google, Bing).

Several open source projects are running on the cloud computing platforms in OGC WPS specification standards for contributing to the society. OpenGeo suite (Internet-2, 2017) is operating in Amazon EC2 and Skygone Cloud platform whereas Quantum GIS (Internet-3, 2017) also deployed in the cloud for various GWS. The flexible, ascendable & sustainable environment in cloud computing construct using Docker which act open-source lightweight technology in cloud computing container with Linux OS. Open-source software deployed are IPython, NumPy, GDAL, and Grass GIS etc (Wang et al.,2015).

In the era of mobile computing, cloud computing and spatial technology, several architectures have been proposed for visualization of spatial data in wide range of application areas with OSS resources (De Abreu et al., 2014; Athanasis et al., 2015). Mobile GIS applications entail an effectual collaboration frame for web services in mobile Spatial Data Infrastructure (SDI) Model. SDI applications in Mobile executed in dispersed environment by support of client-server Model (Hae-Kyong and Ki-Joune, 2005; Stefanakis. and Prastacos, 2008; Fiduccia et al., 2016;Gomes et al., 2017). Mobile Data Service (MDS) is architecture for implementing web services of geo-spatial data in mobile client (Chang, 2003). J2ME and mobile scalable vector graphics (SVG) based light mobile GIS framework proposed and implemented where the specification of mobile SVG, mobile XML and the rules for spatial data organizing is used (Chunling et al., 2009).

Several designing architectures implemented in SDI Model based on mobile cloud computing. Several system architectures has been proposed and implemented in android app to display the spatial data visualisation of spatial and non spatial data and also for the data acquisition (Brinkhoff et al., 2008; Brovelli et al., 2016). It has been perceived that mobile computing with cloud devices have great potential for spatial data visualization and collection in diverse field. In forest inventory management, mobile mapping with high resolution remote sensing images, grid maps and vector maps used for data processing and surveying, those are essential tools for the forestry science workers (Brovelli et al., 2014). Air pollution mobile app has also developed with the help of open source software for air quality management, greenhouse gases and emission of air pollutants by California Environmental Protection Agency and United States Environmental Protection Agency (Dong et al., 2010; Gkatzoflias et al., 2013).

SDI has been one of the important requirements for geo-scientist for accurate raw data collection (Chen and Xiao, 2011). EpiCollect and Creek Watch are two examples of mobile apps and are popular for SDI Models in Mobile in field of environmental monitoring and watershed management (Aanensen et al., 2009). It is benefitted for the general people in health sector. Several prototypes have developed and implemented as a mobile app for public health information collection. With the help of earth observation data, an android based mobile app also developed for the accurate dispersal of rainfall estimations and visualise patterns of growth in floral resources (Liu, 2016; Boonchieng et al., 2014; Mantas et al., 2015; Arundel et al., 2016). There are several case studies and prototype have been designed and implemented for water resources management by using cloud computing and mobile technologies with the help of open source software.

With the help of web services, cloud computing and mobile platform, many researchers have developing various kind of web GIS application in the field of water resources management. The project like Hydrologic Information System and HydroShare are the few which has been developed at Consortium of Universities for the Advancement of Hydrologic Science, Inc. (CUAHSI), United State. The main aims of these systems are designed for accessing and publishing the water data (Delipetrev et al., 2014; Ames et al., 2012; Conner et al., 2013; Peckham and Goodall, 2013). There are also capabilities to collaborative environments to share hydrological data and different associated models with different open source software resources (Swain et al., 2016; Swain et al., 2015).

With the integration of mobile, web and spatial technology, it has been a greater potential for numerous functionality in terms of spatial data sharing over internet. It can offer a dynamic way and real time to represent information through maps. Therefore, there is an urgent need to establish a well organised SDI Model which is a geo portal where each and every stakeholder can use, exchange and access spatial data for economic, social and environmental application. Spatial Web Services are the key technologies which are required for development and implementation of SDI (He et al., 2015; Buyya et al., 2009).

Design and implementation of SDI is used Cloud computing technology which it is used for sharing the information about the Temple City, Bhubaneswar, India. It enables data analyst and naive user to quickly look into the problem and gets the information according to their need. From these several reviewing of the research works, it has been summarized that Cloud SDI Model in thick, thin and mobile users, is the need of the hour.

TOURISM INFORMATION INFRASTRUCTURE: INDIAN PROSPECTIVE

Tourism is driven by the natural urge of human being for the new experiences the motivation for tourism includes business interest and a religious fact. India is endowed with significant tourism sites which includes various Temples, Mosques, Churches, Monuments, Lakes, Mountains, Rivers, Forests etc. which attracts tourists not only from across the country but also from different foreign countries (da Cruz Albuquerque et al., 2014; Boori et al., 2015). Progress in the transport, railway networks, air and the facilities provided for the tourist has been encouraged people to venture and experience new world from the other parts beyond the boundaries. Figure 6 shows the graph during last few years India witnessed a significant growth in the Foreign Tourist Arrivals (FTAs) and Foreign Exchange Earnings (FEEs) (Chaudhuri & Ray, 2015; Internet-4, 2016).

The significance of tourism, as an instrument of employment generation and economic development, has been well recognized in the world over. It is a large-scale service industry globally in terms of gross revenue and FEEs. Tourism can an effective role in achieving the growth with equity and has the high potential growth rate which ensures consequential development in the service sector and provides sustainable models of growth. Tourism sector stimulates other economic sectors like: horticulture, poultry, transport, agriculture, construction, handicrafts etc. In

Figure 6. FTAs in India 1997-2011(Internet-5, 2016)

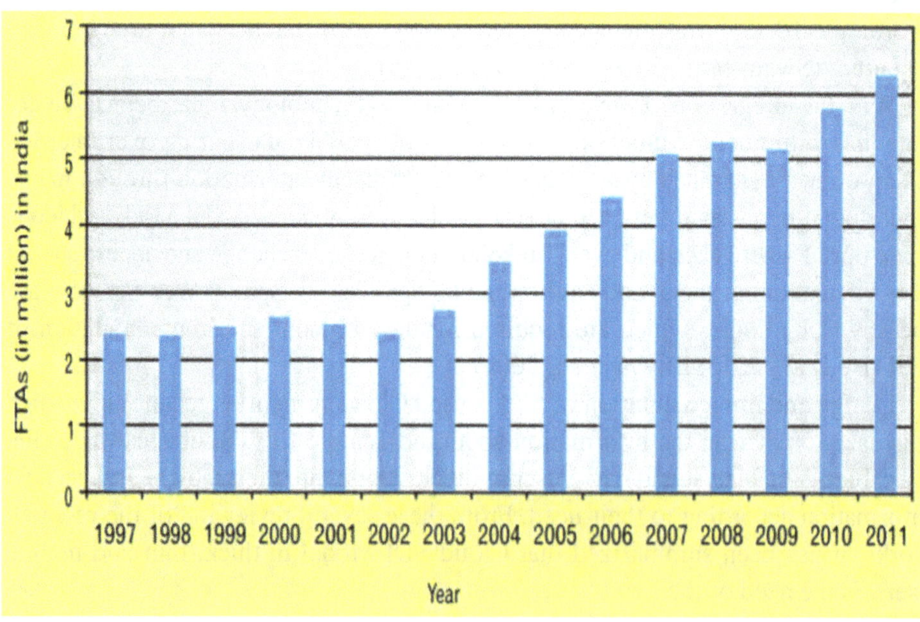

order to improve all the aspects of tourism in the country, it is vital to utilize all the available high-end computing and IT technologies (Lin et al., 2014; Yang et al., 2015). Though various initiatives have been taken by the Ministry of Tourism, Government of India for the rapid development of the concerned sector, still it is a way behind the goal because of the lack of emphasis on a GIS based Tourism Information Infrastructure platform, which enable Tourist to have recommendation and up-to-date information (Sadoun and Al-Bayari, 2009; Baviskar, 2017; McGehee, 2014). Several Models have been implemented in which GIS technology has been taken a prime focus for giving information infrastructure management in tourism sector (Wei, 2011, Singh et al., 2011, Avdimiotis and Christou, 2006 and Gao and Gao, 2011; Ali et al., 2017; García-Palomares et al., 2015). But none of the models have shown cloud computing based with spatial web services. That leads to define a robust cloud computing based SDI Model for tourism information infrastructure management.

OBJECTIVE OF PRESENT STUDY

By going through several literature reviews, the objective of the present study is to development and implementation of prototype based on Cloud SDI Model i.e. TCloud for the temple city, Bhubaneswar, India. It projected the system architecture of TCloud particularly on desktop, web browser and mobile environment. It has also proposed an approach for the development of spatial database for the temple city, Bhubaneswar, India. It also elaborated the comparative analysis of Quantum GIS and Map Window GIS which have been utilized for tourism spatial database creation. Finally, it performed various overlay analysis of tourism spatial database in thick, thin and mobile environments.

Proposed System Architecture of TCloud

The architecture of TCloud classified into two layers. Initial phase of the layer is Cloud SDI layer. In this layer, spatial database stored with the help of Quantum GIS. There are several data providers i.e. Google and Open Street map have been used with the developed database for further analysis. In this layer, catalog services for searching of desired services have been maintained by catalog service. For invocation of different services such as WMS and WFS, cloud server has been maintained thorough the QGIS Cloud Provider. QGIS Cloud Provider is overall responsible for the management of cloud SDI layer. Finally, the client tier layer is the layer where the three categories of users have been used to work on the developed Model. Figure 7 has shown the proposed system architecture of the TCloud.

Figure 7. Proposed System architecture of TCloud framework for processing of tourism spatial data

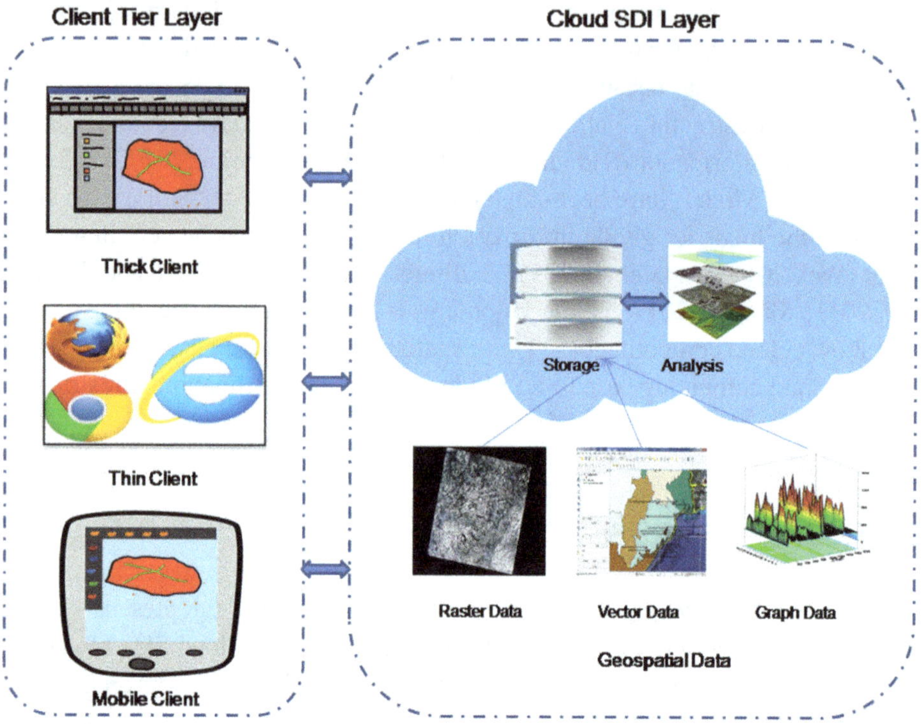

METHODOLOGY ADOPTED

For creation of spatial database, the prime emphasis has been on the real-world approach to discover and spread the thought of spatial database creation in academics sector. The established spatial database has to provide a proficient means of allocation of spatial and non-spatial data in Cloud SDI Model. The prototype is based on the spiral model to combine the time critical nature and strong user focus. Figure 8 represents the fully spiral model procedure for development of TCloud.

In addition, it is expected that each phase would reveal unique features related to the requirements of the underlying infrastructure based web services and enable exploration of the interfaces between SDI components. The flow of processes in the system is captured in the form of state chart diagram that need to be translated into design element of the software. In design phase, sequence diagram describe the workflow behaviour of TCloud and is shown in Figure 9.

TCloud framework consists of three main functional phases. Module I provides the comparison analysis of Quantum GIS and Map Window GIS software and the

Figure 8. Spiral Model approaches for development and implementation of TCloud framework

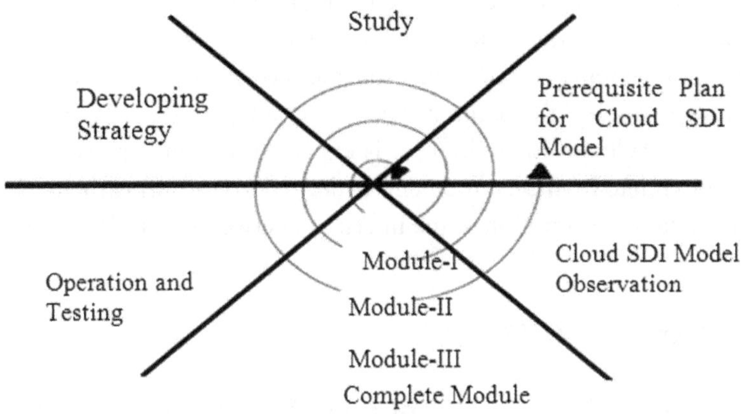

Figure 9. Sequence Diagram for TCloud framework

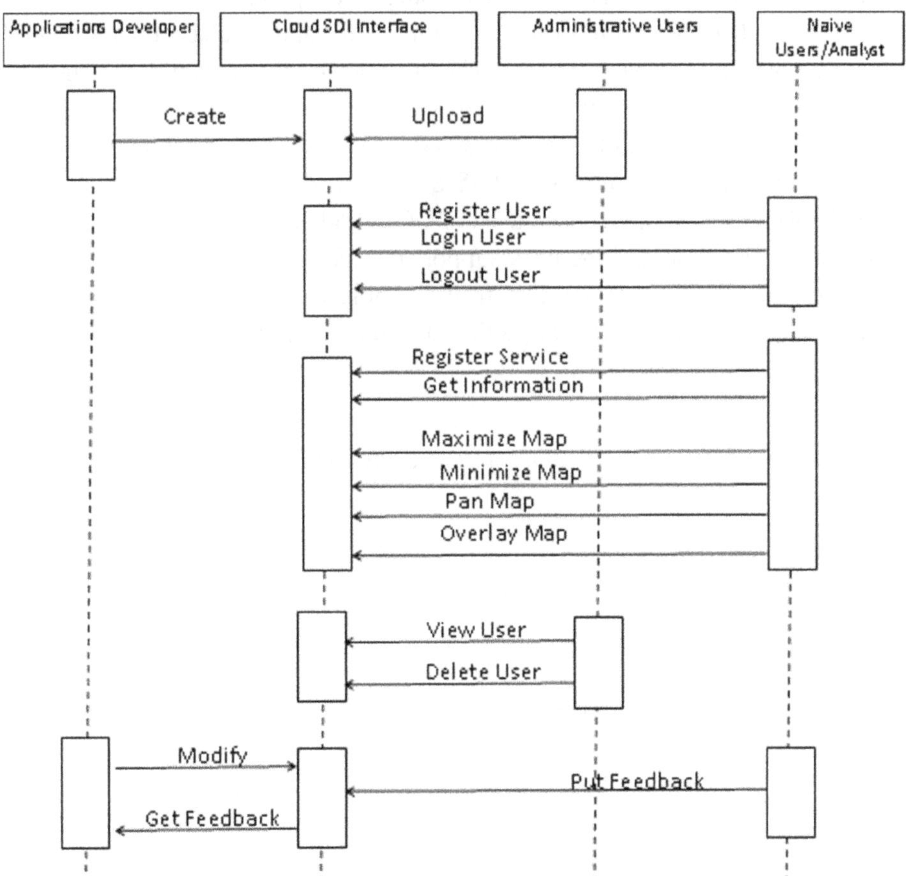

database created for the spatial database for Temple City, Bhubaneswar, India. Module II emphasizes on the development of various OGC compliant services such as WMS and WFS. It also describes the integration of these services with respect to the QGIS Cloud Providers. Module III describes the integration of the developed Model with the different types of users. From the above analysis and designing phase, the core issue is to select the optimized adoption of Open Source GIS for complete development of TCloud. In addition, it is expected that each phase would reveal unique features related to the requirements of the underlying infrastructure based web services and enable exploration of the interfaces between SDI Model components.

RESULTS AND DISCUSSION

Comparison Analysis of Quantum GIS and Map Window GIS

An OS GIS licensed under the GNU General Public License is Quantum GIS (QGIS). The formal assignment of Open Source Geospatial Foundation (OSGeo) is QGIS. It has been supported various raster, vector and database formats. QGIS project was officially released in May of 2002 when coding began. It is a multiple stage application and executed on various OS like UNIX, Linux, Mac OS X, and Microsoft Windows. It can be utilized as GUI to GRASS and having trivial size of file in comparison to commercial GIS Software. SPIT (Shape file to Post GIS Import Tool) plug-in tool has been used in QGIS and can be used to load multiple shape files at one time and includes support for schemas. To use SPIT in QGIS, it needs to Open the plug-in manager from the plug-in menu and check the box to the SPIT plug-in. Figure 10(a) shows the snapshot of SPIT plug-in tool with filed input box (Quantum GIS, 2010).

Figure 10. (a) SPIT Plugin Tool. (b) Spatial Converter Tool

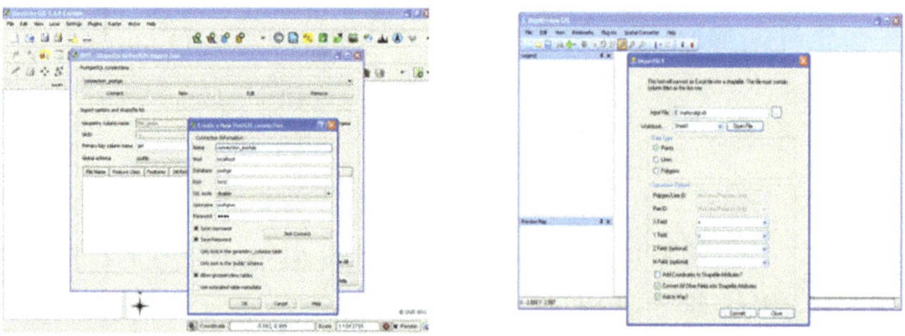

MapWindow GIS is an OSGIS Software which has collections of programmable planned modules. It is adopted as the primary GIS platform for its BASINS (Better Assessment Science Integrating Point and Nonpoint Sources) watershed analysis and modeling software by the United States Environmental Protection Agency. It is purely developed under Microsoft .NET Technology platform. It includes plug-ins for several geo processing jobs like accessing online data sources, buffer and merge, watershed delineation and an experimental geodatabase plug-in. Map Window GIS is provisioned with wide variety of data format and read/ write ESRI outline records. The structures and aspects of spatial have been revised in accordance to the requirements for managing the database. Map Window GIS has also unique plug-ins 'Spatial Converter' to create spatial database from excel datasheet. Spatial converter also has the features to import and export ESRI shape files from different file formats. Figure 10(b) shows the snap shot of spatial converter tool in which Excel file imported to create ESRI shape file in point data (MAPWINGIS, 2007).

The following Table summarized the analysis of two OSGIS in terms of various parameters associated with in the user points of level.

From, the above comparison analysis, it has been considered to be used these two OSGIS software according to the user requirement and the platform which has to be demand for the development of spatial database creation. Thus, the next section describes about the tourism spatial database creation.

Tourism Spatial Database Creation

The creation of GIS digital database is significant & tedious assignment where efficacy in GIS project depends upon. Integrated spatial database creation include stages such as inputs of data on spatial and attribute, its authentication by connecting with same set of data. Spatial database delivers a platform in which organizations interrelate with technologies to nurture actions for spending, handling and generating geographic data (Yang et al., 2011). The development of spatial database supports in various administrative and political levels through these decision-making functions.

Table 1. Comparison Analysis of Quantum GIS and Map Window GIS.

Spatial Database Creation Software	Open source	Free Software	Web	Linux/ BSD/ Unix	Mac	Windows	Other
Quantum GIS	Yes	Yes	Yes	Yes	Yes	Yes	Google Earth Plugin, KML, WMS
Map Window GIS	Yes	Yes	No	No	No	Yes	No

Quantum GIS 1.6.0 and Map Window GIS 4.8 are two OS GIS software selected to examine the competences w.r.t. creation of spatial database.

The procedure model of spatial database creation is recurring or frequent in nature and each operation improves the study and strategy steps through assessment and testing of a completed component. In complete component, Quantum GIS Open Source GIS software has set up toursim spatial database by the help of political map of India. QGIS is also used for integrated spatial database creation. The tourism spatial database has been nominated to illustrate the capabilities of developed framework. Spatial Database for toursim Sector has been prepared by Quantum GIS Ver. 1.6.0 and Map Window GIS Ver. 4.8.1.

Initially, the base image of India has been downloaded from the Google Earth. The downloaded image is geo-referenced with the help of Geo-referencer tool in Quantum GIS Ver. 1.6.0. For geo-referencing the base map of India, 10 numbers of GCPs have been taken. The GCPs have been selected at the intersection of latitude and longitude lines. For universal coordinate system, WGS-84 with EPSG:4326 coordinate reference system has been chosen. Now, the image is ready for geo-referencing. After geo-referencing, the generated image is used to extract the thematic maps. Figure 11 shows the snapshot of geo-referencing of India map in

Figure 11. Geo-referencing: Map of India

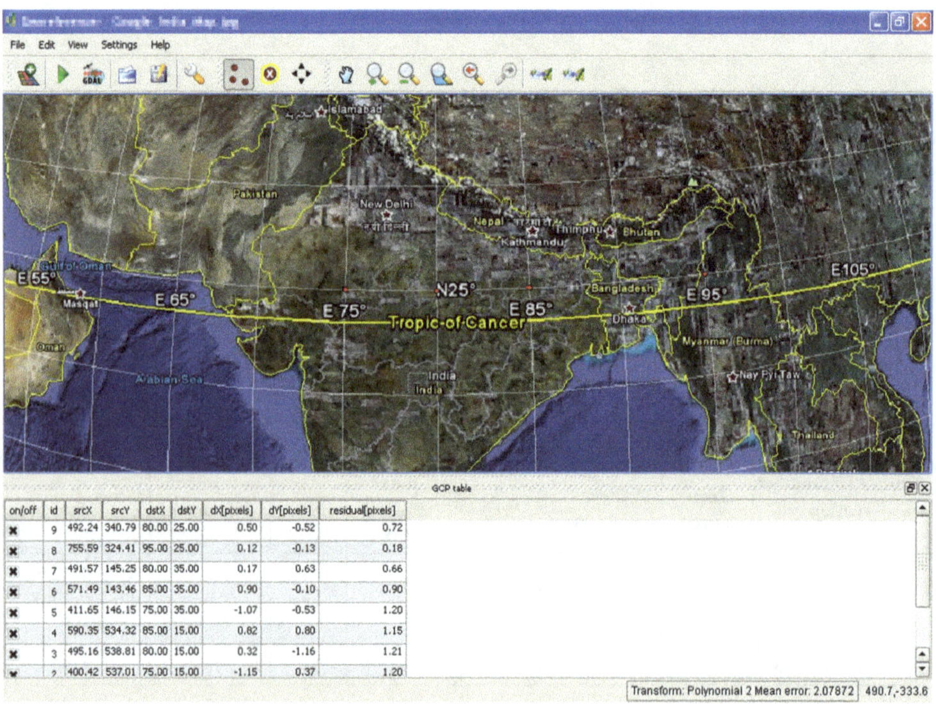

Geo-referencer Tool from Quantum GIS Ver. 1.6.0. In the present application case study, the entire tourism related informations of Odisha have been taken. These have been categorised into the different layers with schema definition.

Spatial Data Visualisation of Temple City, Bhubaneswar

The procedure Model of spatial database creation is recurring or frequent in nature and each operation improves the study and strategy steps through assessment and testing of a completed component. In complete component, Quantum GIS Open Source GIS software has set up the spatial database for Temple City, Bhubaneswar in India. Quantum GIS Ver. 2.14.10 is used for integrated spatial data visualisation of all the seven different shape files related to Temple City, Bhubaneswar, India. For universal coordinate system, WGS-84 with EPSG:4326 coordinate reference system has been chosen in the visualisation of spatial database. In Figure 12, it has been visualisation with google maps and all the shape files related to Temple City, Bhubaneswar are invoked in one platform on Quantum GIS desktop environment.

Overlay Analysis in TCloud

In this section, spatial data analysis particularly overlay analysis has been performed for well location type spatial data of Temple City, Bhubaneswar, India. Overlay Analysis has been one of the important spatial data analyses. It has superimposed various raster and vector spatial data in a common platform. It has been designed

Figure 12. Integrated spatial database of the temple city, Bhubaneswar

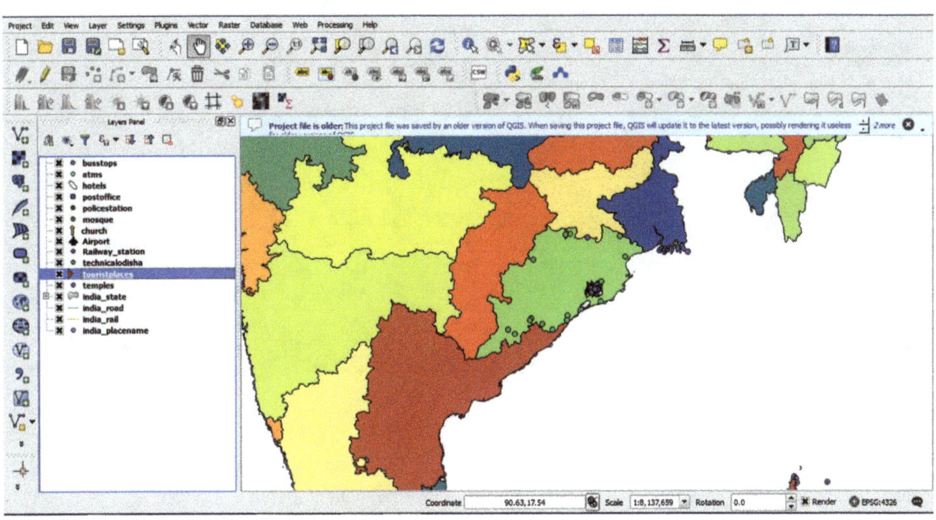

and developed the dataset both raster and vector spatial data with the help of QGIS. It has been found that the raster data in ESRI grid format and 12 number of vector data related to Temple City, Bhubaneswar in shape file formats. Road network, Administrative boundaries, Temple location, ATM location, church, postal and police station in Temple City, Bhubaneswar are the vector data have been used; which are stored in ESRI shape file formats. In the present study, it has been used the vector data for processing in TCloud. The overlay analysis of various vector data and raster data of particular area has been performed. Initially, the developed datasets have been opened with Quantum GIS; desktop based GIS analysis tools, and performed some join operations.

In Quantum GIS, plugin named as QGISCloud has been installed. The said plugin has the capability of storing various raster and vector data set in cloud database for further overlay analysis. After storing in cloud database, it generates the mobile and thin client link for visualization of both vector and raster data set. The overlay analyses on localhost, thin and mobile client have been shown in Figure 13, 14 and 15 respectively. It can observe that the overlay analysis is a useful technique for visualization of spatial data.

Figure 13. Overlay operation on thin client local environment in TCloud

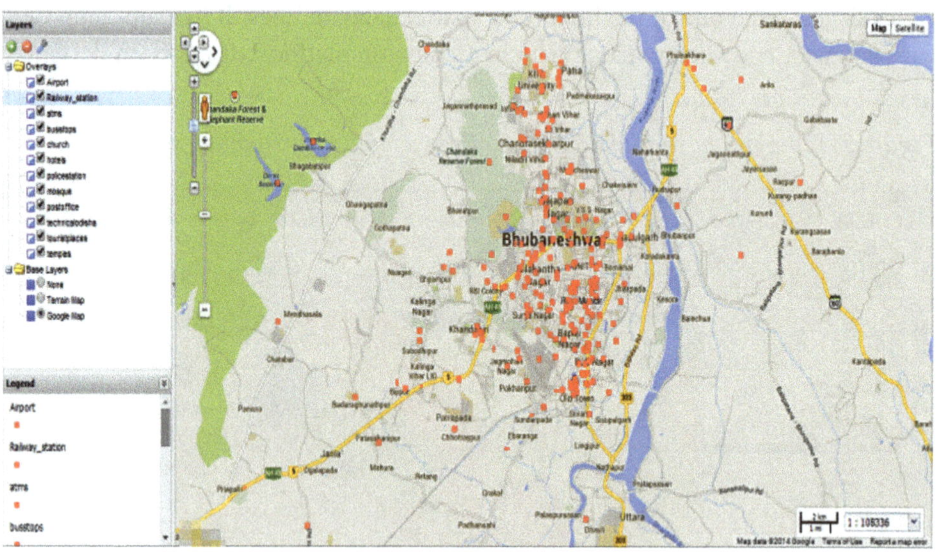

Figure 14. Overlay operation on thin client local environment in TCloud (Internet-6, 2017)

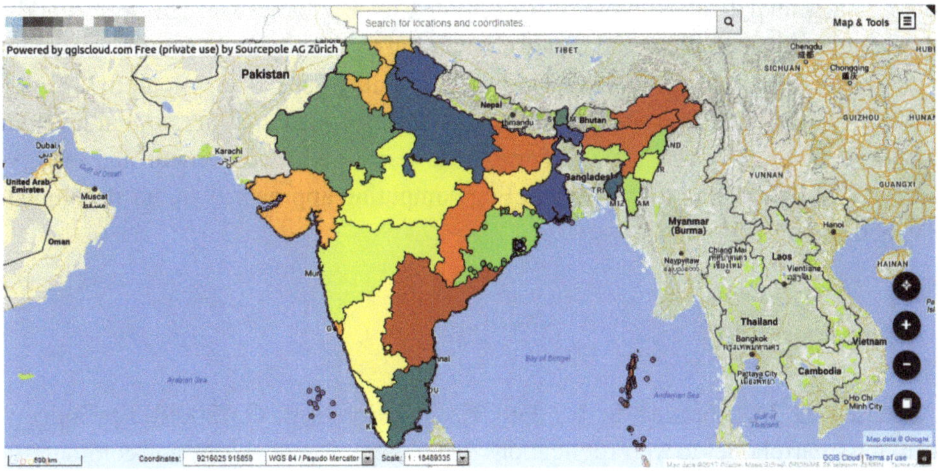

Figure 15. Overlay operation on mobile client environment in TCloud (Internet-7, 2017)

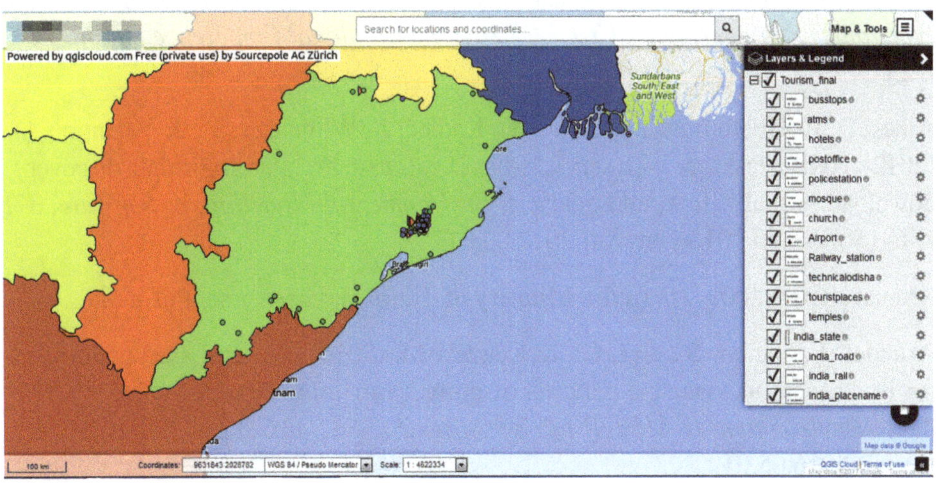

CONCLUSION

In this paper, it has been developed and validated TCloud that employed cloud computing gateway in Spatial Cloud Computing Framework. It leads to overall efficiency of spatial data processing using TCloud framework which is benefit to the numerous decision maker/end users. Spatial database of Temple City, Bhubaneswar has been taken as case study in our developed TCloud framework. Here, it has also introduced the optimum adaptation of open source GIS software for the development of tourism spatial database creation for Temple City, Bhubaneswar. This feature adds the edge intelligence in Spatial Cloud Computing environment. In future, we would like to add more intelligent processing and feasibility aspects on Cloud SDI Layer for achievement of successful SDI Framework implementation at national level in health, education, coastal, natural resource, watershed management and energy sector. It will also plan to implement Fog computing approaches in our developed TCloud framework.

REFERENCES

Aanensen, D. M., Huntley, D. M., Feil, E. J., & Spratt, B. G. (2009). EpiCollect: Linking smartphones to web applications for epidemiology, ecology and community data collection. *PLoS One*, *4*(9), e6968. doi:10.1371/journal.pone.0006968 PMID:19756138

Ali, T. A. T., Saeed, R. A., & Fageeri, S. O. (2017). Web-based GIS Business Hotels Tourism Sites in Khartoum, Sudan. In *Communication, Control, Computing and Electronics Engineering (ICCCCEE), 2017 International Conference on* (pp. 1-5). IEEE.

Ames, D. P., Horsburgh, J. S., Cao, Y., Kadlec, J., Whiteaker, T., & Valentine, D. (2012). HydroDesktop: Web services-based software for hydrologic data discovery, download, visualization, and analysis. *Environmental Modelling & Software*, *37*, 146–156. doi:10.1016/j.envsoft.2012.03.013

Annual Report 2009–10. (n.d.). Ministry of Tourism, Government of India.

Arundel, J., Winter, S., Gui, G., & Keatley, M. (2016). A web-based application for beekeepers to visualise patterns of growth in floral resources using MODIS data. *Environmental Modelling & Software*, *83*, 116–125. doi:10.1016/j.envsoft.2016.05.010

Athanasis, N., Karagiannis, F., Palaiologou, P., Vasilakos, C., & Kalabokidis, K. (2015). AEGIS App: Wildfire Information Management for Windows Phone Devices. *Procedia Computer Science*, *56*, 544–549. doi:10.1016/j.procs.2015.07.249

Avdimiotis, S., & Christou, E. (2006). GIS Applications in Tourism Planning: A Tool for Sustainable Development Involving Local Communities. *23rd Euro Chrie Conference.*

Barik, R. K., Das, P. K., & Lenka, R. K. (2016). Development and implementation of SOA based SDI model for tourism information infrastructure management web services. In *Cloud System and Big Data Engineering (Confluence), 2016 6th International Conference* (pp. 748-753). IEEE. doi:10.1109/CONFLUENCE.2016.7723633

Barik, R. K., & Samaddar, A. B. (2014a). Service Oriented Architecture Based SDI Model for Education Sector in India. *Proceedings of the International Conference on Frontiers of Intelligent Computing: Theory and Applications (FICTA)*, 555-562. doi:10.1007/978-3-319-02931-3_63

Barik, R. K., & Samaddar, A. B. (2014b). Service Oriented Architecture Based SDI Model for Mineral Resources Management in India. *Universal Journal of Geosciences (Prague)*, *2*, 1–6.

Barik, R. K., Samaddar, A. B., & Gupta, R. D. (2009). Investigations into the Efficacy of Open Source GIS Software. *Map World Forum, on Spatial Technology for Sustainable Planet Earth.*

Barik, R. K., Samaddar, A. B., & Samaddar, S. G. (2011). Service Oriented Architecture based SDI Model for Geographical Indication Web Services. *International Journal of Computers and Applications*, *25*, 42–49. doi:10.5120/3017-4077

Baviskar, M. R. (2017). Designing and Developing a GIS Database for Tourism in Aurangabad: A Review. *International Journal of Computers and Applications*, *165*(5).

Boonchieng, E., Boonchieng, W., Senaratana, W., & Singkaew, J. (2014). Development of mHealth for public health information collection, with GIS, using private cloud: A case study of Saraphi district, Chiang Mai, Thailand. *Computer Science and Engineering Conference (ICSEC), 2014 International*, 350-353. doi:10.1109/ICSEC.2014.6978221

Boori, M. S., Voženílek, V., & Choudhary, K. (2015). Land use/cover disturbance due to tourism in Jeseniky Mountain, Czech Republic: A remote sensing and GIS based approach. *The Egyptian Journal of Remote Sensing and Space Sciences*, *18*(1), 17–26. doi:10.1016/j.ejrs.2014.12.002

Brinkhoff, T. (2008). Supporting mobile GIS applications by spatial web services. *The International Archives of the Photogrammetry, Remote Sensing and Spatial Information Sciences*, *37*, 865–870.

Brovelli, M. A., Minghini, M., Moreno-Sanchez, R., & Oliveira, R. (2016). Free and open source software for spatial applications (FOSS4G) to support Future Earth. *International Journal of Digital Earth*, 1–19.

Brovelli, M. A., Minghini, M., & Zamboni, G. (2014). Public participation GIS: A FOSS architecture enabling field-data collection. *International Journal of Digital Earth, 8*(5), 345–363. doi:10.1080/17538947.2014.887150

Brovelli, M. A., Minghini, M., & Zamboni, G. (2016). Public participation in GIS via mobile applications'. *ISPRS Journal of Photogrammetry and Remote Sensing, 114*, 306–315. doi:10.1016/j.isprsjprs.2015.04.002

Buyya, R., Yeo, C. S., Venugopal, S., Broberg, J., & Brandic, I. (2009). Cloud computing and emerging IT platforms: Vision, hype, and reality for delivering computing as the 5th utility. *Future Generation Computer Systems, 25*(6), 599–616. doi:10.1016/j.future.2008.12.001

Chang, Z. (2003). *Design and Implementation of distributed Web Services in Mobile GIS Application* (Master's Thesis). International Institute for Geo-information Science and Earth Observation, The Netherlands.

Chaudhuri, S., & Ray, N. (2015). Application of Web-Based Geographical Information System (GIS) In Tourism Development. In Encyclopedia of Information Science and Technology (3rd ed.; pp. 7026-7036). IGI Global.

Chen, H., & Xiao, K. (2011). The design and implementation of the geological data acquisition system based on mobile GIS. *Geoinformatics, 2011 19th IEEE International Conference* on, 1-6.

Chen, Z., Chen, N., Yang, C., & Di, L. (2012). Cloud computing enabled web processing service for earth observation data processing. *IEEE Journal of Selected Topics in Applied Earth Observations and Remote Sensing, 5*(6), 1637–1649. doi:10.1109/JSTARS.2012.2205372

Chunling, Y., Man, Y., & Jianwei, N. (2009). A Light Mobile GIS Framework Based on Mobile SVG. *Computer Science and Information Engineering, 2009 WRI World Congress on, 7*, 616 – 619.

Coleman, D. J., Rajabifard, A., & Kolodziej, K. W. (2016). Expanding the SDI environment: Comparing current spatial data infrastructure with emerging indoor location-based services. *International Journal of Digital Earth, 9*(6), 629–647. doi:10.1080/17538947.2015.1119207

Conner, L. G., Ames, D. P., & Gill, R. A. (2013). HydroServer Lite as an open source solution for archiving and sharing environmental data for independent university labs. *Ecological Informatics, 18*, 171–177. doi:10.1016/j.ecoinf.2013.08.006

GSDI Cookbook. (2004). *Developing spatial data infrastructure: The SDI Cookbook, 2004*. Available from: http://www.gsdi.org/docs2004/Cookbook/ cookbookV2.0.pdf

da Cruz Albuquerque, H. C., Cardoso Pedrosa Martins, F. M., & Tomás Galiza Cardoso, L. M. (2014). *Geographical information systems, tourism*. Springer International Publishing.

de Abreu, F. (2014). Development of a Mobile Mapping Solution for Spatial Data Collection Using Open-Source Technologies. *Procedia Technology*, *16*, 481–490. doi:10.1016/j.protcy.2014.10.115

Delipetrev, B., Jonoski, A., & Solomatine, D. P. (2014). Development of a web application for water resources based on open source software. *Computers & Geosciences*, *62*, 35–42. doi:10.1016/j.cageo.2013.09.012

Dong, C., Liu, F., Wang, H., & Chen, F. (2010). Application research of mobile GIS in forestry informatization. *Computer Science and Education (ICCSE), 2010 5th IEEE International Conference on*, 351-355.

Evangelidis, K., Konstantinos, N., Stathis, M., & Constantine, P. (2014). Spatial services in the Cloud. *Computers & Geosciences*, *63*, 116–122. doi:10.1016/j.cageo.2013.10.007

Fiduccia, A., Gugliermetti, F., Pagliaro, F., Mattoni, B., Nardecchia, F., & Canu, C. (2016). Web-Based Spatial Decision Support Systems to Monitor and Manage Coastal Environments. *18th Italian National Conference on Photonic Technologies*. doi:10.1049/cp.2016.0920

Gao, S., & Gao, C. (2011). Design and Implementation of Tourism Geographical Information System Based on ArcIMS. *IEEE International Conference on Computer Science and Service System*, 3916-3919.

García-Palomares, J. C., Gutiérrez, J., & Mínguez, C. (2015). Identification of tourist hot spots based on social networks: A comparative analysis of European metropolises using photo-sharing services and GIS. *Applied Geography (Sevenoaks, England)*, *63*, 408–417. doi:10.1016/j.apgeog.2015.08.002

Georis-Creuseveau, J., Claramunt, C., & Gourmelon, F. (2016). A modelling framework for the study of Spatial Data Infrastructures applied to coastal management and planning. *International Journal of Geographical Information Science*, *31*(1), 122–138. doi:10.1080/13658816.2016.1188929

Giuliani, G., Lacroix, P., Guigoz, Y., Roncella, R., Bigagli, L., Santoro, M., ... Lehmann, A. (2016). Bringing GEOSS Services into Practice: A Capacity Building Resource on Spatial Data Infrastructures (SDI). *Transactions in GIS*.

Gkatzoflias, D., Mellios, G., & Samaras, Z. (2013). Development of a web GIS application for emissions inventory spatial allocation based on open source software tools. *Computers & Geosciences*, *52*, 21–33. doi:10.1016/j.cageo.2012.10.011

Gomes, J. L., Jesus, G., Rogeiro, J., Oliveira, A., da Costa, R. T., & Fortunato, A. B. (2017). An Innovative Web Platform for Flood Risk Management. Advances in Network Systems, 217-231. doi:10.1007/978-3-319-44354-6_13

Guan, W. W., Bol, P. K., Lewis, B. G., Bertrand, M., Berman, M. L., & Blossom, J. C. (2012). WorldMap–a spatial framework for collaborative research. *Annals of GIS*, *18*(2), 121–134. doi:10.1080/19475683.2012.668559

Gupta, A., & Merchant, P. S. (2016). Automated Lane Detection by K-means Clustering: A Machine Learning Approach. *Electronic Imaging*, *14*, 1–6. doi:10.2352/ISSN.2470-1173.2016.1.VDA-502

Hae-Kyong, K., & Ki-Joune, L. (2005). A Framework for Dynamic Updates of Map Data in Mobile Devices. *Web and Wireless Geographical Information Systems, Springer*, *3833*, 66–77. doi:10.1007/11599289_7

Hamylton, S. M., & Prosper, J. (2012). Development of a spatial data infrastructure for coastal management in the Amirante Islands,Seychelles. *International Journal of Applied Earth Observation and Geoinformation*, *19*, 24–30. doi:10.1016/j.jag.2012.04.004

Harper, E. (2006). *Open Source Technologies in Web-based GIS and Mapping* (Master's Thesis). Northwest Missouri State University, Maryville, MO.

He, L., Yue, P., Di, L., Zhang, M., & Hu, L. (2015). Adding spatial data provenance into SDI—a service-oriented approach. *IEEE Journal of Selected Topics in Applied Earth Observations and Remote Sensing*, *8*(2), 926–936. doi:10.1109/JSTARS.2014.2340737

Huang, Q., Yang, C., Liu, K., Xia, J., Xu, C., Li, J., ... Li, Z. (2013). Evaluating open-source cloud computing solutions for geosciences. *Computers & Geosciences*, 5941–5952.

Idrees, M. O. I. (2015). Challenges in coastal spatial data infrastructure implementation: A review. *South African Journal of Geomatics*, *4*(4), 369–383. doi:10.4314/sajg.v4i4.2

India Tourism Statistics at a Glance. (n.d.). Market Research Division, Ministry of Tourism, Government of India.

Kim, D.-H., & Kim, M.-S. (2002). Web GIS Service Component based on Open Environment. *IEEE Geoscience and Remote Sensing Symposium, 6*, 3346-3348.

Koswatte, S., McDougall, K., & Liu, X. (2015). SDI and crowdsourced spatial information management automation for disaster management. *Survey Review, 47*(344), 307–315. doi:10.1179/1752270615Y.0000000008

Laura, J. R., Hare, T. M., Gaddis, L. R., Fergason, R. L., Skinner, J. A., Hagerty, J. J., & Archinal, B. A. (2017). Towards a Planetary Spatial Data Infrastructure. *ISPRS International Journal of Geo-Information, 6*(6), 181. doi:10.3390/ijgi6060181

Leidig, M., & Teeuw, R. (2015). Free software: A review, in the context of disaster management. *International Journal of Applied Earth Observation and Geoinformation, 42*, 49–56. doi:10.1016/j.jag.2015.05.012

Li, H., Lu, J., Cai, B., & Yao, S. (2008). Study on SOA-Orient WebGIS framework. *14th IEEE International Conference on Automation and Computing.*

Lin, P. J., Kao, C. C., Lam, K. H., & Tsai, I. C. (2014). Design and implementation of a tourism system using mobile augmented reality and GIS technologies. In *Proceedings of the 2nd International Conference on Intelligent Technologies and Engineering Systems (ICITES2013)* (pp. 1093-1099). Springer. doi:10.1007/978-3-319-04573-3_133

Mansourian, A., Rajabifard, A., Valadan Zoej, M. J., & Williamson, I. P. (2005). Using SDI and web-based system to facilitate disaster management. *International Journal of Computers & Geosciences, 32*(3), 303–315. doi:10.1016/j.cageo.2005.06.017

Matans, V. M., Liu, Z., & Pereira, A. J. S. C. (2015). A web service and android application for the distribution of rainfall estimates and Earth observation data. *Computers & Geosciences, 77*, 66–76. doi:10.1016/j.cageo.2015.01.011

McGehee, N. G. (2014). Volunteer tourism: Evolution, issues and futures. *Journal of Sustainable Tourism, 22*(6), 847–854. doi:10.1080/09669582.2014.907299

Morris Steven, P. (2006). Geospatial Web Services and Geo archiving: New Opportunities and Challenges in Geographic Information Services. *Library Trends, 55*(2), 285–303. doi:10.1353/lib.2006.0059

Mwange, C., Mulaku, G. C., & Siriba, D. N. (2016). Reviewing the status of national spatial data infrastructures in Africa. *Survey Review*, 1–10. doi:10.1080/00396265.2016.1259720

Nasr, M. R. (2007). *Open Source Software: The use of open source GIS software and its impact on organizations* (Doctoral dissertation). Middlesex University.

OGC. (2004a). *Open GIS Web Map Server Cookbook, 2004*. Available from: http://www.opengeospatial.org/standards/wms

OGC. (2004b). *Open GIS Web Map Service Implementation Specification, Version 1.3.0, 2004*. Available from http://portal.opengeospatial.org/files/?artifact_id=4756

OGC. (2005). *OpenGIS Web Feature Service Implementation Specification, Version 1.1.0*. Retrieved from http://portal.opengeospatial.org/files/?artifact_id=8339

Patra, S. S., & Barik, R. K. (2015). Dynamic Dedicated Server Allocation for Service Oriented Multi-Agent Data Intensive Architecture in Biomedical and Spatial Cloud. In *Cloud Technology: Concepts, Methodologies, Tools, and Applications* (pp. 2262–2273). IGI Global. doi:10.4018/978-1-4666-6539-2.ch107

Peckham, S. D., & Goodall, J. L. (2013). Driving plug-and-play models with data from web services: A demonstration of interoperability between CSDMS and CUAHSI-HIS. *Computers & Geosciences*, *53*, 154–161. doi:10.1016/j.cageo.2012.04.019

Pun-Cheng, L. S., Lai, W. W., & Chang, R. K. (2016). Exploring utility system SDI–Managerial and technical perspectives. *Tunnelling and Underground Space Technology*, *54*, 13–19. doi:10.1016/j.tust.2016.01.026

Puri, S. K., Sahay, S., & Georgiadou, Y. (2007). A Metaphor-Based Sociotechnical Perspective on Spatial Data Infrastructure Implementations: Some Lessons from India, Research and Theory in Advancing Spatial Data Infrastructure Concepts. ESRI Press.

Quantum GIS User Guide, Version 1.6.0. (2010). Academic Press.

Raghunathan, S., Prasad, A., Mishra, B. K., & Chang, H. (2005). Open source versus closed source: Software quality in monopoly and competitive markets. *IEEE Transactions on Systems, Man, and Cybernetics. Part A, Systems and Humans*, *35*(6), 903–918. doi:10.1109/TSMCA.2005.853493

Rajabifard, A., Feeney, M. E. F., & Williamson, I. P. (2002). Future Directions for SDI Development. *International Journal of Applied Earth Observation and Geoinformation ITC, 4*(1), 11–22.

Reference Manual.MAPWINGIS (2007). MAPWINGIS.

Sadoun, B., & Al-Bayari, O. (2009). A GIS system for tourism management. *IEEE International Conference on AICCSA*, 226-232.

Schäffer, B., Baranski, B., & Foerster, T. (2010). Towards Spatial Data Infrastructures in the Clouds. *Spatial Thinking Lecture Notes in Geoinformation and Cartography, 0*, 399–418. doi:10.1007/978-3-642-12326-9_21

Senaratne, H., Mobasheri, A., Ali, A. L., Capineri, C., & Haklay, M. (2017). A review of volunteered geographic information quality assessment methods. *International Journal of Geographical Information Science, 31*(1), 139–167. doi:10.1080/13658816.2016.1189556

Singh, S. P., Sharma, J., & Singh, P. (2011). A GeoReferenced Information System for Tourism (GeoRIST). *International Journal of Geomatics and Geosciences, 2*(2), 456–464.

Smith, J., Mackaness, W., Kealy, A., & Williamson, I. (2004). Spatial data infrastructure requirements for mobile location based journey planning. *Transactions in GIS, 8*(1), 23–44. doi:10.1111/j.1467-9671.2004.00166.x

Stefanakis, E., & Prastacos, P. (2008). Development of a coastal SDI using GeoFOSS. *the Proceedings of the 11th AGILE Conference.*

Strain, L., Rajabifard, A., & Williamson, I. (2004). Spatial Data Infrastructure to facilitate coastal zone management. *Coastal Zone Asia Pacific Conference*, 1-6.

Supak, S. K., Devine, H. A., Brothers, G. L., Rozier Rich, S., & Shen, W. (2014). An open source web-mapping system for tourism planning and marketing. *Journal of Travel & Tourism Marketing, 31*(7), 835–853. doi:10.1080/10548408.2014.890153

Swain, N. R., Christensen, S. D., Snow, A. D., Dolder, H., Espinoza-Dávalos, G., Goharian, E., ... Burian, S. J. (2016). A new open source platform for lowering the barrier for environmental web app development. *Environmental Modelling & Software, 85*, 11–26. doi:10.1016/j.envsoft.2016.08.003

Swain, N. R., Latu, K., Christensen, S. D., Jones, N. L., Nelson, E. J., Ames, D. P., & Williams, G. P. (2015). A review of open source software solutions for developing water resources web applications. *Environmental Modelling & Software, 67*, 108–117. doi:10.1016/j.envsoft.2015.01.014

Tang, Z., & Liu, T. (2016). Evaluating Internet-based public participation GIS (PPGIS) and volunteered geographic information (VGI) in environmental planning and management. *Journal of Environmental Planning and Management, 9*(6), 1073–1090. doi:10.1080/09640568.2015.1054477

Vaccari, L., Shvaiko, P., & Marchese, M. (2009). A geo-service semantic integration in Spatial Data Infrastructure. *International Journal of Spatial Data Infrastructures Research*, *4*, 24–51.

Wang, X. Z., Zhang, H. M., Zhao, J. H., Lin, Q. H., Zhou, Y. C., & Li, J. H. (2015). An Interactive Web-Based Analysis Framework for Remote Sensing Cloud Computing, *ISPRS Annals of the Photogrammetry. Remote Sensing and Spatial Information Sciences*, *4*, W2.

Wei, W. (2011). Research on the Application of Geographic Information System in Tourism Management. *International Conference on Environmental Science and Engineering- ICESE 2011.*

Xiaolin, Lu. (2005). An Investigation on Service Oriented Architecture for constructing Distributed WebGIS Application. *IEEE International Conference on Services Computing (SCC'05)*, *1*, 191-197.

Yang, C., Goodchild, M., Huang, Q., Nebert, D., Raskin, R., Xu, Y., ... Fay, D. (2011). Spatial cloud computing: How can the spatial sciences use and help shape cloud computing? *International Journal of Digital Earth*, *4*(4), 305–329. doi:10.1080/17538947.2011.587547

Yang, C., Huang, Q., Li, Z., Liu, K., & Hu, F. (2017). Big Data and cloud computing: Innovation opportunities and challenges. *International Journal of Digital Earth*, *10*(1), 13–53. doi:10.1080/17538947.2016.1239771

Yang, C., Raskin, R., Michael, G., & Gahegan, M. (2010). Spatial cyberinfrastructure: Past, present and future. *Computers, Environment and Urban Systems*, *34*(4), 264–277. doi:10.1016/j.compenvurbsys.2010.04.001

Yang, Y., Tang, J., Luo, H., & Law, R. (2015). Hotel location evaluation: A combination of machine learning tools and web GIS. *International Journal of Hospitality Management*, *47*, 14–24. doi:10.1016/j.ijhm.2015.02.008

Yue, P., Guo, X., Zhang, M., Jiang, L., & Zhai, X. (2016). Linked Data and SDI: The case on Web geoprocessing workflows. *ISPRS Journal of Photogrammetry and Remote Sensing*, *114*, 245–257. doi:10.1016/j.isprsjprs.2015.11.009

Yue, P., Zhou, H., Gong, J., & Hu, L. (2013). Geoprocessing in Cloud Computing platforms–a comparative analysis. *International Journal of Digital Earth*, *6*(4), 404–425. doi:10.1080/17538947.2012.748847

Chapter 7
The Benefits of Applying GIS in Spiritual Tourism Management and Promotion

Farooq Haq
Canadian University Dubai, UAE

Anita Medhekar
Central Queensland University, Australia

ABSTRACT

The main contribution of this chapter is to critically discuss the benefits of applying geographic information systems (GIS) as a tool for management and promotion of spiritual tourism circuits (STC) and ST destinations. This research-based chapter also examines the extent to which GIS can be used in spiritual tourism management and promotion, proposes a model for the use and benefits of GIS in spiritual tourism management and promotion in India and Pakistan, and proposes GIS connected STC. This chapter identifies the socio-economic and business benefits of applying GIS to spiritual tourism circuits (STC). In this research, the spiritual tourism product is exclusively based on spiritual place, sites, or destinations, which is also called spatial or geographical data.

INTRODUCTION

Tourism is one of the fastest growing service sectors of today's global economy. Tourism contributes with over 10% to global GDP and is one of the highest employment generators around the world (UNEP, 2016). Modern economies are

DOI: 10.4018/978-1-5225-5088-4.ch007

depending upon effective information technology (IT) and tourism as the most powerful engine of economic development. Nowadays, tourism and IT are providing strategic opportunities and tactical instruments for economic growth, environmental awareness, redistribution of wealth and cross-cultural harmony. Since tourism involves people travelling by road, rail or air to distance geographical locations, the use of inter-active web-based Geographical Information System (GIS), geo-portals and digital maps on mobile phone and GPS can ease the navigation process. GIS has recently being used for decision making in tourism for collecting, storing, managing, promotion, and inventory mapping of spatial geographical data efficiently (Thecla & Chigozie, 2012; Fajuvigbe, Balogun & Obembe, 2007).

The GIS has been defined by (Grimshaw, 1993, p. 206) as: "is at its simplest level a technology that enables decision-makers to explore the geographical dimension of data". GIS is a database tool which will help in sustainable development of tourism locations, resource planning (Bunruamkaew & Murayama, 2012) and attractions by meeting the competing socio-economic and environmental objectives in a virtual manner. The GIS operations are based on spatial and non-spatial or attribute data, where spatial or geographical data refers to a known location on the Earth's surface (Masron, Mohamed & Marzuki, 2015; Jovanivic' & Negus, 2008), which can also be applied to spiritual locations.

Spiritual Tourism (ST) has been recognized as an emerging type of tourism niche segment that attracts tourists from all social, cultural and geographical human segments (Haq & Medhekar, 2016; Haq, 2015). The ST product and service is based on people, places and events while the social and commercial success of this emerging segment of spiritual tourism depends on planning, development, management and promotion (Haq, 2015; Matheson, Rimmer & Tinsley, 2014). With an increase in per-capita income, ease of travel and ageing population, more and more people are travelling overseas for various types of tourism including spiritual tourism (Bowler, 2014).

This paper identifies the socio-economic and business benefits of applying GIS to Spiritual Tourism Circuits (STC) (Medhekar & Haq, 2012). In this research the spiritual tourism product will be exclusively based on spiritual place, sites or destinations, which is also called as spatial or geographical data. Since it could be better aligned with GIS, rather people or events, which is a part of non-spatial or attribute data. This can be illustrated by (i) creating a spatial GIS locational map of major spiritual destinations, (ii) inventory of heritage and spiritual tourism places, and (iii) providing interactive information on internet and mobile devices regarding access to these spiritual and heritage places by various means of transportation, availability of accommodation, tourists information centres, hospitals, restaurants, shopping, recreation and other local attractions.

With growing urbanisation, it has both positive as well as negative impact on the daily lives of citizens (Shinde, 2016 & 2007; UNEP, 2016; Kundu, 2011), as it is becoming difficult to navigate around towns and cities with development of roads and infrastructure for the locals and tourists. Latest advances in software and hardware technologies have delivered systems which provide a range of searching, querying, and analysing functions in a more user-friendly manner. GIS technology can help the tourist in finding the visual location, road maps, traffic, preserving, managing, marketing and promotion (Masron et al., 2015; Abdel-Fadeel, Saad & Omran, 2013; Fadanhusi, 2011; AlQeed, Bazazo, Hasoneh & Alqaid, 2010; Fajuvigbe et al., 2007; Elliott-White & Finn, 1997) and mapping the route to various cultural heritage and spiritual destinations and circuits.

In this millennium, tourism related activities have developed as a new form of service industry such as spiritual tourism, wellness and medical tourism, adventure and heritage tourism, as an alternative for sustainable socio-economic development of a country. For sustainable tourism planning, management, and promotion, GIS is extensively useful and applied in Canada, USA and UK (Bahaire & Elliott-Ehite, 1999) for sustainable tourism development (Bunruamkaew & Murayama, 2012; Yianna & Poulicos, 2011; Goodchild, 2010). The tourism sector has widely implemented Web based GIS which is integrated with live World-Wide-Web Google map for various spatial and non-spatial use of the tourism related resources, service providers from the supply side for information sharing, and overall tourism and destination development, management and promotion of tourism destinations.

From the demand side, tourists can plan their travel, by visual presentation of the geographical location by GIS of the topographic conditions, and its relationship with transport facilities airport, railway stations and bus terminals, online ticket booking, medical facilities, recreational, cultural, places of worship and tourists information centers, museums, population density (Farsari & Prastacos, 2004). This helps to categorise and preserve a place as a popular tourism destination for various tourism segments such as: heritage, medical, adventure, historical and spiritual tourist destination, which can improve navigation and monitor the travel route to reach a specific spiritual tourist destination in an innovative way, using state-of-the-art GIS technology.

In a competitive global tourism industry, tourism planning and destination management require the use of GIS and World-Wide-Web application to enhance their business operations and adopt innovative business models to be competitive. The proliferation of online services has created a broad spectrum of tourism information shared by several websites. As a result, it is almost impossible for tourists to use accumulated data from different accessible sources, and they are confused with dissimilarities regarding information presented on various websites. Hence, Jovanivic' and Negus (2008) suggested to use maps to present information in an effective way as maps are a natural means of indexing and presenting tourism related information. Nowadays, maps are being used to navigate during travels and maps exploit the two dimensional capabilities of human vision and present the information in a focussed and easy form, to read by the traveller (Jovanivic' & Negus, 2008).

The interactive Web-GIS platform can also be useful where the content of a spiritual destination is developed and managed by spiritual tourists and updated using the web 2.0 related to spiritual tourism destinations and STC all over the world such as Mecca and Medina, Lourdes, Vatican in Rome, Jerusalem, Buddhist trail, Amritsar in India. GIS and interactive web-based geo-portals and maps can help potential tourists with navigation, for example - planning the route for the trip, selecting hotels, restaurants, washroom facilities, online-booking, police stations, hospitals and other attractions. Web-GIS can be used for sustainable planning, management and promotion of the spiritual tourism destinations and places where the visual information in the form of text, images, brief highlighted content and 1-2 minute multi-lingual videos could be presented in an innovative manner. The aim of GIS in ST is that it can be easily understood by spiritual tourists and empowers them to plan and manage their trip from start to finish to a new spiritual tourism destination, as well as contribute to the content on the interactive web about the experience (Chen, 2007).

The main contribution of this chapter is to critically discuss benefits of applying GIS for management and promotion of spiritual tourism circuits (STC) and ST destinations. The purpose of this chapter is: (i) to examine the extent to which GIS can be used in spiritual tourism management and promotion, (ii) to propose a model for the use and benefits of GIS in spiritual tourism management and promotion in India and Pakistan, and propose GIS connected STC. Based on the properties of GIS specified by Bahaire & Elliott-Ehite (1999), this paper will elaborate upon benefits of GIS as a process and a database for managing and promoting spiritual tourism. The process will indicate the support from GIS in collecting, analysing and adopting spatial data from the sites or places associated with spiritual tourism in India and Pakistan. The spiritual tourism database could be supported by GIS in order to record, report and plan the data regarding the environment, culture, people and buying behaviour of spiritual tourists. GIS and Tourism: GIS can be used as a tool for marketing and promotion of tourism destinations, location and facilities used by tourists such as hotels, restaurants, local transport and recreation facilities, and hospital to meet tourists needs (Luberichs & Wachowiak, 2010; Hsu & Powers, 2001).

This chapter introduces spiritual tourism and significance of innovative GIS technology for ST management and promotion. Section two provides the literature review on spiritual tourism in general and in case of key popular spiritual destinations significant to various religious groups in particular. This is followed by the use of GIS in ST management and promotion. Section three critically discusses possible benefits of GIS application in spiritual tourism managerial decision making process for tourism providers and buyers. Section four proposes a model of GIS connected STC (GIS-STC) for India and Pakistan, and identifies the key stake holders and benefits framework for GIS-STC. The final section discusses recommendations for managerial and policy implications for the application of GIS in spiritual tourism destination management, conclusion and future research directions to advance the knowledge in this area.

LITERATURE REVIEW

The literature review presented in this chapter covers three major areas relevant to the study. Firstly, the concept of spiritual tourism and characteristics of a spiritual tourists; secondly, the applications of GIS to spiritual tourism, and; thirdly, advantages of using GIS in tourism in general and spiritual tourism management and promotion in particular in India and Pakistan, which can be generalised for other countries as well.

Spiritual Tourism

The importance of tourism in today's world has been discussed and it has been observed that the tourism industry employs 3% of the global work force and it rises to 8% if indirect jobs are counted (UNEP, 2016). The tourism industry has experienced amazing growth around the world with an increase of 4.3% in 2014, which averages more than 4.5% international tourist per year since 2010 (WTO, 2015). According to World Tourism Organisation, WTO (2009), 689 million people travelled to foreign countries in 2000, out of which nearly 40 million people travelled for spiritual purposes from Christian, Muslim and Hindu families. It is generally accepted that a religion inspires its followers for spiritual tourism and religious/ pilgrimage tourism that are considered as sub-segments of cultural tourism. An interrelating parallels shared by spiritual tourism, religious tourism and pilgrimage is that they are the oldest forms of travel and they are associated with religion since they gift the traveler with hope and inner peace (Bowler, 2014; Matheson et al., 2014; Moal-Ulvoas & Taylor, 2014). It is also recognised that ST is the most considerably growing type of tourism due to its multi-dimensional and non-commercial benefits, such as peace, understanding and harmony (Haq & Medhekar, 2016; Bowler, 2014).

To understand spiritual tourism, it is necessary to conceptualise spirituality. Spirituality is derived from the Latin 'spiritus', meaning 'breath of life', which finds its origins in the Greek word 'pneuma' that was used in the New Testament to describe a person's spirit guided by God (Pricipe, 1983, p. 128). The term spirituality has also been referred to as the personalised and subjective side of religious understanding

(Egresi & Kara, 2016; Matheson, et al., 2014; Kale, 2004). A two dimensional definition of spirituality states that, firstly, it is the desire and need to find meaning and purpose in one's life in order to live an integrated life, and secondly, spirituality includes the belief in a Supreme Power controlling the entire universe (Mitroff & Denton, 1999; Chittick, 1992; Pricipe, 1983).

A non-religious definition of spirituality described spirituality as an unfolding mystery, which realties to harmonious interconnectedness and inner strength (Mitroff & Denton 1999; Kale, 2004). A multi-religious definitional approach to spirituality expands the meaning and application of spirituality to daily life. Spirituality has been explained as an inner truth that is the focal point of all religions as recognised by Chittick (1992). Ibn Al-Arabi's well-known multi-religious definition has been translated as: "my heart has become capable of every form: it is the pasture of gazelles and a covenant of Christian monks and a temple of idols and the pilgrim's Ka'ba (Mecca) and the tables of the Tora and the book of Quran. I follow the religion of love: whatever way love's camel, take that is my religion and faith" (Nicholson, 1978, p.67).

It has been argued that the desire and need to find meaning and purpose in one's life in order to live an integrated life is more applicable to the business scenario but accepting that spirituality includes the belief in a Supreme Power controlling the entire universe may not fit with organisational management theory (Haq & Jackson, 2009; Kale, 2004; Mitroff & Denton 1999). However, this study stresses upon the marriage of spiritual tourism as a spiritual and human product; with management and promotion supported by GIS applications. The concept of spiritual tourism can be understood by following the definition of a spiritual tourist that is adopted in this study: "someone who visits a specific place out of his/her usual environment, with the intention of spiritual meaning and/or growth, without overt religious compulsion, which could be religious, non-religious, sacred or experiential in nature, but within a Divine context, regardless of the main reason for travelling" (Haq & Jackson, 2009, p. 145). The above given definition covers tourists who visit sacred places and events seeking a Divine presence, even without classifying themselves as followers of a religion. Historically, South Asia or the Indo-Pak subcontinent has been administratively secular, but recognized as the home of global spiritual activities attracting religious and spiritual tourists associated with all religions to satisfy their spiritual obligations and aspirations (Shinde, 2016; Haq & Medhekar, 2015a; Sharpley & Sundaram, 2005). Spiritual tourism could be referred to as the last bridge concerning travel and tourism between India and Pakistan, given its common history, aesthetics, socio-cultural heritage, besides Bollywood and sports of cricket and hockey. History has witnessed several oral and written records of people traveling for spiritual experiences and accomplishments to different areas of this special region (Shinde, 2016; Haq & Medhekar, 2015b; Sharpley & Sundaram, 2005).

GIS Applications in Tourism

GIS as a tool has been explained in the introduction. However, a review of literature on GIS applications to tourism in general is pertinent for this study. GIS can also be referred to as the geospatial information system that is being used to store and process data referenced to the Earth's surface (Masron et al., 2015). GIS supports tourism development, management and planning by providing insights on location, site conditions, related trends and usage patterns and routing options. The internet based GIS, or Web-GIS have been successful because it is more mobile, flexible and effective in sharing and communicating geographical knowledge among tourism buyers and sellers. Moreover, the Web-GIS allow tourists to access GIS services without installing the GIS software on their personal machines (Abdel-Fadeel et al., 2013; Fadanhusi, 2011; AlQeed et al., 2010).

GIS has been observed, researched and appreciated as an emerging tool that facilitates the development of applications that use geographic information in combination with other media. Tourism planners and managers are using GIS to deliver digital maps as the base for printed maps, digital files for Internet mapping, digital files for mobile mapping and websites with interactive mapping (Masron et al., 2015; Jovanivic' & Negus, 2008). The core benefit of GIS is the analytical capabilities of the system, specifically, the spatial analysis functions that are useful for tourism buyers and sellers. Spatial analysis helps to pinpoint and forecast trends on the data, creating new relationships from the data, viewing complex relationships between data sets, and making better decisions (Jovanivic' & Negus, 2008). The main purpose of geographic analysis is to transform data into useful information to satisfy the criteria and goals of all stakeholders in the tourism process (Luberichs & Wachowiak, 2010). GIS can also help in the analysis that could predict events in another location or at another point in time (Chen, 2007). It helps tourists not only to plan their trip and identify places and routes but also to change the routes if there is a natural event like cyclone or road blockage.

GIS in Spiritual Tourism Management and Promotion

GIS has been successfully applied for mass tourism marketing (Tremblay, 2005). The success of ST will depend on a country's capabilities to harness the full potential of ST places, people, products and events and promote the same to the spiritual tourists. Spiritual tourism resources include natural resources like spiritual gardens, rivers banks and oceans, ashrams, spiritual places of worships like temples, mosques, churches, shrines and cemeteries of historical figures, including historical forts and architecture. GIS can be used to capture, store ST related spatial and non-spatial data and visualise to integrate it virtually in real time to help in tourism development, planning, management and promotion to organise spiritual tourism related activities in the spatial context (Jovanivic & Negus, 2008). The capabilities of GIS are borderless and significant in terms of map production and spatial analysis for spiritual tourism destination circuits (Bahaire & Elliott-Ehite, 1999).

The map production by GIS will support spiritual tourism by tracking and guiding visitors to specific locations that are considered as sacred and holy treasures but hard to reach. Likewise, spatial analysis by GIS will help spiritual tourism operators to collect, retain and update data to inform spiritual tourists about the spiritual richness or historical significance of a place including attractions related to nature, heritage, culture, spiritual events, and festivals, recreation facilities and other local entertainment and tourist attractions, accommodation, transportation and support services. ST location is not only limited to spiritual temples, churches, mosques, and synagogues, but also ancient heritage sites, historical monument such as forts, cultural attractions, ashrams, spiritual gardens, national parks, forests, mountains, lakes and beaches.

Benefits of GIS in ST Promotion and Management

Potential benefits of the application of GIS in ST are significant, and indirectly help to locate, plan, develop, preserve and promote cultural and spiritual tourism places, destinations and events and provide all the information in an integrated manner on the GIS platform. Business and practice related benefits of GIS for promotion and management of STC are detailed below:

- **User Friendly Tourism Management:** Web-GIS system will be easy to navigate by tourism planners for development, management of tourism resources, marketing and promotion of ST destinations (Bunruamkaew & Murayama 2012; Bahaire & Elliott-White, 1999). It can be used as web 2.0, where tourists can also update the information and experiences for future facilitation.
- **User Friendly Spiritual Tourists:** Spiritual tourists gather information from GIS database to plan their virtual trip on a virtual map effectively. Tourists can visualise spacial relationships, and search for spiritual tourism destination routes, hotels, hospitals, restaurants and other facilities required by the spiritual tourists, during the trip to view the routes on an interactive web-map, videos with hyper-links to find information liked with ST location, using GIS (Kushwaha, Chatterjee & Mandal, 2011). For example in case of Buddhist Circuit (Medhekar & Haq, 2012), GIS system can help to identify the locations along the route and identify all the information about historical, cultural significance, events and services required by the spiritual tourists.

- **Security:** of STC by using GIS information system data is essential, as it is used for collecting, storing, managing and mapping of special data related to ST destination and its facilities. ST data should be backed automatically to prevent any damage or misuse of data by the people.
- **Better Tourism Services:** The tourism package being offered for spiritual tourists could be improved my enhancing the online services based on GIS support. It can be done by offering GIS application and support linked with major religions and STC in India and Pakistan.
- **Users Save Time:** from the supply-side stakeholders, ST planners and also the spiritual tourists from the demand-side can have a single platform to get information about ST location and attraction and other support services with the use of GIS.
- **Socio-Economic Benefits:** from the use of modern GIS technology of information system can improve if applied to determine ST demand and ST operation, planning, management, promotion, and determining ST places and monuments for protection and revenue generation by attracting spiritual tourists.
- **Environmental Benefits:** where the application of GIS technology is useful in locating, preserving and promoting historical, cultural and spiritual sites. Also it will help overtime to monitor changes in biodiversity and environmental impact of tourism development (Bahaire & Elliott-White, 1999) and spiritual tourist's movement on the site and plan for preservation.
- **Planning Spiritual Tourism Circuits/Routes:** GIS helps in planning and developing ST related infrastructure around STC/routes, so that the tourists can visualise their trip virtually along with travel time and visit tourists places along the route, beside the main ST destination, which is helpful in case of an independent spiritual tourists travelling alone (Wei, 2012).
- **Sustainable ST Promotion and Development:** In order to develop sustainably the niche segment of ST at the destination; web-GIS system on mobile phones will be most advantageous in developing countries, to mainly monitor the ST marketing impact; economic and social impact; monitoring movement of spiritual tourists; managing and promoting cultural and annual religious pilgrimage, gatherings and celebrations like Kumbh Mela, Hajj or Umrah.

Proposed Model of GIS –STC Application for India and Pakistan

The aim of this section is to propose a model by identifying possible STC related to major religions places, sites and destinations in India and Pakistan where spiritual destination's development, management and promotion will be based on GIS as a process and database, and for mapping STC locations and spatial analysis.

Given the coexistence of secular multi-faith spirituality which is penetrating amongst the educated, secular and religious population in India and Pakistan; following the trend of unity of all religions and humanity, is being appreciated by majority of population in both the countries. There is a need for peace and economic development in India and Pakistan to harness the potential benefits from developing spiritual tourism circuits (Haq & Medhekar, 2013; Medhekar & Haq, 2012), and GIS application for STC across the length and breadth of the two countries, which integrates India and Pakistan, with its many similarities and shared history, culture food, spirituality for economic development and business opportunities and travel of people between the two countries.

This paper presents a unique study on developing GIS for STC, for management and promotion of spiritual tourism destinations between India and Pakistan. Findings of our previous study reflected high interest in both nations for cross-border spiritual tourism based on the peace brand (Haq & Medhekar, 2013). The different religious circuits by rail and road are suggested and proposed below which can be mapped by GIS-STC for development, managemnrt and promotion of spiritual tourism between India and Pakistan via GIS-STC spatial mapping, to benefit from GIS application to tourism.

Hindu Circuit

Given the numerous historical temples and spiritual gardens which are thousands of years old from Mahabharata period in present day ***Pakistan,*** are concentrated across the main regions of Khyber Pakhtunkhwa, Punjab and Sindh. For example (a) Hinglaj Mata mandir in Balochistan, Sri Swaminarayan mandir, Valmiki mandir, Punchmukhi Hamuman temple, Aditya Sun-temple, Jagannath temple Krishna temple, and Katas-Raj. In the case of ***India*** which also has numerous spiritual tourism circuits from East, West, North and South and Central India. (b) Vaishnodevi in Jammu and Kashmir, Haridwar-Kedarnath, Allahabad, Varanasi, Joshimath and Badrinath, (c) Barsana, Gokul, Nandgoan, Mathura, Vrindavan, Govardhan, and (d) the numerous temples in South Indian states in Kerala, Tirupathi and Mahabalipuram.

Islamic Circuit

India and Pakistan have numerous centuries old heritage Mosques, Durgas, spiritual Gardens, Forts and mausoleums of historical and Islamic heritage and significance, established by Delhi sultanate and Mughal Empire since 10th century across the length and breadth of the country. Various STC can be identified and mapped for spiritual tourists; such as starting from Mohabbat Khan Mosque (670) in Peshawar, Grand

Jamia Mosque (7th largest in the world 2014), Badshahi Mosque and Masjid Wazir Khan (1642) built by Shah Jahan in Lahore, Masjid-e-Tooba in Karachi, Abbassi Mosque in Bahawalpur and National Faisal Mosque in Islamabad covering the Grand Mosques and other spiritual gardens built by the Mughal emperor Jahangir and Shahjahan mosque (1647) in Thatta. Moving on to India, there are various mosques in Jammu Kashmir, Rajasthan, New Delhi, Srinagar, and Bhilai, Nagina Masjid in Agra, Taj-Ul Masjid in Bhopal, Fathepur Sikri, Bara and Chota Imambara in Lucknow, Ajmer in Rajasthan, and Aurangabad. Such as Hazratbal in Kashmir, Jama Masjid built in 1650 in Delhi and Agra. Second circuit could cover Islamic heritage and spiritual Mosques such as Mecca Masjid (1694) and Char-Minar in Hyderabad, mosques and gardens in southern state of India covering the states of Karnataka, Tamil Nadu, Kerala, Andhra Pradesh, and Telangana.

Sikh Circuit

Similar to the Buddhist circuit, Sikh circuit application can also be made available on GIS, showing the road and rail routes along with all road-side spiritual tourist facilities and attractions in Pakistan and India. Starting from Delhi to visit famous Sikh temples/TAKHTs across the country Sri Akal Takht Sahib, and Bhatinda, Sri Patna Sahib, Sri Hazoor sahib at Nanded, Sri Darbar Harmandir Sahib, Sri Keshgarh Sahib, Sri Damdama Sahib situated in Amritsar, Sirhind, and across to Lahore in Nankana Sahib, where the founder of Sikh religion was born, and the end of journey with the visit to Sahib Golden Temple Akal Takht in Amritsar.

Jain Circuit

Jainism is also one of the ancient religions with historical places of worship mainly in Rajasthan can be mapped by GIS as spiritual tourism destinations. They includes (a) Jodhpur, Bikaner, Jaisalmer and Nagaur, (b) Bhavnagar, Mehsana, Patan, Udaipur, Sirohi, Pali and Jaisalmer, (c) Bengaluru, Mandya, Hassan, dashing Kannada, and Udupi, (d) Bhojpur, Vidisha, Damoh and Chattarpur. In Pakistan, the Bhabhra merchant community follow Jainism, which is evident from Indus-Valley civilisation and the memorial shrine of Vijayanandsuri in Gujranwala, Lahore, Nagar bazaar, Taxila, and ancient Jain temples at the foot hills of Karoonjar Mountain's.

Jewish Circuit

Jewish Circuit to include Synagogues of historical significance all over India (a) David Sassoon Synagogues in Pune, Mumbai in Maharashtra, Cochin, Malabar, Calcutta in Bengal and Manipur in East India and Karachi in Pakistan.

Zoroastrian Circuit

Zoroastrians migrated to India to avoid persecution in Persia. It is one of the oldest pre-Islamic religions. Spiritual places include Fire Temples of historical significance all over India (a) Navsari and IranShah Atash Behram in Gujarat established 1742, (b) Mumbai and Pune in Maharashtra (c) Calcutta in Bengal, and (d) in Pakistan, Karachi, Jandial temple which is UNESCO protected in Taxila, Rawalpindi and Lahore has historical Fire-Temples/ Atishgah.

Bahai' Circuit

Baha'i, which teaching unity of humanity Circuit to include Baha'i, temples such as Lotus Baha'i Temple in New Delhi and other Baha'i temples in Bangalore, Bombay, and Bihar and Karachi including the key religious places of worships from the list above as Baha'i religion is inclusive of teachings of key major religious faiths.

Sufi Circuit

Since Sufism can be considered as separate from mainstream Islam, various Sufi shrines and places that hold events and spiritual gatherings under various Sufi traditions can be specified as Sufi Sites: (a) Islamic places in Delhi, Agra, Fatehpur-Sikri and Nizamuddin (b) Bijapur, Shridi, Gulbarga, Aurangabad in Maharashtra, (c) Awadh- Lucknow region and the (d) Sufi Saints' shrines, Hazratbal shrine in Jammu and Kashmir, Ajmer Sharif in Rajasthan, Salim Chisti Dargha in Fatehpur Sikri built in 1570, Haji Ali in Mumbai and durgha Kutub-shahi in Delhi, and (e) In Pakistan, Multan is known as the city of saints and the shrine of Lal Shahbaz Qalandar in Sindh province. The Sufi shrines also attract devotee and spiritual visitors for Sufi-music festivals associated with various new and old religions.

Christian Circuit

Numerous churches were built by the Portuguese, Dutch and the British during the colonial times in cantonment cities and hill stations of Shimla, Kashmir, Murree and Darjeeling which still exist in India and Pakistan. They include the Churches from (a) Maharashtra and Goa such as Mount Mary's church, Se Cathedral Goa, Lady of Immaculate Conception, the Chapel of St. Sebastian, Basilica of Infant Jesus, Saint Francis of Assisi (b) Tamil Nadu such as Velankanni, St. John's, Saint Thomas, and Saint George Cathedral; (c) in Kerala-Malayatoor, Santacruz Basilica, and Church of Saint Francis; (d) Sacred Heart Church in Delhi- Nizamuddin, Saint Michael's cathedral in Shimla; Anglican Cathedral in Allahabad (e) St Mary's, Malayatoor Church, Valiya Palli Church, St. Francis Church, Santa Cruz Basilica in Kerala, and Christ Church.

Buddhist Circuit

With the development of Buddhist circuit and proposed circuits for other spiritual tourism circuits by Medhekar & Haq (2012), GIS tools can be adopted for all these circuits, where spiritual tourists can get immediate information on their 'Spiritual Tourism Applications' related to the specific spiritual circuit (Buddhist, Islamic, Hinduism, Sikh, Jain, Jewish, Christian) with drop-down menu choice of option on their mobile phones and web-based software, regarding the circuit route by rail and road including other attractions of cultural and heritage significance and tourists related road-site services and facilities. This will not only encourage potential spiritual pilgrims to visit these placed but also increase the economic value of these spiritual places by preserving and enhancing economic significance in terms of tourists demand, employment, small business related to spiritual tourism, road-side amenities for the tourists such as clean rest-rooms, food stalls, first-aid, gardens, and local micro enterprises related to handicrafts and spiritual souvenirs generating income and inclusive development and growth, along with attracting foreign tourists.

India and Nepal provide key destinations related to the Buddhist circuit. For example in case of Buddhist circuit, starting from place of birth Lumbini and Kapilvastu in Nepal, to the place of enlightenment in Bodhgaya in Bihar, to where the first sermon was delivered in Sarnath, Bodh-Gaya in Uttar Pradesh, and to where he lived and taught at Nalanda University the centre of Buddhist learning and teaching and Vaishali the place of last sermon in Bihar and in Kushinagar where he died, besides other places where Buddha stayed and preached such as Kausambi and Sarasvati in Bihar (World Bank, 2014), including all the other Buddhist monasteries and ashrams which his followers built. Similarly, Pakistan also has its share of Buddhist heritage and relics in Taxila and Swat and in the North West of the country that can also be included in the Buddhist circuit.

Table-1 provides the flow between the GIS-STC application and its user's from the demand side and supply side stakeholders.

Table 2, indicates GIS-STC applications to identify the ST locations, monuments and other cultural and heritage attractions which are spiritual in nature. This is followed by various, support services required by spiritual tourists on route to these spiritual

Table 1. Users of GIS-STC Applications

<table>
<tr><td rowspan="2">GIS-STC
Application</td><td rowspan="2">GIS data + STC database storage</td><td>1. Spiritual tourists</td></tr>
<tr><td>2. Supply-side stakeholders
• B to B
• B to C
• C to C</td></tr>
</table>

Table 2. GIS Model for Spiritual Tourism information systems (where and how to travel and ST support services for stakeholders)

Spiritual Tourism Locations	*ST Support services*	*Stakeholders*
Temples, churches & mosques	Transport	Spiritual tourists
Ancient heritage sites	Accommodation	Government
Historical monument	Local attractions	Religious organisations
Cultural attractions	Recreation	Tour operators
Ashrams and spiritual retreats	Hospital	Transport-rail, road, air
National parks and forests	Shopping	Hotel & Hospitality
Snow peaked mountains	Food	Heritage department
Lakes, rivers and beaches	ST related services	Local communities
	ST souvenirs	

destinations and the key stakeholders who are the users of GIS-STC information. The support for tourism providers is further elaborated as different from business to business, including government agencies, business to customers or the tourists, and from customers who are also engaged in tourism business to other tourists. The benefits of GIS are further illustrated in Table 2.

Managerial and Policy Implications

With economic development and growth of the economies in Asia and growing interest in spirituality and spiritual tourism places, monuments, sites and events by the tourists, various managerial and policy implications of application of GIS in mapping STC and ST destination management and promotion will be presented in this section. The discussion on the managerial and policy implications will start with the identification of various challenges or limitations that could be faced in the process, as discussed below:

- **Management and Protection - of ST Destination Data:** requires information technology expertise from the supplier side for planning, managing and updating STC related data base.
- **Education, Training and Skills:** For operation of web-GIS for ST destination location, development and marketing are essential due to the new technology. There is lack of GIS qualified human resources who can apply the technology to tourism.

- **Expensive Internet and GIS Software Technology:** Can prevent the adoption and use of such advanced technology for STC development, management and promotion.
- **Continuous Updating of GIS-STC Information for the Spiritual Tourists:** Related to ST destination routes and geographical changes to realise the full potential of use of GIS for STC management and promotion. It is essential that the information on the virtual GIS system is state-of-the-art, interactive, accurate and current quality of data to increase the attractiveness of the ST location.
- **User Friendly Web Based GIS Technology:** Can provide a common platform on mobile devises in a timely manner, as demanded by the spiritual tourists on the move, with the choice of language, information regarding the ST location and spiritual tourists related services.
- **High Cost of Implementing GIS Technology:** By tourism operators and mangers
- **Lack of or Inconsistent Spiritual Tourism Digital and Spatial Database:** Related to ST location, sites, monuments, facilities, number of visitors, origin of visitors, spiritual tourists types of expenditure patterns, spiritual travel motivation makes the application of GIS limited in ST development and promotion.
- **Lack of Motivation for Implementing GIS Technology:** For sustainable development and effective promotion of ST destinations, annual pilgrimage, events and ST circuits.
- **Poor Adoption of GIS - by the Key Stakeholders and the Lack of Public Awareness:** In context of the economic benefits and innovative opportunities from the use of GIS in attracting spiritual tourists and planning spiritual pilgrimage/travel, similar to the use of Uber–taxi service.
- **Environmental Impact and Sustainability:** Of the spiritual destinations, cultural and historical monuments which are spiritual in nature is a challenge for ST preservation, development and promotion. Application of GIS could help to manage this challenge and see the changes in ST locations bio-diversity, historical, cultural and spiritual assets due to ST development at local and national level.

The managerial and policy implications could be explained by presenting a framework connecting benefits of GIS for various functions and properties of the spiritual tourism management and promotion. The benefits of GIS are illustrated separately for tourism providers and buyers in Table 3 given below.

Table 3. Benefits and functions of GIS Spiritual Tourism Providers and Tourists

Benefits of GIS as	*Benefits for*	*Properties of GIS for ST*	*GIS Analytical Functions for ST*
A process	Providers	A process for capturing, storing, checking, analysing and displaying data for better service management and promotion for company and customers	Presenting and thematic mapping for promotion
	Tourists	A process of identifying the right place with right services for spiritual wealth, including the budgeting, transportation and accommodation	Better time, place, activities and people planning
A toolbox	Providers	Containing tools for collecting, storing, retrieving, and displaying spatial data for better management and promotion for customers and government agencies	Data query, retrieval and statistical support
A database	Providers	Spatially referenced objects for customers' preferences for place & events	Spatial queries
	Tourists	Spatial references for preferences based on various religious and spiritual beliefs	Spatial info and spiritual guidance
An application	Providers	Cadastral information system for digital cartographic data for related attributes and characteristics, to support marketing and planning information systems for effective tourism management	Database integration
A decision support system	Providers	Integrating spatial data within a tourism management and promotion problem-solving environment	Route finding, alternative sites and events
	Tourists	Integrating spatial data for better planning and budgeting	Route finding, time and cost management

Source: Adapted from Jovanivic' & Negus (2008)

The Table 3 presents the benefits of GIS for STC as a process, toolbox, database, application and a decision support system. These benefits are further related with the support for STC providers, buyers or the tourists. It was observed in this research study that as a toolbox and an application, GIS is more beneficial for the providers only. As a toolbox, it helps to collect, store and retrieve spatial data for tourism suppliers. As an application, GIS helps to develop a cadastral information system for digital cartographic data, which supports the customer service and future advertising and communication strategies. GIS performs as a process, a database and a decision support system for both buyers and sellers of STC. As a process, it facilitates capturing, storing, checking, analysing and displaying data from both dimensions of the supply-side and demand-side. Similarly, as a database and a decision support system, GIS helps to integrate all spatial data to be used for various purposes by both players.

CONCLUSION

In order to realise the full potential from any niche spiritual tourism segment and spatial information, the success of ST segment, depends upon abilities of a country to plan, develop, manage, preserve, and market the ST location, spiritual and historical monuments, annual pilgrimage and ST related people, places, products, cultural events and facilities. Application of GIS in STC is a powerful tool, which will help to create a virtual inventory and database of ST locations and monuments of cultural, heritage and spiritual significance in context of where and how to get their and other tourists related services. With the collaboration of internet, GIS technology spiritual tourism related information will be effectively transmitted for not only planning a trip from the demand perspective, but also for managing and promoting ST destinations along the STC for a particular religious circuit to the pilgrims, from the supplier perspective.

For national spiritual tourism information management with the use of internet and GIS-STC system to be effective, public-private-partnership between the key supply-side stakeholders is essential such as the governments of India and Pakistan, tourism authority, railways, airways and transport department, hospitality sector, national heritage councils, and leaders of religious bodies. Web based GIS application is useful for all supply-side stakeholders in effective management and promotion to attract spiritual tourists. On the other hand GIS-STC also helps the spiritual tourists in planning their spiritual journey, locating maps, destination attractions and other support services.

Future research directions will focus on how to meet the increasing demand for ST (domestic and foreign spiritual tourists) and the quality of virtual GIS database for users of ST. Another area to investigate is to compare with other developed and developing countries from the supply perspective that are using GIS for spiritual tourism circuit development, planning and promotion. Similarly, qualitative and quantitative data can be collected from ST at location regarding the usefulness of GIS-STC in planning their trip and during the travel to spiritual destinations.

Governments need to plan with the private sector key stakeholders such as spiritual organisations, infrastructure and transport ministry, hospitality sector and small businesses in public private partnerships between the countries such as India and Pakistan to employ the services of GIS expert consultants to build the GIS-STC network. This will not only promote packaged STC tours within and between the two countries to the religious places, but also help in the development of hospitality industry, accommodation, rail-roads, and other infrastructure facilities and services required by spiritual tourists, on-route to their spiritual destination using the GIS-STC applications.

REFERENCES

Abdel-Fadeel, M.M., Saad, S.K., & Omran, S. (2013). Opportunities and challenges of using GIS in sustainable tourism development: The case of Egypt. *Conference of Egypt-Thailand Relations Chiang Mai Rajabbat University*.

AlQeed, M. A., Bazazo, I. K., Hasoneh, A. I., & Alqaid, B. A. (2010). Using Geographic Information System to Visualize Travel Patterns and Market Potentials of Petra City in Jordan. *International Journal of Marketing Studies*, *2*(2), 144–159.

Bahaire, T., & Elliott-White, M. (1999). The Application of Geographical Information Systems (GIS) in Sustainable Tourism Planning: A Review. *Journal of Sustainable Tourism*, *7*(2), 159–174. doi:10.1080/09669589908667333

Bowler, B. (2014). How Spiritual Tourism Might Change the World. *The Huffinton Post*, *2*, 13.

Bunruamkaew, K., & Murayama, Y. (2012). Land use and natural resources planning for sustainable ecotourism using GIS in Surat Thani. *Thailand Sustainability*, *4*(3), 412–429. doi:10.3390/su4030412

Chen, R. J. C. (2007). Geographic information systems (GIS) applications in retail tourism and teaching curriculum. *Journal of Retailing and Consumer Services*, *14*(4), 289–295. doi:10.1016/j.jretconser.2006.07.004

Chittick, W. (1992). Notes on Ibn al-`Arabi's influence in the Subcontinent. *The Muslim World*, *82*(1), 218–241.

Egresi, I. O., & Kara, F. (2016). Residents' attitudes to tourists visiting their mosques: A case study from Istanbul, Turkey. *Journal of Tourism and Cultural Change*, 1–21. doi:10.1080/14766825.2016.1192182

Elliott-White, M. P., & Finn, M. (1997). Growing in Sophistication: The Application of Geographical Information Systems in Post-Modern Tourism Marketing. *Journal of Travel & Tourism Marketing*, *7*(1), 65–84. doi:10.1300/J073v07n01_05

Fadanhusi, J. T. (2011). Application of Geographical Information System (GIS) Technology to Tourism Management in Ile-Ife, Osun State, Nigeria. *Pacific Journal of Science and Technology*, *12*(2), 274–283.

Fajuvigbe, O., Balogun, V. F., & Obembe, O. M. (2007). Web-Based Geographic Information System (GIS) for Tourism in Oyo State, Nigeria. *Information Technology Journal*, *6*(5), 613–622. doi:10.3923/itj.2007.613.622

Farsari, Y., & Prastacos, P. (2004). GIS Applications in the planning and management of tourism. In A Companion to Tourism. Blackwell Publishing. doi:10.1002/9780470752272.ch47

Goodchild, M. (2010). Twenty years of progress: GIScience in 2010. *Journal of Spacial Information Science*, *1*, 3–20.

Grimshaw, D. J. (1993). *Bringing GIS into Business Geoinformation*. London: Longman.

Haq, F. (2015). A Framework to Compare Historical and Modern Islamic Spiritual Tourism: Study of Algerian and Malaysian tourism. *European Journal of Business and Management*, *3*(1), 105–118.

Haq, F., & Jackson, J. (2009). Spiritual journey to Hajj: Australian and Pakistani experience and expectations. *Journal of Management, Spirituality & Religion*, *6*(2), 141–156. doi:10.1080/14766080902815155

Haq, F., & Medhekar, A. (2013). Branding Spiritual Tourism as an Innovation for Peace between India and Pakistan. *International Journal of Social Entrepreneurship and Innovation*, *2*(5), 404–414. doi:10.1504/IJSEI.2013.059312

Haq, F., & Medhekar, A. (2015a). Islamic Tourism Destinations in India and Pakistan. In Islamic Tourism Trends, Challenges, and Opportunities in the Global Economy (pp.190-216). IGI Global.

Haq, F., & Medhekar, A. (2015b). Spiritual Tourism between India and Pakistan: A Framework for Business Opportunities and Threats. *World Journal of Social Sciences*, *5*(2), 190–200.

Haq, F., & Medhekar, A. (2016). Is Spiritual Tourism an Innovation in Tourism for India and Pakistan? Business, Economics. In Leadership, Innovation and Entrepreneurship as Driving Forces of the Global Economy. Springer.

Hsu, C., & Powers, T. (2001). *Marketing Hospitality*. Hoboken, NJ: John Wiley & Sons.

Jovanivic', V., & Negus', A. (2008). The Applications of GIS and its Components in Tourism. *Yugoslav Journal of Operations Research*, *18*(2), 261–272. doi:10.2298/YJOR0802261J

Kale, S. H. (2004). Spirituality, religion, and globalization. *Journal of Macromarketing*, *24*(2), 92–107. doi:10.1177/0276146704269296

Kundu, A. (2011). *Trends and processes of urbanisation in India.* Population and Development Branch, United Nations Population Fund, UNFPA, September.

Kushwaha, A., Chatterjee, D., & Mandal, P. (2011). *Potentials of GIS in Heritage and Tourism.* Geospatial World Forum. Retrieved on March, 8, 2017, from: http://www.geospatialworldforum.org/2011/proceeding/pdf/Avantika.pdf

Luberichs, J., & Wachowiak, H. (2010). GIS supported segmentation study of visitors to Majorca Island. *Advances in Hospitality and Leisure*, (2), 1-37.

Masron, T., Mohamed, B., & Marzuki, A. (2015). GIS base tourism decision support system for Langkawi Island, Kedah, Malaysia. *Theoretical and Empirical Researches in Urban Management, 10*(2), 21.

Matheson, C. M., Rimmer, R., & Tinsley, R. (2014). Spiritual attitudes and visitor motivations at the Beltane Fire Festival, Edinburgh. *Tourism Management, 44*, 16–33. doi:10.1016/j.tourman.2014.01.023

Medhekar, A., & Haq, F. (2012). Development of Spiritual Tourism Circuits: The Case of India. *Journal of Global Business Review*, 2(2), 212–218.

Mitroff, I., & Denton, E. A. (1999). *A spiritual audit of corporate America: A hard look at spirituality, religion, and values in the workplace* (Vol. 140). Jossey-Bass.

Moal-Ulvoas, G., & Taylor, V. A. (2014). The spiritual benefits of travel for senior tourists. *Journal of Consumer Behaviour, 13*(6), 453–462. doi:10.1002/cb.1495

Nicholson, R. A. (1978). *The Tarjuman al-Ashwaq: A Collection of Mystical Odes by Ibn Al-Arabi, translated from original.* London: Theosophical Publishing House.

Principe, W. (1983). Toward defining spirituality. *Studies in Religion/Sciences Religiousness, 12*(2), 127-141.

Sharpley, R., & Sundaram, P. (2005). Tourism: A sacred journey? The case of Ashram tourism, India. *International Journal of Tourism Research, 7*(3), 161–171. doi:10.1002/jtr.522

Shinde, K. A. (2007). Visiting Scared sites in India: Religious Tourism or Pilgrimage? In R. Raj & N. D. Morpeth (Eds.), *Religious Tourism and Pilgrimage Management: An international Perspective* (pp. 170–183). CAB International. doi:10.1079/9781845932251.0184

Shinde, K. A. (2016). Planning for urbanisation in religious tourism destinations: Insights from Shirdi, India. *Planning Practice and Research*, 1–20.

Thecla, A., & Chigozie, O. (2014). Designing and Developing a GIS Database for Tourism in Nigeria: The Case of Anambra State. *Journal of the Humanities and Social Sciences*, *19*(10), 109–120.

Tremblay, P. (2005). GIS techniques in tourism and recreation planning: Application to wildlife tourism. In B.W. Ritchie, P. Burns, & C. Palmers (Eds.), Tourism Research Methods: Integrating theory with practice (pp. 163-178). Academic Press.

UNEP. (2016). *Sustainable Tourism Gateway. Environmental Impact of Tourism.* Retrieved on August 18, 2016, from: http://www.gdrc.org/uem/eco-tour/st-documents.html

Wei, W. (2012). Research on the Application of Geographic Information System in Tourism Management. *Procedia Environmental Sciences*, *12*, 1104–1109. doi:10.1016/j.proenv.2012.01.394

World Bank. (2014). *Investing in the Buddhist Circuit.* International Finance Cooperation, World Bank. From: http://documents.worldbank.org/curated/en/911791487142754943/pdf/112816-WP-IN-Buddhist-Circuit-Tourism-Strategy-PUBLIC.pdf

World Tourism Organization (WTO). (2015). *Facts and Figures, The UNWTO World Tourism Barometer.* Retrieved 1 June, 2017, http://media.unwto.org/press-release/2015-07-08/international-tourist-arrivals-4-first-four-months-2015

World Tourism Organization (WTO). (2009). *Facts and Figures, The UNWTO World Tourism Barometer.* Retrieved 10 May, 2017, http://unwto.org/facts/eng/barometer.htm

Yianna, F., & Poulicos, P. (2011). *GIS Contribution for the Evaluation and Planning of Tourism: A Sustainable Tourism Perspective.* Retrieved on August 18, 2016, from: http://www.gipsynoise.gr/HellasGI/KoinonikesPraktikes/papers/Farsaki.pdf

ADDITIONAL READING

Jiang, Y., Ramkissoon, H., & Mavondo, F. (2015). Destination marketing and visitor experiences: The development of a conceptual framework. *Journal of Hospitality Marketing & Management*, 1–23.

Mullins, P. (2009). Tourism Urbanization. *International Journal of Urban and Regional Research*, *15*(3), 326–342. doi:10.1111/j.1468-2427.1991.tb00642.x

Nyaupane, G. P. (2008). Heritage complexity and Tourism: The case of Lumbini, Nepal. *Journal of Heritage Tourism*, *4*(2), 157–171. doi:10.1080/17438730802429181

R. Raj, & N. D. Morpeth (Eds.). (2007). *Religious Tourism and Pilgrimage Festivals Management: An International Perspective*. Oxford: CABI Publishers. doi:10.1079/9781845932251.0000

Singh, U. (2010). Exile and Return: The Reinvention of Buddhism and Buddhist sites in modern India. *South Asian Studies*, *26*(2), 193–217. doi:10.1080/02666030.2010.514744

Spoon, J. (2012). Tourism, Persistence, and Change: Sherpa Spirituality and Place in Sagarmatha (Mount Everest) National Park and Buffer Zone, Nepal. *Journal of Ecological Anthropology*, *15*(1), 41–57. doi:10.5038/2162-4593.15.1.3

Zamani-Farahani, H., & Henderson, J. C. (2010). Islamic tourism and managing tourism development in Islamic societies: The cases of Iran and Saudi Arabia. *International Journal of Tourism Research*, *12*(1), 79–89.

KEY TERMS AND DEFINITIONS

Geographical Information System (GIS): A technology that enables decision makers to explore the geographical/spatial dimension of data.

Spiritual Tourism: A type of tourism where someone visits a specific place out of his/her usual environment, with the intention of spiritual meaning and/or growth, without overt religious compulsion, which could be religious, non-religious, sacred, or experiential in nature, but within a Divine context, regardless of the main reason for traveling.

Spiritual Tourism Circuits (STC): The spiritual locations, destinations, heritage buildings, places of worship (e.g., temples, mosques, churches, and shrines) of a particular religion such as Buddhist or Sufi circuit along with all the essential facilities and services (shopping, restaurants, transport, first-aid, washroom) needed by the spiritual tourists and other touristic attractions along the way on their STC journey.

Spiritual Tourism Management: The process to manage the resources and capabilities in order to satisfy needs of the spiritual tourism business.

Spiritual Tourism Marketing: The process to manage the elements of the tourism marketing mix in order to offer sustainable value to customers or spiritual tourists.

Tourists: A tourist is defined as a person who buys a tourism product and travels at least 80 kilometers away from his/her usual environment, for the purposes of recreation, medical treatment, religious observances, family matters, sporting events, conferences, study, or transit to another country.

Chapter 8
The Relationship Between the Geographical Information Systems and Rural Development:
Turkey Example

Hüseyin Çeken
Muğla Sıtkı Koçman University, Turkey

Emine Yılmaz
Muğla Sıtkı Koçman University, Turkey

ABSTRACT

Geographical data systems (GIS), which enable the creation of entities that are the subject of tourism, tourism planning, the identification of impact areas of tourism values, and the sustainable utilization of tourism entities, are data-collection and decision-making systems that can contribute to tourism planning and the development of tourism Recently, tourists' interests have shifted from sea-, sand-, and sun-focused tourism to alternative tourism types; one of which is rural tourism. With rural tourism, it is expected that economical and socio-cultural opportunities for people who live in rural areas can be improved. The realization of this provides the concept of rural development. Today, in Turkey, 36.40% of the whole population still live in rural areas and agricultural activities are the most important income sources. Therefore, it is possible to investigate rural tourism and rural development in the Turkish context. This chapter deals with suggestions in terms of rural tourism and rural development by using GIS.

DOI: 10.4018/978-1-5225-5088-4.ch008

INTRODUCTION

The objectives of this chapter are:

- Giving information about geographical information systems,
- Explaining the rural tourism concept
- Giving informations about rural tourism potential of Turkey
- Explaining the relationship between rural tourism and rural development
- Giving strategies about rural tourism examples that were done or will be done with GIS in Turkey.

BACKGROUND

There are some researches about GIS or rural tourism and rural development concepts separately. But the there is no any research as book chapter or review, etc. So the main motive when preparing this chapter was fill the gap in the literature. So, this is a conceptual study.

Conceptual Analysis of Geographical Information Systems

Together with the developments in computer technologies and virtual world, at the beginning of the 21st century information systems started to be used to gather data related to the physical and cultural worlds, to investigate these, to solve the practical problems or use them today, available digital and analogue electronic devices facilitates to enable the inventories of the resources and the utilization of resources and facilitate the arithmetic or logical processes quickly. These data systems continue to develop fast and can form data, process them, store and use them in a far faster way when compared other ways (Raju, 2004).

Bringing together the computer hardware, software, personnel and methods in an organized way with aim of obtaining all kinds of data related to GIS, storing and updating them using, analyzing and monitoring them is called as Geographical Data System. The Primary aim of GIS is to convert the raw data into new that support the decision-making processes by means of location sharing and other analytic processes (Jensen and Christensen, 1986).

According to Fitzpatrick and Maguire (2000), GIS is a computer system which is designed to collect the geographical data related to human, location and facility together with their real references on earth within a database; carrying out several analyses in line with the aim; and the presentation of the results as maps, tables and graphs.

GIS is carried out by managing the data which forms a large part of the collected and produced data in our technological world. GIS is a powerful data management system of our age that can divert many developments in not only technical area but economical, managerial, social and cultural areas as well (Nişancı, Yıldırım and Çolak, 2010). In short, GIS, which is utilized by more than 100 disciplines such as transportation, shopping, cargo and traffic services, selection of most appropriate place, tourism information, questioning maps and geographical data, not only provides improvable usage opportunities but also has a qualification that can be used in all places (Kervankıran and Çuhadar, 2014).

The conponents of GIS are: computer system (hardware and operating system), software, location-based data, data management and analysis procedures and human to start the GIS. In addition, GIS cannot work outside the application area which has idea and knowledge tradition and requires a comprehensive point of (Heywood, Cornelius and Carver, 2006).

The most important characteristics of GIS that differs it from traditional data management is the possibility of managing the locational and table data. The basic functions are as in the following (Yomralıoğlu, 2005):

- **Integration of Data:** GIS has the ability to function in an integrated way with numerical and verbal data together in different environments.
- **Location-Based Questioning:** It is possible with CBS it is possible to see graphical and ungraphical information.
- **Automation:** GIS offers ease to users in issues that requires calculation and measurement thanks to its graphical function.
- **Display:** It is possible to display many things with GIS such as tables, graphs, videos, sound, photographs and statistical data etc.
- **Manipulation:** GIS makes information transfer to different systems possible by obtaining new data from available data and producing data in demanded format.
- **Location-Based Analysis:** GIS makes carrying out analyses such as forming graphical and non-graphical data according to the aim, investigation of results and interpretation.
- **Decision Making Analyses:** GIS plays significant role to better understand location-based data which will be graphically supported as a result of the collection of statistical data and categorization of these data according to predetermined characteristics.
- **Model Analyses:** GIS can carry out simulation processes with real models which it will create in computer environments by taking the relationships between geographical entities and their environments.

A general advantage of using GIS can be explained with the following example. For a planning, generally different maps and other data sources are required. Before starting the analyses, all the required maps and documents need to be collected. These maps and data should be collected from different places and sources (Ex. From cartography agency, geological surface research, sand examination, forestry research, census office etc.). to combined the data in the maps, the same measurement and projection need to be converted to worksheets. Making this process manually requires a lot of money and time. However, by using GIS, maps can be preserved in a database digitally in digital coordinates (meter or feet). This makes the measurement conversions redundant and conversions between map projections can be carried out easily via software. To carry out the planning tasks, location-based analysis function of GIS is applied. This process can be accelerated and it allows an easy change in analytical approach (Raju, 2004).

Since wrong and insufficient data will result in irrational use of resources, low productivity, unconscious and unplanned production, marketing problems and low competition power, loss of income and problems with environment it is highly important to provide data from right resource and within a certain system. For a sustainable development, it is necessary to create an information network based on database understanding across Turkey. Only in this way it would not be wrong to argue that an improvement can affect development positively.

THE DEVELOPMENT OF GEOGRAPHICAL DATA SYSTEMS

Maps were used to obtain information about earth surface for centuries. Sailors, cartographers and military staff used maps to show the distribution of important geographical characteristics. For the first time in history, Roman government supported the gathering the geodetic surveys and mapping techniques together and Roman Empire maintained this tradition. Kaşgarlı Mahmud (1076) and Piri Reis (1470-1554) produced maps which were almost impossible to make in those times. The most important ones of these was the map of North America (1528) and map of West Africa, Portugal, Spain and America (1513) (İnan and İzgi, 2011: 3).

After 1940s', the computer technology which has been developed as the result of personal efforts and the unity of ideas and later developments have led the birth of GIS. The developments in hardware and software in computer technology; the development of three dimensional processes' theories in human geography, anthropology, regional science and economy fields; the environmental problems' being up-to-date, recognition of social realities, increase in education are among the various factors resulting in changes in cartographic analyses. Soon after the emergence of GIS in 1960s, it is counted among the most effective 25 developments that have affected human beings lives (Cook et.al., 1994). In their study Birking et.al. (1996), argued that the collection of these systems that played a significant role in the investigation of earth's surface in detail and that has reorganized the methods and dimensions of location-based analyses resulted in reform-like developments in the research methods of many scientific fields in 80s and 90s, especially in geography. First studies were carried out by organizations such as USA census office in North America and Harvard University Laboratories and Environment Systems Institute (Commercial). Then, Canada and England got involved in these early developments. With its studies, New York State University recognized across world. Commercial agent started to develop and offer GIS software (Raju, 2004:110). In those years, Canada started to use GIS mostly for military and strategical goals. Thanks to the developments of functions that allowed the three-dimensional understanding of locations and their analyses, geographers started to recognize GIS and use this technology in their studies in 1980s (Phoenix, 2000). In 1990's, witnessed the developments in system and database and they can be called as location-based data theory for GIS basically. However, the potential of GIS has been realized recently and ıt has become popular among many users for various applications (Raju, 2004:110).

In academic circles, the spread of GIS dates back 1970s. The first GIS symposium was done between 28th September – 2nd October, 1970in Ottawa in Canada. In 1974, the first conference on computer assisted cartography, Autocarto was carried out. Then, internationally autocarto conferences started to be carried out. In 1984, "Basic Reading in Geographic Information Systems" was published by Marble, Calkins and Peuguet. In 1986 "Principles of Geographical Information System for Land Resources Assessment" by Peter Burrough was published. This book is accepted as the fundamental book on GIS theory. In 1987 "International Journal of Geographic Information Systems" in which scientific research and examination articles on GIS are published started to be published. This journal is the first reviewed research-investigation journal. In 1988, the first GIS/LIS conference was carried out. In this conference, is the first one that included the topics Geographical Data Systems and Field Data Systems. In 1989 Stan Aranoff wrote "Geographical Information Systems: A Management Perspective". In 1991, Maguire, Goodchild and Rhind wrote "Geographical Information Systems: Principles and Applications" which was the most comprehensive book on GIS and issued in two volumes (http://www.urisa.org/). Since these dates, studies on GIS have been published.

GIS arrived in Turkey and recognized in 1990s. In those days, GIS was used in various governmental organizations such as land office, State Institute of Statistics, State Meteorology General Directorate, Mineralogy and Research General Directorate and the State Water Supply Administration (Yomralıoğlu, 2002). GIS General Directorate which has been established in Environment and Urbanization Ministry in 2011 pioneered the establishment of Turkey National Geographical Information

Systems which enabled recordings of the management of geographical data obtained from various sources by means of information technologies and which provides good quality and up-to-date electronic services and as a result which increases the eligible data in technical, social and economic fields (Yomralıoğlu ve Aydınoğlu, 2014).

THE RELATIONSHIP BETWEEN GIS AND TOURISM

GIS offers a usable structure in all areas where geographical data can be found. If all the data on earth accepted as geographical data, it could be concluded that there would be a long list for GIS applications. In other words, it can be said that 'there would be GIS application areas as many as the number of its users' (İnan and İzgi, 2011). Today new generation Web based GIS applications like Google Maps, Yahoo Maps or GlobeXplore provide detailed location information to their users across the world (Chang and Caneday 2011). Since GIS includes lots of activities which are location-based, one of its latest usage area is tourism.

Tourism sector is one of the fastest growing and developing economies of the global economy. Depending on the globalization process, speed and comfort in transportation, increase in the amount of leisure time resulted from the technological developments, the utilization of common languages and currencies, the spread of package tour usage and increased momentum in international peace accelerated the development of tourism sector in global terms. The development of tourism also depends on the diversity of tourism data and wealth that countries and regions possess. Especially in recent years the emergence of different demands of people taking part in international tourism activities depending on their changing desires and needs have resulted from the vast diversity of touristic offers. At this point, GIS could be used to promote the less frequently visited places in rural areas each of which is a kind of touristic offer. Moreover, it can help to reduce the density in popular attraction areas, raising awareness in terms of historical values and thematic attractions in a region and their use for touristic purposes.

Tourism and geography are intertwined, and due to its nature tourism bears powerful geographical characteristics. GIS is an information system which offers geographical investigation and decision making services and these are important for tourism. Especially, with its data storage, data processing and location-based functions GIS directly provides services for tourism (Erşen et al., 2016).

Information systems are needed in some circumstances such as providing all kinds of inventories and producing maps and guide books so that tourists can use them. The system is a database in which all kinds of offers that related destination are available according to the tourists' demands. It also allows several analyses to determine services to be offered and the strategies to be used for these services.

GIS technology is made use of for several purposes such as regional development, regional planning and management, preparation of tourism zones and facilities and archeological plans (Nişancı, Yıldırım and Çolak, 2010).

Generally, GIS can be seen as a computer software by which CBS the data uploaded into the system are processed. However, GIS merges common database systems, developed for query and statistical analyses, with maps. And it can be said that this is what differentiates GIS from other traditional computer systems. A traditional computer system can store various data related to tourism like facility, room and bed numbers, staying overnight and accommodation numbers, average stay time for tourists, occupancy rates of facilities, tourism attraction according to cities and countries, and the number of museums. However, from the information given in the form of tables it is almost impossible to really learn where tourism facilities and attraction centers are. By using GIS, it is possible to develop many applications which may help the development of tourism and its planning such as the location based spread of tourism facilities across years, places where tourist flow is piled, the location based distributions of tourists' profiles, the illustration of touristic attractions according to their locations, the distributions of museums, mapping of the tourism routes, the proximities of tourism facilities or tourism organizations to tourism routes, the analyses of inhabitants in close places (Kervankıran and Çuhadar, 2014).

The importance of GIS in tourism sector is related to the possibility of preserving the local touristic products in this system and to the possibility of using these as a means of development tool in local economic development. Thus, GIS can be used to prepare the inventories of entities that are subject to tourism, tourism planning, the identification of impact areas of tourism marketing and values and making use of the touristic entities in a sustainable way. Due to this, GIS can be seen as a trend that is being developed mainly, in Europe and throughout the world. It is still up-to-date in terms of preserving and enabling the sustainability of offer values both dependent and independent from tourism

IMPORTANCE OF RURAL TOURISM

Recently the interests of tourist have shifted from sea, sand and sun based tourism to alternative tourism types (Çeken et. al., 2007:6). The attractiveness of rural life and the lack of limitations and obligations that can create stress and pressure and the existence of opportunities that can make tourists feel free resulted in increased visits to these places in today's holidays. Thus, a new kind of tourism which is called as rural tourism has emerged (Soykan, 1999: 68). Turnock (1999) defines the rural tourism as the activities including all suitable leisure time in rural areas. In most general terms, rural tourism is a type of tourism depending on the agricultural activities in rural places and the use of natural sources (Uçar, Çeken and Ökten, 2017: 5). Since it sometimes gets complicated while trying to define what rural tourism is, below several definitions made by various researchers can be found:

Rural tourism which is among alternative tourism and merge with rural culture, rural inheritance, natural environment and agriculture depending on the demand for tourism which increase fast today and the diversity in touristic supply data, and which can be integrated with other tourism types easily gain more importance day by day since it is complementary of or alternative to available types. Thus, rural tourism is important both in developing and developed countries together with the changes in tourism demands.

The support of rural tourism in countries or areas where touristic supply resources are dense. While it primarily plays a significant role in reducing the density in certain months resulted from the intersection of supply and tourism demands, it balances

Table 1. Definitions of Rural Tourism

Researcher	Definition of Rural Tourism
Lane, 1994: 9	Is a kind of tourism which not only include agricultural tourism but also special interest holidays and eco-tourism, trekking, hiking and horse riding holidays, adventure sports and health tourism, hunting and fishing, educational travels, art and history tourism and ethnic tourism in some regions.
Küçükaltan, 1997: 212	Whole events and relationships resulted from people's visits to the rural regions other than their own residences and their demands for local agricultural producers products and services in locations that suit the regions natural characteristics and their temporary accommodations in which they minimized their desire to save money.
Yağcı, 2002:4	The whole activities carried out in small settlements in which small scale enterprises which provides accommodation according to tourists' expectations, food and beverages and other services to the tourists whose aims are to spend good time while interacting local values.
WTO; 2003: 3	Activities related to the visitors who want to get into meaningful and sincere interaction with rural environment and hosts away from general tendencies of the society and places where there are dense touristic activities.
Soykan, 2004: 11	Is a kind of tourism in which it is aimed to have a rest in natural environments and be together with different cultures going to a rural region and stay there and follow the activities special to that region, and learn about them in an integrative way.
López ve García, 2006: 86	An alternative type of tourism to the intraregional restrictions that are exploited by mass tourism and to the regions.
Çeken et al., 2007	A type of tourism in which agricultural and animal activities are used as supply sources in regions where these activities are carried out densely.
Özkan, 2007: 86	A type of tourism in which tourists go to and stay in a rural area in order to have their vacations in natural environments and to meet different cultures and monitor the activities special to those areas.
Torre ve Gutiérrez, 2008: 114	Touristic activities of sustainable organizations in rural environment.

unequal tourism distribution across geographies of countries and it frees tourism from sea, sand and sun tourism. It enables more balanced distribution of tourism demands. Rural tourism is the most adapted tourism type with sustainable tourism understanding. So, rural tourism has an important function for the preservation of natural and cultural inheritance and presentation of these to the world. Moreover, rural tourism meets the needs of national and international tourists by not damaging touristic supply resources and also play a crucial role to leave these resources to next generations. In this way, demands for areas in which rural tourism is developed gains a sustainable characteristic.

The importance of an alternative tourism to nature, history and socio-cultural touristic values are seen crucial and/or rural tourism can be seen as complementary of alternative tourism. It is possible to show the importance of rural tourism as in the following (Uçar, Çeken and Ökten, 2017: 8-9):

- Tourism is an activity of consumption. So, depending on the additional demand that rural tourism brings about in rural areas, demands for the products that are grown and produced in rural area increase. That is, increase in demand increase the local production and as a result the product gets its real value. In this way, thanks to the products that are sold within their real values the both income and welfare levels of the producers in that rural region increase as well.
- Development of rural tourism in the region result in developments in 54 other sectors that are either directly or indirectly related to tourism.
- Since tourism sector is a labor-intensive sector, it potential to realize the automatization is limited. So, development of tourism in rural areas creates employment opportunities. In this respect, the migration from rural areas to cities and unemployment in rural areas are prevented.
- There are important changes for women in rural tourism. When industries that depend on traditional sources turn into labor-intensive ones like rural tourism, positive changes for women can be seen such as their contributions to their family income, labor status, the flow in their lives etc. (Ritchie and Inkari, 2006: 32).
- As a result of the development of rural tourism in rural areas, some of the money earned in big cities or town is spent in these areas. This narrows the gap between rural areas and cities in terms of their income levels.
- Touristic supply resources in rural areas can be utilized rationally thanks to rural tourism.
- Thanks to rural tourism, more cost effective and enjoyable vacation opportunities can be created. In this way, it is possible to provide vacation opportunities for people with low purchase power.

- Rural tourism raise people's organization awareness in rural areas by activating them.
- Rural tourism enables the reduction of infrastructure inadequacy in rural areas in time.
- Rural tourism rives the local tourism and as a result reduces the dependency to external sources.
- Rural tourism helps preserve the natural and structural resources, their restructuring and strengthening.
- Rural tourism creates income as protection for next generations and economic value for structures which will not have direct economic contributions to local people. It creats opportunities for small scale enterprises as well.
- Rural tourism has good relationships with rural enterprises (ex. farms) and create important amount of income for people working in these enterprises. In addition, another capability of rural tourism is that local people can create small scale projects in which they can use their talents and develop them (Ashley, Roe and Goodwin, 2001).
- In rural tourism, tourism demand to local areas depending to the rural tourism supply sources. As a result, it provides multipurpose activities. This lighten the load of traditional holiday destinations which has started to lose their relaxing effects and reached their saturation points.
- Among rural tourism activities there are village wedding ceremonies, local crafts, traditional competitions (like bull and camel fighting) and festivals. These are highly important for not only the development of rural tourism but the advertisement of region (Köroğlu and Köroğlu, 2006: 238).
- Rural tourism helps decrease the economic, social and psychological problems resulted from people's concentration indoors. Moreover, it frees tourism from the hegemony of traditional tourism and provides opportunities to benefit from relaxing features of rural regions.

Due to all those above-mentioned issues, rural tourism has gained popularity among alternative tourism and it has almost become one of the most widespread type of tourism.

The Effect of Rural Tourism on Development

Before mentioning what rural development is it would be beneficial to briefly mention about the issue of development which is broader. The concept of development is a multifactorial and has dynamic characteristics. That is, it changes according to the contemporary and changing conditions and is a changeable and dynamic phenomenon that focuses on the solutions resulted from those changing conditions. Thus, there are differences while defining the concept and this is seen as normal.

In general, development is defined as a process which includes changes occurred in economic, social and political structure of a country depending on the steady increase in national income levels (Clark, 1996: 34). As can be understand from this definition, development does not only mean increase in national income per capita but means changes in economic, socio-cultural and political structures of a country.

The rural development concept, which is included in the development concept and is more specific or address more specific section, has become a commonly accepted concept by both developing and developed countries. Rural development is the process of merging the efforts of small communities to enhance the economic, social and cultural conditions with the efforts made by government, unifying these communities with the whole nation and enabling their full contributions to national development efforts (Geray, 1974: 13).

First of all, with rural development it is tried to provide more humanistic conditions to communities with less income and flourish the hope for future. Each country can be successful in terms of their efforts to unite the nation to the degree they achieve to provide economical welfare. One of the possible ways to ensure is rural development (Gülçubuk, 1997).

Rural development is a multifaceted development approach which take agricultural development, rural industry, rural tourism and environmental conditions into consideration. The topic of this section deals with the concept of rural tourism which a type of rural development. The development of tourism means the development of other sectors operating in rural areas. Rural tourism has got a leading role. Since rural tourism create economic values for the entities which previously do not have any economical values previously, it increases the level of welfare of local people. Thanks to rural tourism it is possible to overcome the infrastructural problems (road, water, electricity, communication) in rural areas and it positively effects the socialization and the development of tourism awareness. The dependency of rural tourism to natural and socio-cultural sources in rural regions brings about a sustainable environment mentality.

Due to economic factors such as extra employment in rural areas, income effect, sectoral development etc. rural tourism will enable the rural development as well. Because there is a linear relationship between rural tourism and rural development. Rural tourism creates extra income for local people by enabling the rational usage of many geo-economic and socio-cultural sources in that region which do not have economic value. Rural tourism, in this respect, is a success symbol of non-agricultural diversification. The development of rural tourism means the development of other two sectors as well. The goal of rural development is to develop the economic and socio-cultural opportunities of people who live in rural areas. The way of realizing this is only possible by means of rural tourism. Therefore, one of the leading and most significant sectors in rural development is rural tourism.

It is a realty that in Turkey 36,4% of the total population still live in rural areas (TÜİK, 2014) and agricultural activities are still the most important income source in rural areas. Recently in Turkey's socio-cultural and economic development process, the developmental differences between rural areas and cities have become explicit as the results of modernization efforts, industrialization and socio-economic transformation. Due to their socio-economic structures, rural areas have not been able to achieve the same development levels with cities. Although some improvements have been made in terms of increasing people's quality of life living in rural areas and development in rural areas thanks to several policies aimed at decreasing these differences have recently been released in Turkey, these improvements have not been reached to the desired levels so far (DPT, National Development Strategy, 2006:6). Another way of diminishing these development differences is the rural tourism which one of the leading sectors of rural development.

In short, the development of rural tourism has significant positive effects on the development of rural area and local people. The start of growth together with rural tourism is an important step in terms of rural development. On the other hand, rural development will provide significant positive outcomes for countries thanks to multiplier effect.

The Relationship Between GIS and Rural Tourism and Rural Development

When economic, environmental and social changes in the world are taken into consideration, diversifying the one-way usages in rural areas by tourism activities will have effects on local economy, the use of resources in more rational and sustainable ways, and community health. And this fact is being accepted by more and more people. Therefore, the number of studies on identification of usage types suitable to locations is increasing day by day and planning approaches with different analysis techniques are gaining popularity (Topay, 2003).

Latest technological developments affected rural development planning positively. This planning area is highly significant for an agricultural country like Turkey which has various geographical shapes, cultural events and so on. Planning is a commonly accepted way of dealing with resource distribution and complicated problems in terms of decision making procedures and managing these problems. Geographical information system has become compulsory for better planning and development and providing more effective and meaningful direction (Scaria and Vijayan, 2012).

GIS has an important role in terms of demand increase for rural tourism and development of rural area. Turkey has rich rural supply capacity because of her own conditions. Since it is possible to experience the four seasons in Turkey, there are seven regions and each region has unique rural tourism supply potential, the importance of GIS increase day by day. What is important is to transform the potential rural tourism demand into effective demand. GIS has effects on products in rural regions, developing the tourism potential of that region and attracting tourists. Another aim of rural tourism is that freeing tourism from the dominance of sea tourism and divert tourism demand to inner regions. There are different components and values in rural areas and regions depending on the rural tourism. What is important is to advertise these, create a brand value and increase the demand to these. One way of realizing these is to develop GIS and generalize across regions. In this respect, identifying the potential of rural tourism and in which places it occurs, the market opportunities of factories that operate in rural areas, the potential customers for any products and the socio-economic structure to develop marketing strategies for any products are among the ways of various use of GIS (Akça and Esengün, 2003).

Generally, the most frequent use of GIS is the location analysis related to transportation and accommodation. In terms of rural tourism GIS can be used to identify;

- Important and necessary places that can be suitable for rural tourism.
- The potential tourism resources in rural areas.
- The most suitable accommodation facility.
- The most appropriate plan for excursion points.
- The shortest points between any predetermined points.

It can be said that GIS can be used for above-mentioned purposes and desired productivity can be obtained.

One of the most beneficial outcome of GIS in terms of rural tourism is the ease that GIS provides in land evaluation. In land evaluation studies, real estate maps can be obtained via GIS and land evaluation processes can be carried out faster and this can facilitate the jobs done by institutions. In addition, it is possible to obtain more accurate and objective results with GIS and to avoid the value differences that emerge in the same region. As a result, it is possible to save time and reduce the cost of trials and expenditures (Coşar and Engindeniz, 2011).

In short, in case of using GIS successfully in rural tourism applications, it is possible to mention about rural development.

Turkey's Rural Tourism Potential and Rural Development Studies

Turkey is a country which can realize rural tourism best on account of its unique geography, climate in which the four seasons can be experienced and its unique natural beauties. Especially, in regions where rural areas are frequent it is possible to use rural areas for both agricultural and touristic purposes by encouraging the agricultural organizations. This may make significant contributions to the economic development and growth of rural areas. For the social and economic functions of rural tourism should be assessed carefully while determining tourism plans and policies (Çeken et al., 2007).

Turkey has a lot of superior characteristics when compared to other rival countries in terms of history, culture and natural values. Turkey is a country where both summer and winter tourism are lived due to her geographical characteristics. Turkey provides many appropriate characteristics for various alternative tourism activities with her unique bays and sandy beaches. This shows that in Turkey not only the traditional vacation understanding but also rural tourism exists (Çeken, 2003).

Plateaus, mostly in Black sea region and partially in Aegean and Mediterranean regions of Turkey, bears importance not only as temporary accommodation places and pastures for rural communities but also as places sustaining some historical events and values. Moreover, when some activities such as line-fishing, grass skiing, paragliding and trekking are adapted to rural tourism, for national and international tourists who want to live in nature and meet with new people and cultures Turkey can become an important rural tourism destination. For example, paragliding and rafting are encouraged activities in Erzincan (T.B.M.M., 2006). Local craft which reflect Turkish culture from different regions, Anatolian folklore, traditional village weddings, traditional sports (javelin, camel and bull fighting), its natural flora and variable fauna, its plateaus and forests with incredible views constitute Turkey's rural mosaic (Çeken, 2003).

Together with the increase in demands for rural tourism in Turkey, activities such as river sports in rural areas, trekking, cycling tours, cave tourism, sportive line fishing, bird watching plateau tourism are carried out (Kozak et al., 2001).

If Turkey's natural and historical beauties are utilized with rural tourism, there can be a big leap in tourism. When the development of tourism activities in rural areas in Turkey come into question, one of the first things that come into mind is plateaus and plateau tourism. However, many villages in addition to mountain and plain villages have potential in terms of their historical and socio-cultural values. Especially North Anatolian region and Erzurum which is an important plateau city is important for rural tourism with its Palandöken Mountains, Tortum Falls, Uzundere Resting Valley. In addition, Kaçkar Mountains in Rize, Artvin and its proximities are important rural tourism areas as well. It is observed that such kind of rural tourism is being developed in the proximities of big cities such as İstanbul and İzmir, touristic cities such as Muğla and Antalya, cities that has cultural and historical richness such as Bursa and Kastamonu. Also, it can be seen in villages which are close to ancient cities (Çeken, 2003).

Since 1963 there has been a planned period in Turkey and development plans have been prepared. By 2018 the eleventh of these plans, which are made for five-year period, will have been released. However, the concept of "rural development" was first mentioned in the Eighth Five-Year Development Plan (Uçar, Çeken and Ökten: 2017). In this plan, continuing developing the regions that have first priority to enable the rural development; within sustainable development to increase income and employment by activating the local potentials, developing the industry that depends agriculture; supporting the entrepreneurs to increase income of population living in rural areas and conducting physical planning studies to decrease the developmental differences between regions where regional planning is done and so on are aimed (Çelik, 2005).

"Rural Development Projects" were applied in various regions both to better make use of the natural resources in Turkey's various regions and to decrease the socio-economic differences observed in rural sections. The distinctive feature of these projects is the improvement of rural infrastructure. In addition, by making use of the available resources in region in most appropriate ways, they dealt with increase in herbal and animal products and measurements that could bring about income increase integratively. In this respect, first rural development project was conducted for Çorum-Çankırı in 1976. It is followed by the Erzurum rural development project. Then, in 1990, Bingöl-Muş, in 1991 Yozgat, Ordu-Giresun in 1998 and Erzincan-Sivas development projects in 2005 were carried out. After these projects were carried out not city based but regionally. One of the most notable regional projects not only in Turkey but in the world was the South Anatolia Project. Other projects that aimed regional development and carried out in integrative way was East Anatolia Project and East Black sea Regional Development Project. All of these projects were not only regional but also projects that were depending on sustainable humanistic development (Uçar, Çeken and Ökten, 2017).

Rural Tourism Examples That Were Done or Will Be Done With GIS in Turkey

In above section of this study, shortly, information about GIS is given, rural tourism and rural development topics are examined and Turkey's rural tourism potential is put forward. Finally, in this section it would be appropriate to mention about rural tourism studies carried out or could be done will be mentioned.

When required literature review is done, it is found out that there are no GIS-based studies in Turkey. Although there are some studies dealing with the use of rural areas, it is seen that the topic has not been investigated in terms of tourism. However, Topay and Parladır (2015) investigate the topic within the alternative tourism types. They have found out that the failure of Isparta's integration to tourism results from the under advertisement. In this vein, there is a need for a planning study to be done as soon as possible within Isparta borders. The study not only meet the needs of local people in terms of economic development but also it will be an important step toward sustainable use of resources and contribute to the recreation-tourism need of the region. It is thought that this can be done in most planned way via GIS. So, they investigate the convenience analysis of alternative tourism activities in Isparta by means of GIS. In their study, they identify the most suitable areas for alternative tourism activities such as horse riding, mountain cycling, camping, skiing and trekking. The collected data are transferred into digital environment by means of GIS tools. Data of research area are questioned in computer environment in line with method and the most appropriate locations for each activity is identified and mapped.

Since there are not enough studies on this topic, suggestions were developed for potential studies. In this respect, rural tourism studies that make use of GIS in Turkey can be realized as in the following ways:

- First of all, it should be noted that the main goal is rural development and workshops and seminars should be carried out the raise awareness of local people.
- Turkey has been divided into seven geographical regions and each region has different physical characteristics, climate and supply resources. Therefore, it can be useful to evaluate each region individually. For rural development goal, it is important to embrace a holistic view across Turkey is needed.
- For a productive study, each rural tourism components (trekking, climbing, agro-tourism, camping and so on) should be dealt individually according to the region.
- While carrying out all these studies, regions need to be analyzed in terms of whether they are defined as rural area and a SWOT analysis should be done. This can facilitate the identification of most appropriate strategies by making better inferences.
- While carrying out the studies, people working in tourism area and professional GISS web operators should cooperate.
- To ensure that studies on this topic keep up with international standards, programs and studies should be written not only in Turkish but in other languages.

- When rural tourism is mentioned in Turkey, first nature walk and nature sports come into mind. In addition, village visits including shopping and food and beverage on the way can be understood (Soykan, 2002). The trekking routes for these places can be identified by GIS, the villages on these roads can be shown and shopping and food/beverage locations can be identified. Also, by pointing out the possible nature sports that can be done on the route, contribution to the rural development of the region can be done.
- Rural tourism activities that are programmed by GIS should be supported by the government.
- By hosting experts both national and international, media and important people, advertisement studies should be strengthened.

FUTURE RESEARCH DIRECTIONS

People call the age that we live, information age. So, information technologies are really important. As we explain before, this chapter is a conceptual study. There is no other studies about GIS and rural tourism. Thus, for future researches, it is recommended that researchers can develop a method and develop this conceptual study. Maybe tourism and engineering professors can work together. This can be more productive.

CONCLUSION

GIS is a computer system which is designed to collect the geographical data related to human, location and facility together with their real references on earth within a database; carrying out several analyses in line with the aim; and the presentation of the results as maps, tables and graphs. Tourism sector is one of the fastest growing and developing economies of the global economy. Depending on the globalization process, speed and comfort in transportation, increase in the amount of leisure time resulted from the technological developments, the utilization of common languages and currencies, the spread of package tour usage and increased momentum in international peace accelerated the development of tourism sector in global terms. Tourism and geography are intertwined, and due to its nature tourism bears powerful geographical characteristics. GIS is an information system which offers geographical investigation and decision making services and these are important for tourism. Especially, with its data storage, data processing and location-based functions GIS directly provides services for tourism.

Recently the interests of tourist have shifted from sea, sand and sun based tourism to alternative tourism types. Rural tourism which is among alternative tourism and merge with rural culture, rural inheritance, natural environment and agriculture depending on the demand for tourism which increase fast today and the diversity in touristic supply data, and which can be integrated with other tourism types easily gain more importance day by day since it is complementary of or alternative to available types. Thus, rural tourism is important both in developing and developed countries together with the changes in tourism demands.

GIS has an important role in terms of demand increase for rural tourism and development of rural area. Turkey has rich rural supply capacity because of her own conditions. Since it is possible to experience the four seasons in Turkey, there are seven regions and each region has unique rural tourism supply potential, the importance of GIS increase day by day. What is important is to transform the potential rural tourism demand into effective demand. GIS has effects on products in rural regions, developing the tourism potential of that region and attracting tourists.

REFERENCES

Akça, H. & Esengün, K. (2003). Coğrafi Bilgi Sistemlerinin Tarım Ekonomisi Alanında Kullanım Olanakları. *TKK Ekin Dergisi*, 25.

Ashley, C., Roe, D., & Goodwin, H. (2001). *Pro-poor Tourism Strategies: Making Tourism Work for the Poor. A Review of Experience*. London: Overseas Development Institute.

Birking, M., Clarke, G., Clarke, M., & Wilson, A. (1996). *Intelligent GIS: Location Decisions and Strategic Planning*. Cambridge, UK: GeoInformation International.

Çeken, H. (2003). *Küreselleşme, Yabancı Sermaye ve Türkiye Turizmi*. İstanbul: Değişim Yayınları.

Çeken, H., Karadağ, L., & Dalgın, T. (2007). Kırsal Kalkınmada Yeni Bir Yaklaşım Kırsal Turizm ve Türkiye'ye Yönelik Teorik Bir Çalışma. *Artvin Çoruh Üniversitesi Orman Fakültesi Dergisi*, *8*(1), 1–14.

Çelik, Z. (2005). *Planlı Dönemde Türkiye'deki Kırsal Kalkınma Politika ve Uygulamaları Üzerine Bir Değerlendirme, Planlama, TMMOB*. Ankara: Şehir Plancıları Odası Yayını.

Chang, G., & Caneday, L. (2011). Web-based GIS in tourism information search: Perceptions, tasks, and trip attributes. *Tourism Management*, *32*(6), 1435–1437. doi:10.1016/j.tourman.2011.01.006

Clark, J. (1996). *Kalkınmanın Demokratikleşmesi. (Çev.) Serpil Ural.* Ankara: Türkiye Çevre Vakfı Yayını.

Cook, W. J., Collins, S., Flynn, M. K., Guttman, M., Cohen, W., & Budiansky, S. (1994). 25 Breakthroughs That Are Changing the Way We Live and Work. *U.S. News & World Report*, *2*(May), 46–60.

Coşar, G. Ö., & Engindeniz, S. (2011). Tarım Arazilerinin Değerlemesinde Coğrafi Bilgi Sistemlerinden Yararlanma Olanakları. *Ege Üniversitesi Ziraat Fakültesi Dergisi*, *48*(3), 283–290.

DPT Ulusal Kalkınma Stratejisi. (2006). Retrieved April 5, 2017, from www.ekonomi.name

Erşen, G., Yüncü, H. R., Metin, T. C., & Çabuk, A. (2016). Gastronomi Turizminde Yeni Açılımlar: Karaburun Zeytin Rotalarının Coğrafi Bilgi Sistemleri İle Analizi. Balıkesir Gastronomi Turizmi Kongresi.

Fitzpatrick, C., & Maguire, D. J. (2000). GIS in Schools: Infrastructure, Methodology and Role. In GIS: A Sourcebook for Schools. Taylor & Francis.

Geray, C. (1974). *Planlı Dönemde Köye Yönelik Çalışmalar*. Ankara: Sevinç Matbaası.

Gülçubuk, B. (1997). *Kırsal Kalkınmada TKV'nin Yeri*. Ankara: Yayınlanmamış Doktora Tezi, Ankara Üniversitesi Fen Bilimleri Enstitüsü.

Heywood, I., Cornelius, S., & Carver, S. (2006). *An Introduction To Geographical Information Systems*. Pearson, Prentice Hall. Retrieved from http://www.urisa.org/

İnan, A., & İzgi, E. (2011). *GIS (Coğrafi Bilgi Sistemi)*. Retrieved April 12, 2017, from http://cbs.ormansu.gov.tr/cob2011/wp-content/uploads/2011/05/cbs1.pdf

Jensen, J. R., & Christensen, E. J. (1986). Solid and Hazardous Waste Disposal Site Selection Using Digital Geographical Information Systems Techniques. *The Science of the Total Environment*, *56*, 265–276. doi:10.1016/0048-9697(86)90331-1

Kervankıran, İ., & Çuhadar, M. (2014). Turizm Rotalarının Oluşturulmasında Coğrafi Bilgi Sistemlerinin Önemi. III. Disiplinlerarası Turizm Araştırmaları Kongresi, 576-589.

Köroğlu, A. & Köroğlu Ö. (2006). Kırsal Turizmin Yöre Kalkınmasındaki Rolü: Buldan Yöresi Kırsal Turizm Potansiyeli. *Buldan Sempozyumu*, 233-242.

Kozak, N., Kozak Akoğlan, M., & Kozak, M. (2001). *Genel Turizm İlkeler- Kavramlar*. Ankara: Detay Yayıncılık.

Küçükaltan, D. (1997). Trakya Ekonomisi İçin Bir Bölgesel Kalkınma Modeli: Kırsal Turizm. Ulusal Bölge Bilimi Bölge Planlama Kongresi 20-22 Ekim. İzmir: ProOfset Matbaacılık.

Lane, B. (1994). What is Rural Tourism? *Journal of Sustainable Tourism*, *2*(1-2), 7–21. doi:10.1080/09669589409510680

López, E. P. & García, F. J. C. (2006). Agrotourism, Sustainable Tourism and Ultraperipheral Areas: The Case of Canary Islands. *PASOS Revista de Turismo Y Patrimonio Cultural*, *4*(1), 85-97.

Nişancı, R., Yıldırım, V., & Çolak, H. E. (2010). Coğrafi Bilgi Sistem Uygulamaları. *Bilim ve Teknik*, *514*, 58–63.

Özkan, E. (2007). *Türkiye'de Kırsal Kalkınma Politikaları ve Kırsal Turizm*. Ankara: Yayınlanmamış Yüksek Lisans Tezi, Ankara Üniversitesi Sosyal Bilimler Enstitüsü.

Phoenix, M. (2000). *Learning with GIS, Arcuser Online*. Retrieved May 8, 2017, from http://www.esri.com/news/arcuser/0700/umbrella11.html

Raju, P. L. N. (2004). *Fundamentals Of Geographical Information System*. World Meteorological Organisation.

Ritchie, B., & Inkari, M. (2006). Host Community Attitudes Toward Tourism and Cultural Tourism Development: The Case of the Lewes District, Southern England. *International Journal of Tourism Research*, *8*(1), 27–44. doi:10.1002/jtr.545

Scaria, R., & Vijayan, P. K. (2012). Sustainable Rural Development With The Application Of Spatial Information Technology And Mahatma Gandhi National Rural Employment Guaranteed. *International Journal of Geomatics and Geosciences*, *2*(4), 1048–1061.

Soykan, F. (1999). Doğal Çevre ve Kırsal Kültürle Bütünleşen Bir Turizm Türü: Kırsal Turizm. *Anatolia Turizm Araştırmaları Dergisi*, *10*(1), 67–75.

Soykan, F. (2002). *Kırsal Turizmin Sosyo-Ekonomik Etkileri ve Türkiye*. Türkiye Dağları I. Ulusal Sempozyumu, Haziran, Kastamonu.

Soykan, F. (2004). *Kırsal Alanların Turizm Potansiyelinin Saptanması ve Şirince Köyü'ne (İzmir) Uygulanması*. Bornova-İzmir: Ege Üniversitesi Edebiyat Fakültesi Yayınları.

T.B.M.M. (2006). *T.B.M.M. Genel Kurul Tutanağı, 22.Dönem, 3. Yasama Yılı.* Retrieved April 26, 2017, from www.tbmm.gov.tr

Topay, M. (2003). *Bartın-Uluyayla Peyzaj Özelliklerinin Rekreasyon-Turizm Kullanımları Açısından Değerlendirilmesi Üzerinde Bir Araştırma.* Ankara: Yayınlanmamış Doktora Tezi, Ankara Üniversitesi Fen Bilimleri Enstitüsü.

Topay, M., & Parladır, M. Ö. (2015). Isparta İli Örneğinde CBS Yardımıyla Alternatif Turizm Etkinlikleri İçin Uygunluk Analizi. *Journal of Agricultural Sciences, 21*, 300–309.

Torre, G. M. V., & Gutiérrez, E. A. (2008). The Demand of Rural Tourism in a Natural Park in Southern Spain. In P. M. Burns & M. Novelli (Eds.), *Tourism Development: Growth, Myths and Inequalities* (pp. 113–125). London: CAB International. doi:10.1079/9781845934255.0113

TÜİK. (2014). *Türkiye 2014 Nüfus Yoğunluğu Gridi.* Retrieved May 14, 2017, from http://www.resmiistatistik.gov.tr/?q=tr/content/18-kırkent-tanımının-revizyonu

Uçar, M., Çeken, H., & Ökten, Ş. (2017). *Kırsal Turizm ve Kırsal Kalkınma.* Ankara: Detay Yayıncılık.

WTO. (2003). *Conclusions-Seminar on Rural Tourism "Guidelines for Development".* Retrieved April 16, 2017, from http://www.unwto.org/regional/europe/PDF/Conclusions%20_Ukraine.pdf

Yağcı, Ö. (2002). *Turizmde Çeşitlendirme, Sürdürülebilirlik ve Kırsal Turizm.* Retrieved May 01, 2017, from http://www.emyo.ankara.edu.tr/iktisad/TURKONF/web/ozcan_yagci.doc

Yomralıoğlu, T. (2002). GIS Activities in Turkey. *Proceedings of International Symposium on GIS*, 834-840.

Yomralıoğlu, T. (2005). *Coğrafi Bilgi Sistemleri: Temel Kavramlar ve Uygulamalar.* Trabzon: Akademi Kitabevi.

Yomralıoğlu, T., & Aydınoğlu, A. Ç. (2014). *Türkiye Ulusal Coğrafi Bilgi Sistemi (TUCBS) ve BB'ler İçın Coğrafi Veri Altyapısı.* Retrieved May 01, 2017, from http://arifcagdas.com/wp-content/uploads/2014/11/AYOP2014-tucbs.pdf

ADDITIONAL READING

Lane, B. (1994). What is Rural Tourism? *Journal of Sustainable Tourism*, *2*(1-2), 7–21. doi:10.1080/09669589409510680

López, E. P. & García, F. J. C. (2006). Agrotourism, Sustainable Tourism and Ultraperipheral Areas: The Case of Canary Islands. *PASOS Revista de Turismo Y Patrimonio Cultural*, *4*(1), 85-97.

Scaria, R., & Vijayan, P. K. (2012). Sustainable Rural Development With The Application Of Spatial Information Technology And Mahatma Gandhi National Rural Employment Guaranteed. *International Journal of Geomatics and Geosciences*, *2*(4), 1048–1061.

KEY TERMS AND DEFINITIONS

Rural Development: It is a versatile development approach that considers agricultural development, rural industry, rural tourism, and environmental conditions.

Rural Tourism: It is a type of tourism that is carried out by tourists away from the sea and aiming to live the tourism movements brought in and by rural life.

Chapter 9
The Value and Scope of GIS in Marketing and Tourism Management

Mertcan Taşçıoğlu
Istanbul Medeniyet University, Turkey

Dursun Yener
Istanbul Medeniyet University, Turkey

ABSTRACT

Geographical information systems (GIS) are the systems that store location-based data and analyze them. GIS originated from the conventional cartographic techniques of simply drawing maps with a pencil and board. Following the adoption of computer technology, GIS further evolved as a geo-referenced dynamic information system, which can today be considered as a multi-disciplinary instrument that links different disciplines like geography, computer science, remote sensing, civil engineering, statistics, marketing, and other social and behavioral sciences. In this chapter, GIS is defined and its importance and functions are described in detail. Usage of GIS in marketing is explained, especially its development in marketing theory. Also, the value and scope of GIS in tourism management with a view to understand the spread of GIS applications in tourism is explored.

INTRODUCTION

A geographical information system is at its simplest level a technology that enables decision-makers to explore the geographical dimension of data (Grimshaw, 1993). However, it was only after the emergence of GIS software in the late 1970s and

DOI: 10.4018/978-1-5225-5088-4.ch009

1980s that GIS started to be adopted as a useful tool employed by a wide range of disciplines such as environmental planning, property management, automobile navigation systems, urban studies, market analyses, business demographics (Chen, 2007) and quite recently, in tourism industry. Information that is required in marketing activities is gathered through some tools and geographic information systems are one of them. Its usage raises as its importance are realized in marketing activities (Yener, 2017). By means of geographic information system (GIS), collecting tremendous data, storing them in database, updating when required, analyzing, drawing visible maps based on analyses, accessing the necessary information obtained as a result of analyses are possible. A GIS is essentially a digital map linked to a database management system which can be used for the purposes of displaying and querying information, carrying out spatial analysis and assisting in the decision-making process. Although GIS's emergence is not so old, its usage spreads rapidly in many different areas such as geography, criminology, archeology, education, tourism, management and marketing. With combination of geography and marketing disciplines result in emergence of a study area which is called geomarketing. Geomarketing is a recent discipline that combines the geographic visualization and analysis with marketing techniques, aiming at more efficiently attaining the ultimate goal of organizations. Important marketing activities such as segmentation, positioning and research can be performed more effectively through geomarketing.

GEOGRAPHIC INFORMATION SYSTEMS

Geographical information systems (GIS) are defined as an integrated collection of software and data that is used to visualize and organize location-based data for the purposes of performing geographic analyses and creating maps (Wade and Sommer, 2006). GIS is capable of assembling, storing, manipulating, and displaying geographically referenced information (Coyle, 2011). It combines maps, tabular data and analysis capabilities. It allows users to take information, view that information spatially and analyze that information so users can reach conclusions through correlations (Elliot, 2014). As with most technologies, the effective use of GIS depends on the person using it, rather than the technology itself (Fung and Remsen, 1997). GIS is analytical so that able to answer such questions as: "what is adjacent to this intersection?", "what are the conditions like within 15 km. of this point?", "how far is the nearest water line?" (Black, Powers and Roche, 1994).

The first modern GIS system was used in 1962 in Ottawa, by the Canadian Department of Forestry to map land use and data about farming and wildlife. The designer of the system was Roger Tomlinson, who was asked by the Canada Land Inventory project to pursue his idea of using computers to combine maps and statistics (Coyle, 2011). The high cost of computer hardware limited the widespread use of GISs. As software and hardware have become more affordable, available and powerful, more researchers are realizing the potential of GISs (Elliot, 2014). In 2007, Tomlinson stated that he thought GIS would be implemented by just about every modern business, if the business was not using GIS it was going to be considered old-fashioned (Coyle, 2011). In its present form, GIS technology represents a software/hardware system designed to capture, manage, manipulate, and analyze geographic data (Star and Estes, 1990; Fung and Remsen, 1997) GIS has applications in many different areas such as geology, defense, population, environment and local governments (Bensghir and Akay, 2006). GIS is now a tool that finds application in most fields of human endeavor at local, regional and global levels (Musyoka et al, 2007).

GIS is an information technology that has experienced an expansion out of mainly the public sector into the competitive world of the private sector (Maguire, 1995). As businesses, such as large retail companies and advertising agencies, begin to see the benefits of this technology that assists in spatial-based decision making (Smith, 2005). GIS software continues to be popular as it allows users to manipulate free downloadable data and to create their own data in an unlimited way, as opposed to a web mapping application where the user is limited to what the designer provides (Donnelly, 2010). ESRI (Environmental Systems Research Institute) software is used by 300,000 organizations worldwide, including 7,000 colleges and universities. There is currently a growing movement in the development and adoption of free and open source software (FOSS) for GIS (Donnelly, 2010). GIS has three distinct and unique advantages for economic development (Black, Powers and Roche, 1994).

1. GIS permits the disaggregation and re-aggregation of sub-county data, which offers a level of customization and accuracy not otherwise possible.
2. The technology is designed to overlay and compare different variables or types of data so that the relationships between important decision-making factors can be understood.
3. Because of the spatial capabilities of GIS, its map display features are particularly well suited for economic development marketing.

Definitions of GIS are differentiated according to its user's purposes. Some of them define GIS as a concept that analyzes geographical information that covers all geographical information systems, and others define it as a computer-based tool that converts geographical data to digits. Some managers define GIS as database management system that assists to organizations. GIS is related with computer aided designs and database management systems, however it has capability to produce new data and ability to analyze geographically as a difference (Bensghir and Akay, 2006).

A GIS is essentially a digital map linked to a database management system which can be used for the purposes of displaying and querying information, carrying out spatial analysis and assisting in the decision making process (McMahon, 1997). Although various GIS packages will have different features and levels of sophistication, all systems must offer a basic level of functionality. These basic features of the software include a *database management system*, tools that support geographic *query*, *analysis* and finally a *graphical user interface* (GUI) for easy interaction (Burrough, 1986).

- **Database Management System:** With GIS all the data can be stored in a structured and appropriate format, making access easier and less frustrating.
- **Query Function:** The technology allows the user to generate enquiries by retrieving any selected subset of the database and then apply spatial analysis routines to data using a standard SQL (Structured Query Language).
- **Data Analysis:** It will allow all the different sources of data to be integrated and analyzed. Geographic information can be easily combined and processed to consequently bring about benefits in the form of generating new types of information.
- **Graphical Interface:** GIS allows data that has been stored only once to be output from different devices and in different formats. Different spatial information can be output together (layered on top of each other) to give the user a full picture of what is happening in different areas (Anumba et al, 2005).

Since geographic information system output is most frequently displayed as maps, GIS and computer-generated maps may appear identical. However, their outputs are totally different. While computer-generated maps are static, a GIS is as dynamic as its underlying database. The essential difference lies in the ability of a GIS to analyze the data behind the map and to focus on the relationships among the data items lying in tables in a relational database that (Brody, 1999).

GIS is a form of computerized data visualization. It is a collection of computer hardware, software, data, processes, and people designed to efficiently and effectively capture, store, update, manipulate, analyze, and display forms of geographically referenced information (Laudon & Laudon, 2003). GIS is a powerful database system that "uses geo-references as the primary means of storing and accessing information" (Payton, 2010).

Using GIS as a planning tool will benefit the industry in the future. Having the information pictorially on a map shows which areas have a hub of activities going on, and which areas have moderate activities happening or none at all (Anumba et al, 2005). A GIS should be able to generate a map output. However, it is able to offer much greater power than a simple automated mapping system for two reasons: because of its analytic functions, and because of the importance of relationships between objects, which are not significant in simple mapping (Goodchild, 1989).

GIS, like all information systems, consists of five key components: hardware, software, data, process and people (Tomlinson, 2003). These independent components are all important but must be synchronized in order to function efficiently and effectively (Payton, 2010).

There are two different types of GIS data: spatial and attribute. Spatial data identify the geographic location of features. It is the “where”. Spatial data come in three types of formats: vector, raster and semantic (Payton, 2010).

Raster data consist of images, such as satellite imagery or aerial photography. Raster data represent geographic features through pixels; each pixel can be referred to by its column and row position. Rasters are geo-referenced, positioned to a specific location according to a coordinate system (Elliot, 2014). The advantage of using raster data is that all layers will be aligned with the layer under or over it. However, according to Reynolds (1997) it does have its disadvantages. Increasing the resolution will exponentially increase the size of the data set, thus creating a storage problem.

Vector data is referenced according to a coordinate system. A structure location placed on the map in the form of a point or polygon, based on information collected with a global positioning system (GPS) collection unit, is an example of vector data. Vector data represents features in GIS application, such as trees, rivers, buildings and boundaries (Elliot, 2014). The advantage of using vector data is that it can accurately define a small object more efficiently than raster data. This allows more data to be stored in less space (Payton, 2010).

The third form, semantic or “geo-database model” (Martin et al., 2005), is an object-oriented model. It is a hierarchical structure where subclasses inherit propertied of a higher class. This form of data tends to look more natural then raster or vector and enables full and complex descriptions (Reynolds, 1997; Tomlinson, 2003).

Attribute is the second classification of data needed in GIS. Attribute data provides information about each feature. Attribute data describe and classify the features. It is the “what, where and why”. Expanding on the vector data example, attribute data provides information such as: the owner of the structure, the year it was built, the assessed value, and physical address (Marakas, 2003). For each feature, descriptive information would be stored in an attribute table. Data could include street address, latitude and longitude, color, owner, age, architectural style (Elliot, 2014).

GIS IN MARKETING

Geo-Marketing

GIS has rapidly expanded into many different business areas including marketing (Hess, Rubin, and Lawrance, 2004). Many marketing departments collect information on their current and potential customers including addresses and the products they purchase. The literature on GIS in sales and marketing focuses mainly on marketing for retailers. Because retailers have recognized that geography is one of the most important factors in their activities (Smith, 2005). The choice of geographic location is a major differentiating decision in retailing in general and in grocery trade in particular (Orhun, 2013). GIS techniques would be beneficial for marketing. Location planning would allow the company to decide the best location or venue to hold a specific event. Customer or prospect profiling would allow marketers to gain knowledge of who their customers are and what types of events they are interested in. Once a sales and marketing department better understands their customer base, the company will make more educated decisions on the types of events they should select. Sale territory planning would allow marketers to visually see their market which would assist in the area and extent of where to advertise each event (Smith, 2005)

The marketing decision support system is designed to help marketing managers in the analysis and decision making processes. In this system, statistical techniques and decision models are used together (Türk, Kitapçı and Dörtyol, 2014). Information systems are widely used tools to collect and to interpret data. One of these information systems is called GIS. At present, GIS is used in Marketing Information Systems as a tool for the decision making process. Kotler (1997) defined a marketing information system as, a structure consisting of people, equipment, and procedures to gather, sort, analyze, evaluate, and distribute needed, timely, and accurate information to marketing decision makers. A GIS is an information system with geographical access; a system for input, storage, analysis and output of geographically referenced data. It is clearly something more than a mapping system, because of the emphasis on analysis rather than display (Goodchild, 1989). It supports logistic and marketing managers to evaluate placement options for warehouse stores, based on costs (transport, labor, utilities, etc.), and on customers and competitors demographics (Vlachopoulou, Silleos and Manthou, 2001). GIS's present popularity is based on its powerful visual presentation capabilities and spatial relationship analysis. Xia has listed these advantages as follows (Xia, 2004):

- GIS has the ability to perform spatial analysis,
- GIS can categorize and place objects into individual themes and thus has the ability to treat these objects differently,

- GIS has the ability to store data related to the visual presentation over the spatial database,
- GIS can be implemented online, thus control of the system can be easy and remotely conducted,
- The cost of the implementation of a GIS is considerably low.

Geomarketing is a recent discipline that combines the geographic visualization and analysis with marketing techniques, aiming at more efficiently attaining the ultimate goal about selling products, services, or ideas (Freire and Santos, 2012). Marketing geography provides the most basic approach to store location strategy (Applebaum, 1965). Before the 1950s, economic geography was focused on production and transportation of goods to distributors. Applebaum (1954) argued that the function of marketing was the transfer of goods from distributors to consumers. Throughout the next decade, marketing geography became more focused on measuring market areas of existing stores and estimating market areas of potential store locations (Berry 1967). Marketing is concerned with what is known as the four P's: product, price, place, promotion. Location analysis is often associated with place which refers to location of customers and the connections between geography of production and patterns of consumption (Berry 1967; Parrish, 2009). Until recently the concept of geomarketing had a rather narrow meaning in traditional (business) marketing aimed at geographic segmentation (geo-segmentation) of the marketplace and geographic positioning (geo-positioning) of the products (goods and services) (Anderson, 2004). The primary marketing research data collection activities in a personal interview, telephone, mail, with tools such as e-mail is made based on the survey. Sampling activities can be done easily with GIS. Research of the geographical planning, organizing settlements to be reached more easily by researchers on the theme of GIS (Fidan, 2009). Concerning their geographical scale or scope, analyses can be conducted at four levels (Freire and Santos, 2012);

- Continental
- National
- Regional
- Local

The main focus of geomarketing is how to best locate a commercial facility in order to maximize the potential customer coverage and therefore profit. Conducting such an analysis requires as a minimum data on population/potential customer distribution and information concerning the business. However, population density is not constant due to human activities; population counts and their distribution vary widely from day to day and may be misrepresented by census data. In geomarketing, this information is especially useful for retail sales, banking, insurance, lodging, real estate, and franchising (Freire and Santos, 2012). Geomarketing information enables the user to take better and faster decisions about marketing and sales activities. The main sources of information are geographic, demographic, and statistical data. Geomarketing information is gathered from internal company data, which are combined with external demographic, statistic and geographic data (Krek, 2006). Traditional geomarketing can;

- Help firms determine which products and promotions match the lifestyles and buying patterns of their customers in geographic perspective.
- Delimitate of the trade catchment areas, identify of retail sites, to make spatial competitor analysis.
- Create a multidimensional snapshot of trends to plan trade areas, predict sales spatially, and design sales territories, plan media and advertising geographically by using GIS (Anderson, 2004).

GIS provide value for marketing decision making through two mechanisms (Hess et al, 2004);

1. GIS provide a way to analyze internal or external marketing intelligence data in a format particularly suited to marketing decision making; and
2. GIS provide the ability to integrate both internal and external marketing intelligence data to greatly improve the effectiveness of these marketing decisions.

Geo-demographics have also made a major contribution to improving survey design and sampling. An area-based classification which can be linked to other survey data is a very powerful tool in the hands of a designer of samples. They can be used to create matched samples or sub-samples; they are also useful in probability sampling, hall-testing and minority product sampling. To summarize, the main strengths of geo-demographics are (Mitchell and McGoldrick, 1994);

- Ability to link different data sets which have been geo-coded;
- Coverage of all addresses;
- Ease of use, being linked to the postcode system;
- Ability to link "above the line" and "below the line" marketing activities together, e.g. door-to-door promotions.

In marketing literature, there are some theories using geographical data in order to predict the preferences of customers. These theories and models are;

1. Central Place Theory
2. Trading Areas
3. Gravity Models
4. Analog Modeling

Central Place Theory

In the first half of the twentieth century, marketing science was emerging in the academic environments of business and economics, but it was largely ignored by geography (Berry, 1967) except German geographer Walter Christaller, and his colleague, economist August Lösch. They provided fundamental theoretical concepts of Central Place Theory which states that the most efficient place to locate a new store is exactly between existing stores. By following this pattern, all trade areas will become equal in size and shape. CPT has some assumptions;

- All consumers have equally wealth and are rational,
- Cost of entry into the market is equal for businesses,
- Travel costs are the same for all consumers,
- Consumers only purchase one product.

In this environment, a trade area for a store is a perfect circle. Demand reaches zero at the point to where consumers are unwilling to travel to purchase the product because the total cost (product price plus travel cost) is more affordable elsewhere. CPT does not take into account social or political factors that influence consumer behavior. In the real world, for example, there are multiple goods which may be priced differently for various reasons that CPT cannot take into account (Parrish, 2009).

Trading Areas

One of the most important ideas in marketing geography is a trade area. The American Marketing Association defined a trade area as a district whose size is usually determined by the boundaries within which it is economical in terms of volume and cost for a marketing unit to sell and/or deliver a good or service (Huff 1964). The total number of potential customers in the trade area is equal to the sum of the expected number of consumers in each of the demand zones. According to Berry (1967), an ideal theoretical trade area is a perfect circle drawn around the store. Demand for a good is calculated as the price consumers are willing to pay for the good plus the cost of traveling. At a certain point, consumers are no longer willing to travel resulting in zero demand. The distance from a store to where demand is equal to zero is used to compute the radius of the circle. Once the area is found, the total quantity of the demanded product or service can be derived by multiplying the area of the circle by population density. In the most basic method of representing a trade area, concentric circles can be used to tally customers and sales. As an example, if 20 percent of a store's spotted customers come from any zone and the store has $30,000 sales each week, then $6,000 sales are assumed to come from this zone (Parrish, 2009).

Gravity Models

The retail gravity model was intended to determine trade areas for two competing cities. The gravity models from which he worked included both Reilly (1931) and Converse (1943). William Reilly designed the model to express the proportion of business each city would attract from the population center of a smaller town located between the two competing cities. Reilly's formula is shown in Equation 1 (Parrish, 2009);

$$\frac{Ba}{Bb} = \left(\frac{Pa}{Pb}\right) * \left(\frac{Db}{Da}\right)^2$$

where

B_a = the proportion of the retail business from an intermediate town attracted by city A;

B_b = the proportion of the retail business from an intermediate town attracted by city B;

P_a = the population of city A;

P_b = the population of city B;

D_a = the distance from the intermediate town to city A; and,

D_b = the distance from the intermediate town to city B.

The formula reflects distance and population as influential factors. Any small town between two major cities of the same size will be more influenced by the city to which it is closer. Further, proximity is juxtaposed by population size, thus any city with a larger population will have a stronger influence on the small town (Reilly 1931).

Reilly's gravity model was later modified by P.D. Converse (1943) to identify a breakpoint between the cities. When traveling from one city toward the other, the breakpoint represents the distance between the cities at which the influence of one city ends and the influence of the competing city begins. The breakpoint formula derived by Converse is shown in Equation 2.

$$Db = \left(Dab\right) / \left(1 + \sqrt{Pa / Pb}\right)$$

where;

D_b = the breaking point between city A and city B in miles from B;

D_{ab} = the distance separating city A from city B;

P_b = the population of city B; and

P_a = the population of city A.

A firm could be surrounded by any number of competitors and its trade area could be defined as the distance between the firm and each of its breakpoints. Huff (1964) proposed an alternative model that focuses on the consumer rather than the firm. This model described the process by which consumers choose a firm for specific services. The expression is shown in Equation 3;

$$Pij = \left(\frac{Sj}{Tij^{\lambda}}\right) / \left(\sum_{j=1} Sj / Tij^{\lambda}\right)$$

where;

P_{ij} = the probability of a consumer at a given point of origin i traveling to a particular shopping center j;

S_j = the size of a shopping center j (measured in terms of the square footage of selling area devoted to the sale of a particular class of goods);

T_{ij} = the travel time involved in getting from a consumer's travel base i to a given shopping center j; and

λ = a parameter which is to be estimated empirically to reflect the effect of travel time on various kinds of shopping trips.

One important difference between Huff's approach and Reilly's model is that consumer spatial and non-spatial behavior can be explained by the Huff model (Haines et.al. 1972). The model captures spatial behavior which can be used to predict consumer behavior within the context of multiple store locations. The parameter λ is not considered to be constant and varies according to the type of product. High order goods, such as automobiles, will have a smaller λ than low order goods, such as groceries. Lastly, the model allows the data to be visually represented on a map (Huff, 1964). Later research argued that Huff's model was not accurate in predicting sales for a single location. Studies such as those by Stanley and Sewall (1976) assume that consumers and stores are spatially independent in the Huff model (Parrish, 2009).

Analog Modeling

Another prominent method of estimating trade areas is the analog method. Thompson (1982) defines analogs as snapshot of a store's performance at a given point in time. These snapshots are useful for identifying historical trends in the market area for a particular site and predicting future sales. Applebaum developed the analog model as a means to compare potential sites to existing ones. Analog models allow the analyst to predict what may happen at a proposed site, based on knowledge of existing sites (Munroe, 2004). To analyze the trade area, it can be subdivided into primary, secondary and tertiary trade areas after performing customer spotting (Applebaum, 1966). The primary trade area is where the store gets the majority of its business. The breakpoint is arbitrary, but Applebaum suggested that 60 to 70 percent of a site's sales come from the primary trade area. The secondary trade area should account for 15 to 25 percent of sales, and the tertiary trade area represents the remaining customers and sales (Parrish, 2009).

THE ROLES OF GIS IN TOURISM

Tourism is defined by Abomeh and Nuga (2013) as the temporary movement from home to unfamiliar place for business, recreate, sightsee, visit or points of interest purposes whether near or far. According to this definition the points of interest can vary from person to person. It can range from a hotel, a night club, a restaurant a park to historic sites, conservation sites and other outdoor places within a city or urban centre. Tourism is an activity highly dependent on environmental resources and it also involves geographical attributes. However, if it lacks of planning and management its environmental base will be worn out rapidly (Abomeh and Nuga, 2013).

Although GIS is rarely discussed in the context of tourism, it is now recognized widely as a valuable tourism management instrument due to tourism's strong ties to geographical tools. Information technologies and the tourism share a common characteristic that they are two important areas for the economic growth. Tourism has various geographical features and GIS can play an important role in tourism since it offers functions such as data collection, data processing and storage for planning a trip or making a tourist investment. Most of the tourist information is difficult to use because the information is widespread and scattered through different areas and traditional maps are difficult and expensive to keep to date and they are static and complex so tourists are not able to find a particular data that they needed. On the other hand, if tourists and businesses use GIS, questions about locations, modeling, conditions, and trends are easier to answer. GIS provides the linkage between the spatial data and semantic information while making a combination between database operations and geographic analysis benefits that are offered by maps (Shyti and Kush, 2012). GIS can help to categorize arrival of tourists and the tourism generating areas while providing information about other competing destinations. This system contributes to geographical distribution for tourism offices and branches as well as the locations of tourism boards. In order to meet tourist needs effectively it is important to classify markets. GIS can identify tourism markets whether they are local, regional or global and can divide those markets on a geographical basis as well as other factors such as demographic, psychological and behavioral, in order to make a perfect map of the market to assist in decision-making (Bazazo and Hasoneh, 2010). GIS can also be used at a strategic level to support tourism development by helping in decision-making processes and how, when and where tourism development should go, accessing politicians and managers in tourism development, and also by giving integrated and spatial touristic information. With the help of GIS tools it is possible to make touristic resources inventories, to identify the best places for new destinations, and to evaluate tourism impacts in the environment in order to protect natural and cultural resources (Albuquerque et al. 2016).

Many institutions use GIS in their operations. For example National Park Service sees GIS as a necessary tool for upholding the mandate of the National Park Service to manage parks for future generations. The department benefit from GIS to analyze location, features or objects such as a stream, and feature characteristics such as water quality, direction of flow. It has also been used to investigate location, area condition, new trends and changes, and resource usage related to tourism development, to determine the best areas for tourist destinations while maintaining sustainable natural zones, to identify ecotourism areas, and to deliver tourism information (Dye and Shaw 2007).

GIS also made big contributions to various areas of recreational resource management such as resource inventory and building a spatial decision system. GIS-based inventories make managing the spatial database of recreational resources more flexible, more objective, and more efficient. Walking tracks, waterfalls, historical and cultural sites of interest, and other recreational features can be stored as spatial objects, and data can be stored in GIS as waterfalls representing point feature, a walking track representing line feature, and a picnic ground representing area feature (Chhetri and Arrowsmith, 2008).

In a tourism management context, GIS can be used by a tour operator. Within a specific area, GIS can show all roads represented as lines and hotels represented as points. Then GIS can be used to identify all hotels that match search criteria such as those providing facilities of interest to the visitors within this specific area. Another important function of GIS is data integration, where two different map layers of the same zone can be viewed together as a new layer. For example a buffer zone operation for identifying hotels' distance from a trekking route can be combined with hotel data. With the help of improved visualization users can remove all other features and show only trekking routes and villages offering hotel accommodation. In short GIS helps to manipulate data in order to identify various patterns or relationship based on specific needs which enables reaching value-added information for decision-making process (McAdam, 1999).

Use of GIS in Tourism Planning

Tourism planning refers to the integrated planning of natural or cultural attraction, services such as accommodation, restaurants, shops, visitor information, tour and travel operations, and transportation facilities (Jovanovic, 2016). Tourism implies travelling between close and far places and to determine the location, maps are an important tool; as a result tourism is more concerned with maps and thematic information than other industries. When looking at maps, tables, charts, and reports it is sometimes difficult to pull out required information and make sense out of them. GIS opens new perspectives for people on using spatial data to make decisions and tourism planning. GIS usually operates on spatial and attribute data. Spatial or geographical data can be expressed as a grid coordinate or in degrees of latitude and longitude. GIS contribute to tourism planning by manipulating data and spatial attributes in order to provide required value added information, identifying patters and spatial relationships. GIS also has the ability to work with other technologies such as remote sensing, GPS, CAD. These developed systems provide a wide range of searching and querying analytical functions which could further facilitate more tools to tourism planning (Shah and Wani, 2015).Tourism planning requires all locations and their interrelations to be defined and analyzed within a spatial context so much more spatial data collecting and processing is needed. In this context GIS can describe and identify tourism infrastructure elements geometrically, thematically, and topologically while dealing with both object (e.g., visitor centers, trails, forest patches) and field data (e.g., humidity, slope, altitude) (Patwal, 2013).

Applications of GIS in Tourism

There are various usages of GIS in tourism. GIS applications in tourism range from facility inventory, tourism-based land management, visitor impact assessment, recreation related wildlife conflicts, wilderness perception mapping, to tourism information system management and decision support systems (Raghuvanshi et al. 2006). In order to highlight wilderness areas and differentiate them from other protected areas, a wilderness continuum map was developed in the UK. With the help of GIS, this map can identify potential risk areas from tourism activities. A GIS supported landscape management system is developed and used in the Cairngorm Mountains to describe environmental and cultural resources and landscape designations. The system was very useful for managing high level of scenic interest and where visitors are attracted to that scenic amenity. In Red Stone Park, GIS was used in the landscape planning. GSI helped to protect bio-diversity, landscape and visual quality of the area by enabling process simulation, impact analysis, and landscape change model development. Another application of GIS was in the northern Nepal and in the Siberian Mountains to monitor the impact of tourism. The collected data was used to analyze the number of tourists which is generally very low, and to formulate a cultural resource management framework (McAdam, 1999).

The Recreation Opportunity Spectrum (ROS) (1982) is a classification tool related with the recreational resource management which is developed by The US Department of Agriculture and Forest Services. The ROS classifies forest service areas into six management categories: urban; rural; roaded natural; semi-primitive motorized; semi-primitive non-motorized; and primitive. There are three setting characteristics: physical, social, and managerial. Physical setting is related with type of access, remoteness, and area size, social setting includes the number of people encountered, and managerial setting is based on visitor management, level of development, and naturalness. Similar to the ROS approach, Australian Department of the Environment and Energy (2008) created the National Wilderness Inventory. There are four wilderness quality indicators: Remoteness from Settlement remoteness from places of permanent occupation; Remoteness from Access remoteness from established access routes; Apparent Naturalness the degree to which the landscape is free from the presence of permanent structures associated with modern technological society; Biophysical Naturalness the degree to which the natural environment is free from biophysical disturbance caused by the influence of modern technological society.

The ROS and the National Wilderness Inventory try to identify the wilderness quality across the landscapes which creates base for various GIS applications (Chhetri and Arrowsmith, 2008).

CONCLUSION

Today the most important asset of organizations is information. They require information to make more efficient decisions to reach their ultimate goals. Organizations have different information sources such as their client, competitors or even themselves. However, managing information is more important than getting it. For this purpose, they use some systems such as decision support systems and information systems. The decision support system is designed to help managers in the analysis and decision-making processes and information systems are widely used tools to collect and to interpret data. These systems have some component and geographic information systems (GIS) are an important part of information systems.

A GIS is essentially a digital map linked to a database management system which can be used for the purposes of displaying and querying information, carrying out spatial analysis and assisting in the decision-making process. Although output of GIS is displayed as maps, they are totally different. While computer-generated maps are static, a GIS is as dynamic as its underlying database. The essential difference lies in the ability of a GIS to analyze the data behind the map and to focus on the relationships among the data items lying in tables. The first modern GIS system was used in 1962 in Ottawa, by the Canadian Department of Forestry to map land use and data about farming and wildlife. The designer of the system was Roger Tomlinson, who was asked by the Canada Land Inventory project to pursue his idea of using computers to combine maps and statistics (Coyle, 2011). Today more than 300,000 companies use the GIS. By means of GIS, collecting tremendous data, storing them in database, updating when required, analyzing, drawing visible maps based on analyses, accessing the necessary information obtained as a result of analyses are possible. Its usage spreads rapidly in many different areas such as geography, criminology, archeology, education, tourism, management and marketing. With combination of geography and marketing disciplines result in emergence of a study area which is called geomarketing. Geomarketing is a recent discipline that combines the geographic visualization and analysis with marketing techniques, aiming at more efficiently attaining the ultimate goal of organizations. In marketing literature, there are some models and theories that tried to explain consumer behavior when they were dealt with geographical based decision makings. These theories are central place theory, trading areas, gravity models and analog modeling

Developments such as GIS and associated technologies such as GPS, remote sensing, etc. are considered as a means of improving decision-making in tourism planning, in the way that they make information more accessible and usable (Bahaire & Elliott-White, 1999). As tourism is a spatial phenomenon at its very core, which basically refers to travel from one place to another, activities related to tourism planning requires spatial data collecting and processing. Tourism industry is using GIS since the early 1990s in the way that they are very useful tools for the evaluation and planning of different aspects of tourism like natural and cultural resources, facilities, activities, services and so on. However, there are also various critics for GIS systems. GIS is considered as a quite complicated technology necessitating more advanced, than simple word processing, skills. It can be easy to manipulate the results to support the interests of particular groups, and it does not by itself ensure fairness, equity and compatibility with sustainability principles (Avdimiotis et al., 2006; Bahaire and Elliot-White, 1999; Farsari, 2003). With these warnings in mind, GIS can be used effectively by tourism managers but not as a solely final decision-making tool. Tourism managers can use GIS to identify where tourists are coming from, to plan for new site selections, to analyze the transportation infrastructure, and to analyze the necessary space. Moreover, tourists can use GIS to plan routes and accommodation, to access interactive maps, cultural events, and to see other needed information in the area they want to travel.

REFERENCES

Abomeh, O.S., & Nuga, O.B. (2013). Utilization of GIS technology for tourism management in Victoria Island Lagos. *European Scientific Journal, 9*(3).

Albuquerque, H. C. D. C., Martins, F. M. C. P. F., Raposo, R. M. D. A., Cardoso, L. M. T. G., Beça Pereira, P. M. D. S., & Dias, P. (2016). Construction of a web-based geographical information system–the case of "Ria de Aveiro" region. *Anatolia, 27*(1), 71–81. doi:10.1080/13032917.2015.1083210

Anderson, V. M. (2004). *Developing Integrated Object-Oriented Conception of Geomarketing as a Tool for Promotion of Regional Sustainable Development: The Case Study of Ukraine*. Research Paper, 2004-7.

Anumba, C., Dainty, A. R. J., Ison, S. G., & Sergeant, A. (2005). The application of GIS to construction labour market planning. *Construction Innovation, 5*(4), 219–230. doi:10.1108/14714170510815267

Applebaum, W. (1954). Marketing Geography. *American Geography: Inventory and Prospect*, 245-251.

Applebaum, W. (1965). Can store location research be a science? *Economic Geography*, *41*(3), 234–237. doi:10.2307/141900

Applebaum, W. (1966). Guidelines for a Store-Location Strategy Study. *Journal of Marketing*, *30*(4), 42–45. doi:10.2307/1249498

Australian Department of the Environment and Energy (2008). *National Wilderness Inventory*. Department of the Environment and Energy.

Avdimiotis, S., Mavrodontis, T., Dermetzopoulos, A.S., & Riavoglou, K. (2006). GIS applications as a tool for tourism planning and education: A case study of Chalkidiki. *Turizam: znanstveno-stručničasopis, 54*(4), 405-413.

Bahaire, T., & Elliott-White, M. (1999). The application of Geographical Information Systems (GIS) in sustainable tourism planning: A review. *International Journal of Sustainable Tourism*, *7*(2), 159–174. doi:10.1080/09669589908667333

Bazazo, I. K., & Hasoneh, A. I. (2010). Using Geographic Information System to visualize travel patterns and market potentials of Petra City in Jordan. *International Journal of Marketing Studies*, *2*(2), 144.

Bensghir, T. K., & Akay A. (2006). Bir Kamu Politika Aracı Olarak Coğrafi Bilgi Sistemleri (CBS): Türkiye'de Belediyelerin CBS Uygulamalarının Değerlendirilmesi. *Çağdaş Yerel Yönetimler, 15*(1), 31-46.

Berry, B. J. L. (1967). *Geography of Market Centers and Retail Distribution*. Englewood Cliffs, NJ: Prentice Hall.

Black, S., Powers, G., & Roche, M. (1994). A Gis-based approach to community analysis for targeted marketing. *Economic Development Review (Schiller Park, Ill.)*, *12*(2), 69–73.

Brody, R. (1999). Geographic Information Systems. *Journal of Business & Finance Librarianship*, *5*(1), 3–18. doi:10.1300/J109v05n01_02

Burrough, P. (1986). *Principles of geographical information systems for land resources assessment*. Oxford, UK: Clarendon Press.

Chen, R. J. C. (2007). Geographic information systems (GIS) applications in retail tourism and teaching curriculum. *Journal of Retailing and Consumer Services*, *14*(4), 289–295. doi:10.1016/j.jretconser.2006.07.004

Chhetri, P., & Arrowsmith, C. (2008). GIS-based modelling of recreational potential of nature-based tourist destinations. *Tourism Geographies*, *10*(2), 233–257. doi:10.1080/14616680802000089

Converse, P. D. (1949). New Laws of Retail Gravitation. *Journal of Marketing*, *14*(3), 379–384. doi:10.2307/1248191

Coyle, A. (2011). Interior library GIS. *Library Hi Tech*, *29*(3), 529–549. doi:10.1108/07378831111174468

Donnelly, F. P. (2010). Evaluating open source GIS for libraries. *Library Hi Tech*, *28*(1), 131–151. doi:10.1108/07378831011026742

Dye, A. S., & Shaw, S. L. (2007). A GIS-based spatial decision support system for tourists of Great Smoky Mountains National Park. *Journal of Retailing and Consumer Services*, *14*(4), 269–278. doi:10.1016/j.jretconser.2006.07.005

Elliott, R. (2014). Geographic information systems (GIS) and libraries: Concepts, services and resources. *Library Hi Tech News*, *31*(8), 8–11. doi:10.1108/LHTN-07-2014-0054

Farsari, Y. (2003, April). GIS-based support for sustainable tourism planning and policy making. In *Proceedings of the XII International Leisure and Tourism symposium* (pp. 3-4). Academic Press.

Fidan, H. (2009). Marketing Information System (MIS) and USAGE of Geographic Information System (GIS) in Marketing. *Journal of Yaşar University*, *4*(14), 2151–2171.

Freire, S., & Santos, T. (2012). Advancing GeoMarketing Analyses with Improved Spatio-temporal Distribution of Population at High Resolution. In *Proceedings of the European Conference on Information Management* (pp.100-108). Academic Press.

Fung, D. S., & Remsen, A. P. (1997). Geographic Information Systems Technology for Business Applications. *Journal of Applied Business Research*, *13*(3), 17–24. doi:10.19030/jabr.v13i3.5748

Goodchild, M. F. (1989). Geographic Information Systems and Market Research. *Papers and Proceedings of Applied Geography Conferences* (*vol. 12*, pp. 1-8). Academic Press.

Grimshaw, D. J. (1993). *Bringing GIS into Business Geoinformation*. London: Longman.

Haines, G. H., Leonard, S. S., & Marcus, A. (1972). Maximum Likelihood Estimation of Central-City Food Trading Areas. *JMR, Journal of Marketing Research*, *9*(2), 154–159. doi:10.2307/3149948

Hess, R. L., Rubin, R. S., & Lawrence, A. W. Jr. (2004). Geographic information systems as a marketing information system technology. *Decision Support Systems, 38*(2), 197–212. doi:10.1016/S0167-9236(03)00102-7

Huff, D. L. (1964). Estimating a Trading Area. *Journal of Marketing, 28*(3), 34–38. doi:10.2307/1249154

Jovanović, V. (2016). The application of GIS and its components in tourism. *Yugoslav Journal of Operations Research, 18*(2).

Kotler, P. (1997). *Marketing Management: Analysis, Planning, and Control* (9th ed.). Upper Saddle River, NJ: Prentice-Hall.

Krek, A. (2006). Efficient Pricing of Geomarketing Internet Services: European vs. American Approach. *International Journal of Business and Management, 1*.

Laudon, K. C., & Laudon, J. P. (2003). *Essential of management information systems* (5th ed.). New Jersey: Prentice Hall.

Lee, B., & Graefe, A. (2004, October). GIS: A Tool to Locate New Park and Recreation Services. *Parks & Recreation*, 34–41.

Maguire, D. (1995). Implementing Spatial Analysis and GIS Applications for Business and Service Planning. In G. Clarke & P. Longley (Eds.), *GIS for Business and Services Planning* (pp. 171–191). New York: Pearson Professional.

Marakas, G. M. (2003). Modern data warehousing, mining, and visualization: Core concepts. In *Data mining and data visualization* (pp. 97–101). Upper Saddle River, NJ: Prentice Hall, Pearson Dedication, Inc.

Martin, W. E., Brown, C. V., DeHayes, D. W., Hoffer, J. A., & Perkins, W. C. (2005). *Managing information technology* (5th ed.). Prentice Hall.

McAdam, D. (1999). The value and scope of geographical information systems in tourism management. *Journal of Sustainable Tourism, 7*(1), 77–92. doi:10.1080/09669589908667327

McMahon, P. (1997). *Geographic Information System in the construction industry – an underused resource?* The Chartered Institute of Building, No.82.

Mitchell, V.-W., & McGoldrick, P. J. (1994). The Role of Geodemographics in Segmenting and Targeting Consumer Markets: A Delphi Study. *European Journal of Marketing, 28*(5), 54–72. doi:10.1108/03090569410062032

Munroe, K. (2004). Meaningful Metrics: Site Evaluation and Market Optimization Models. In The Power of Place: Advanced Customer and Location Analytics for Market Planning (pp. 127-156). San Diego, CA: Claritas.

Musyoka, S. M., Mutyauvyu, S. M., Kiema, J. B. K., Karanja, F. N., & Siriba, D. N. (2007). Market segmentation using geographic information systems (GIS): A case study of the soft drink industry in Kenya. *Marketing Intelligence & Planning*, *25*(6), 632–642. doi:10.1108/02634500710819987

Orhun, Y. A. (2013). Spatial differentiation in the supermarket industry: The role of common information. *Quantitative Marketing and Economics*, *11*(1), 3–37. doi:10.1007/s11129-012-9123-x

Parrish, J. S. (2009). *Geographic Information Systems and Spatial Analysis of Market Segmentation for Community Banks* (Unpublished master dissertation). University of North Carolina.

Patwal, A. (2013). Sustainable Tourism Infrastructure Planning: A GIS Based Approach. *International Journal of Hospitality and Tourism Systems*, *6*(1).

Payton, K. F. (2010). *GIS-Based Market Analysis: A Model for Implementing a Geographic Information System as a Market Analysis Tool* (Unpublished master dissertation). The College of St. Scholastica, Duluth, MN.

Raghuvanshi, T. K., Belwal, R., & Solomon, N. (2006). An Approach to Develop Tourism Potential in Ethiophia Through Geographic Information System. *The Consortium Journal of Hospitality and Tourism*, *11*(1).

Recreation Opportunity Spectrum. (1982). Retrieved July 10, 2017, from https://www.fs.usda.gov/detail/tonto/landmanagement/planning/?cid=stelprdb5412121

Reilly, W. J. (1931). *The Law of Retail Gravitation*. New York: John Wiley & Sons.

Reynolds, H. (1997). *An introduction to geographic information systems (GIS)*. Retrieved July 17, 2015 from http://badpets.net/IntroGIS/

Shah, S. A., & Wani, M. A. (2015). Application of Geospatial Technology for the Promotion of Tourist Industry in Srinagar City. *International Journal of u-and e-Service Science and Technology*, *8*(1), 37–50.

Shyti, B., & Kushi, E. (2012). The impact of GIS application in the tourism development of Elbasan region. *The Romanian Economic Journal*, *15*(45), 189–210.

Smith, M. K. (2005). *Factors In the Adoption of Geographic Information Systems in Sport Marketing* (Unpublished master dissertation). California State University.

Stanley, T. J., & Sewall, M. A. (1976). Image Inputs to a Probabilistic Model: Predicting Retail Potential. *Journal of Marketing*, *40*(3), 48–53. doi:10.2307/1249994

Star, J., & Estes, J. (1990). *Geographic Information Systems: An Introduction*. Englewood Cliffs, NJ: Prentice-Hall.

Thompson, J. S. (1982). *Site Selection*. New York: Lebhar-Friedman Books.

Tomlinson, R. (2003). *Thinking about GIS*. Redlands, CA: ESRI Press.

Türk, T., Kitapçı, I. O., & Dörtyol, T. (2014). The Usage of Geographical Information Systems (GIS) in the Marketing Decision Making Process: A Case Study for Determining Supermarket Locations. *Procedia: Social and Behavioral Sciences*, *148*, 227–235. doi:10.1016/j.sbspro.2014.07.038

Vlachopoulou, M., Silleos, G., & Manthou, V. (2001). Geographic information systems in warehouse site selection decisions. *International Journal of Production Economics*, *71*(1-3), 205–212. doi:10.1016/S0925-5273(00)00119-5

Wade, T., & Sommer, S. (2006). *A to Z GIS: An Illustrated Dictionary of Geographic Information Systems*. Redlands, CA: ESRI Press.

Xia, J. (2004). Library space management: A GIS proposal. *Library Hi Tech*, *22*(4), 375–382. doi:10.1108/07378830410570476

Yener, D. (2017). Geographic Information Systems and Its Applications in Marketing Literature. In S. Faiz & K. Mahmoudi (Eds.), *Handbook of Research on Geographic Information Systems Applications and Advancements* (pp. 158–172). IGI Global. doi:10.4018/978-1-5225-0937-0.ch006

KEY TERMS AND DEFINITIONS

Decision Support Systems: Systems that are designed to help managers in the analysis and decision-making processes.

Geographical Information Systems: An integrated collection of software and data that is used to visualize and organize location-based data for the purposes of performing geographic analyses and creating maps.

Geomarketing: A recent discipline that combines the geographic visualization and analysis with marketing techniques, aiming at more efficiently attains the ultimate goal of organizations.

Gravity Models: Models that were intended to determine trade areas for two competing cities.

Information Systems: Systems are widely used tools to collect and to interpret data. GIS is a kind of information systems.

Marketing: An organizational function and a set of processes for creating, communicating, and delivering value to customers and for managing customer relationships in ways that benefit the organization and its stakeholders.

Spatial Data: Data identifies the geographic location of features. It has three types of formats: vector, raster, and semantic.

Tourism: The temporary movements from home to unfamiliar place for business, recreate, sightsee, visit, or points of interest purposes whether near or far.

Trading Areas: A geographic area within which a business enterprise or center of retail or wholesale distribution draws most of its business.

Chapter 10
Toward Geospatial Collaborative Tourism Recommender Systems

Zahra Bahramian
University of Tehran, Iran

Rahim Ali Abbaspour
University of Tehran, Iran

Christophe Claramunt
Naval Academy Research Institute, France

ABSTRACT

Tourism activities are highly dependent on spatial information. Finding the most interesting travel destinations and attractions and planning a trip are still open research issues to GIScience research applied to the tourism domain. Nowadays, huge amounts of information are available over the world wide web that may be useful in planning a visit to destinations and attractions. However, it is often time consuming for a user to select the most interesting destinations and attractions and plan a trip according to his own preferences. Tourism recommender systems (TRSs) can be used to overcome this information overload problem and to propose items taking into account the user preferences. This chapter reviews related topics in tourism recommender systems including different tourism recommendation approaches and user profile representation methods applied in the tourism domain. The authors illustrate the potential of tourism recommender systems as applied to the tourism domain by the implementation of an illustrative geospatial collaborative recommender system using the Foursquare dataset.

DOI: 10.4018/978-1-5225-5088-4.ch010

1. INTRODUCTION

Tourism as a social, cultural, and economic phenomenon entails the movement of people to countries or places outside their living environment. Tourism activities are highly dependent on spatial information such as places of interest and attractions that are likely to present an interest for potential visitors. Planning a trip and finding the most appropriate travel destinations and Points of Interests (POIs) are still considered as open research issues in GIScience research applied to the tourism domain. A POI can represent a specific tourist attraction that a user may find interesting to visit such as a museum, historical and cultural buildings. Nowadays, the Internet and the World Wide Web offer huge amounts of free information on travel destinations that may be useful for users wishing to plan a visit. However, the large extent of information available is often overwhelming and time consuming for most users as selecting the most appropriate POIs and planning a tourism trip according to their own preferences is far from being a straightforward task. In order to overcome this information overload problem, Tourism Recommender Systems (TRSs) have been recently introduced in order to suggest items that are most likely of interest to a given user. The main principles behind TRSs are that items of interest should be identified and recommended by taking into account the user's profile and preferences. In other word, TRSs infer users' needs by considering either explicit or implicitly given users' profiles and even by comparing such profiles to users with similar preferences.

This chapter provides a survey of TRSs, its main objectives are as follows:

1. To briefly review current tourism recommendation approaches as well as pros and cons. We introduce different classes of tourism recommendation approaches including from content-based, collaborative, knowledge-based, demographic, to hybrid TRSs. Content-based TRSs suggest items that are similar to the ones that the user liked before, while collaborative TRSs take into account items that similar users liked in the past. Knowledge-based TRSs suggest items based on specific domain knowledge and user's preferences. Demographic TRSs suggest items based on some basic user demographic. Finally, hybrid TRSs take into account the combination of the above mentioned techniques.
2. To review current approaches oriented to the representation of user profiles in TRSs. We will introduce user profile representations such as the ones based on a list of keywords (i.e., as categories associated to the users), vectors (i.e., where each vector value denotes a degree of interest associated to a given item), semantic models (i.e., based on domain-based ontologies) and uncertainty-based models (i.e., to take into account the uncertainty associated to the information stored in the user profile).

3. To present the different functionalities so far provided by current TRS. Such functions are classified according to the services delivered such as personal interest estimation, travel destination and tourist packs recommendation, ranked list of POIs recommendation, route or itinerary, recommendations based on Location-Based Social Networks (LBSN), and group-based recommendations.
4. To provide an insight regarding some additional influential factors that play a role in TRSs including geographical, temporal and social influences.
5. To introduce the different Computational Intelligence (CI) techniques applied in TRSs including statistics, fuzzy sets, artificial neural networks, evolutionary computing, and swarm computing.
6. To provide an insight to the different evaluation metrics currently applied to TRSs including quantitative and qualitative metrics either explicit or implicit.
7. To introduce a case study that shows the potential of TRS. The experiment developed uses Foursquare LBSN data including both their geographical and social dimensions. The objective is to take into account several factors such as social relations and similarities while making some choices. The objective is to recommend some optimal routes that take into account some recommended POIs. The whole approach is quantitatively evaluated. The implementation presents some preliminary findings and future works.

The rest of the chapter is organized as follows: Section 2 introduces a series of TRS approaches and the main user profile techniques applied so far. Sections 3 and 4 present the main functionalities provided by TRSs. Section 5 discusses the potential of LBSN when combined with TRSs. Section 6 provides a review on the different computational intelligence techniques applied in TRSs, while Section 7 present a few metrics applied to TRSs. Section 8 develops a case study realized from Foursquare LBSN. Finally, Section 9 summarizes the chapter and outlines future works.

2. TOURISM RECOMMENDER SYSTEMS

In order to overcome the mentioned information overload problem, Tourism Recommender Systems (TRSs) have been recently introduced as information filtering and decision support tools that suggest items that are most likely of interest to a given user (Aggarwal, 2016c; Borràs et al., 2014; Colombo-Mendoza et al., 2017; Gavalas et al., 2014; Moreno et al., 2015; Ricci et al., 2015b). Most of these approaches make a difference between novel and experienced users, and this at different degrees. TRSs generally make a difference between three different and complementary modelling abstractions, that are, items, users, and transactions (Ricci et al., 2015a). Items are the objects that may be recommended. The value of an item may be positive or negative depending on the item is useful to the user, or the item is not appropriate to the user, respectively. Users may have different goals and characteristics. User data form the user model and has a central role to generate personalized recommendations according to the user preferences and needs. Transactions are relations between the users and the items such as selected item by the user, and its related explicit or implicit rating. Ratings can take on a variety of forms (Schafer et al., 2007):

- Numerical ratings such as the 1 to 5 stars.
- Ordinal ratings, such as "strongly interested, interested, neutral, uninterested, strongly uninterested".
- Binary ratings that model choices in which the user is simply asked to decide if a certain item is interested or uninterested.
- Unary ratings can indicate that a user has selected an item, or rated it positively. The absence of a rating indicates that there is no information relating the user to the item.

User can also evaluate items by assigning tags to the items that the system presents. In conversational TRS, the system collects the various requests-responses to provide the user with more refined results (Mahmood and Ricci, 2009). Let us introduce in the next sub-sections the different classes of tourism recommendation approaches including content-based TRSs, collaborative TRSs, knowledge-based TRSs, demographic TRSs and hybrid TRSs (Figure 1).

2.1. Content-Based TRSs

Content-based TRSs (Figure 2) recommend items that are similar to the ones that the user liked in the past (Lops et al., 2011). They utilize the information about an item and user's profile for recommendations and calculate the similarity of item features the user's profile. The system then selects the items which have the highest similarity

Figure 1. Different classes of tourism recommendation approaches

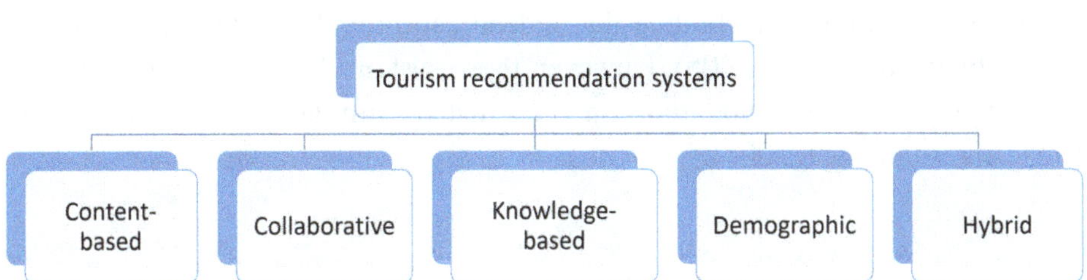

Figure 2. Content-based TRS

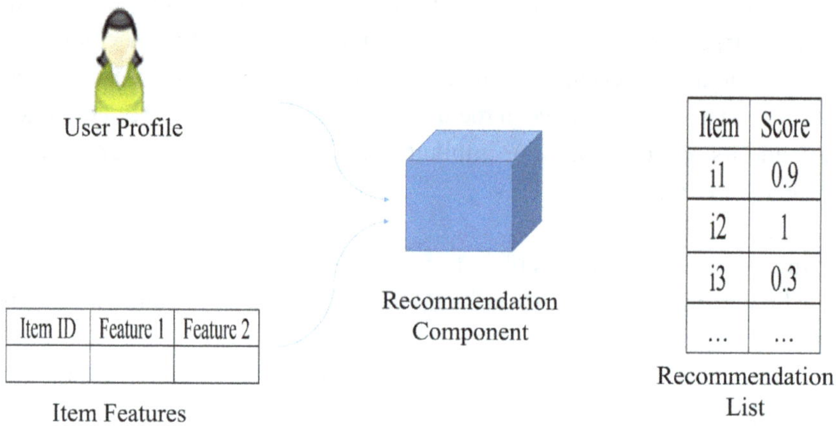

with the user's preferences, and proposes the items similar to their past experiences and preferences (Balabanović and Shoham, 1997; Srisuwan and Srivihok, 2008).

Content-based TRS have the following properties. They could serve the person who has some unique interests (i.e., unlike the other users) and who doesn't need feedbacks from similar users (Husain and Dih, 2012). Moreover, they have the advantage of immediate recommendations, i.e. no feedback from the other users about the items is necessary (Hinze and Junmanee, 2005). Another advantage is that they are not restricted to suggest options previously rated by users. This means that the task of the algorithm is to classify a potential item (an option the user has not expressed any opinion about), as something relevant or irrelevant for the user (Savage et al., 2012). In addition, they can provide the user with a better explanation as to why some option was suggested (Savage et al., 2012). Nevertheless, content-based tourism recommendation systems have some limitations. They need to be learned in order to be efficient with a sufficient amount of data (Hinze and Junmanee, 2005). Therefore, they face the cold start problem since the recommendations lack accuracy at the start, when a user first joins, until the interaction between user and system can enrich the data stored in the user's profile (Pazzani and Billsus, 2007). Automated content-based recommendations are mostly based on the application of some user clustering technique that indeed can provide a solution to the cold start problem (Jian et al., 2005). However, these systems have some limitations as they can only get recommendations on items that are similar to some visited before (Hinze and Junmanee, 2005).

2.2. Collaborative TRSs

Collaborative TRSs (Figure 3) give recommendations to a given user based on items that other users with similar tastes liked in the past (Koren and Bell, 2015; Ning et al., 2015). They compute the similarities among the users based on their rating histories to find users that present preferences similar to the given user. An average of the preferences of similar users is computed to recommend items to the given user (Baltrunas, 2011; García-Crespo et al., 2009; Hinze and Junmanee, 2005; Savage et al., 2012).

Techniques used in collaborative filtering include memory-based collaborative filtering and model-based collaborative filtering (Aggarwal, 2016b, c; Gong, 2009; Horozov et al., 2006; Su and Khoshgoftaar, 2009). Memory-based collaborative filtering, also called user-based collaborative filtering, uses statistical techniques on users' or rating patterns and the correlations between them, and forecasts results according to those users who have similar preferences. In order to deal with the scalability issue, a model-based collaborative filtering technique is widely used. Clustering techniques are algorithmically applied to cluster the users into highly similar groups and makes recommendations based on the group into which the target user is clustered. This resolves the scalability problem as the recommendation process seeks the user rating only from the clustered group rather than the whole database.

One of the advantages of collaborative TRSs is that the user is not restricted to recommend items similar to the one he preferred in the past. However, collaborative

Figure 3. Collaborative TRS

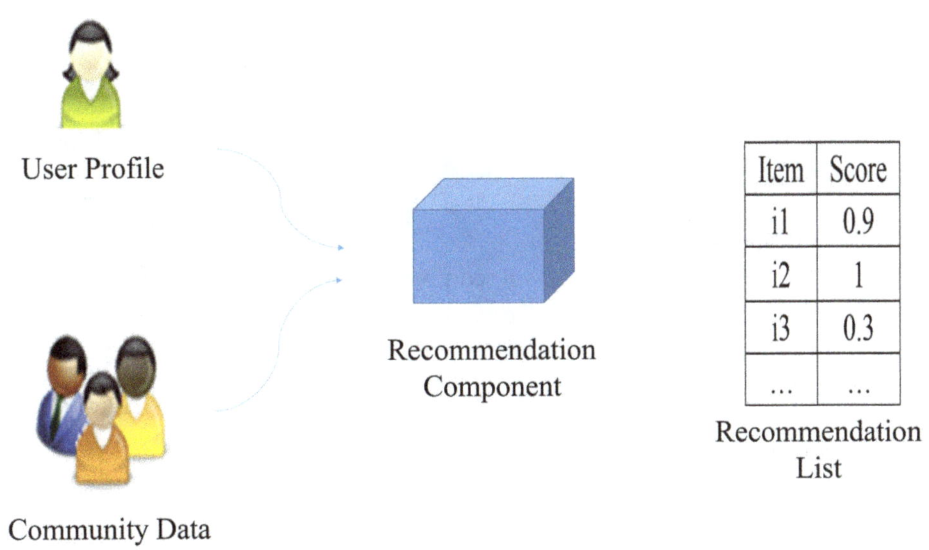

TRSs also have some limitations. They need a large amount of user-related data for delivering appropriate recommendation. Moreover, their accuracy depends on the number of items that are associated with the user. In addition, they neglect new item added to the database as there is no initial rating for them (Hinze and Junmanee, 2005; Husain and Dih, 2012; Kabassi, 2010).

2.3. Knowledge-Based TRSs

Knowledge-based TRSs (Figure 4) suggest items to the user according to both the user's preferences and characteristics of the required item (Aggarwal, 2016c; Hinze and Junmanee, 2005). They collect data explicitly during each user-system interaction. Knowledge-based TRSs store them in the form of statements, rules or ontologies using for example case-based and rule-based TRSs (Berka and Plößnig, 2004; Jadhav and Sonar, 2009).

Case-based tourism recommendation system (Aggarwal, 2016c; Gong, 2009; Husain and Dih, 2012; Jadhav and Sonar, 2009; Zhu et al., 2009) employs past history (cases) in order to solve new cases. It applies an inductive approach to propose solution for new cases. The required knowledge is drawn from the case database, meaning that this method requires less knowledge input from the user in order to propose a solution.

Figure 4. Knowledge-based TRS

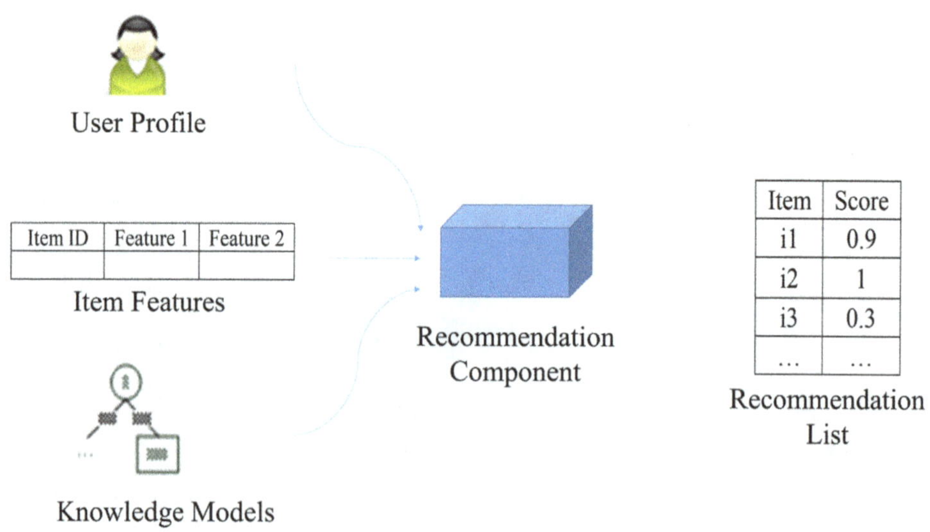

Rule-based tourism recommendation system (Aggarwal, 2016c; Jadhav and Sonar, 2009; Okabe et al., 2008) addresses the recommendation problem in the form of rules applicable to the tourism domain.

2.4. Demographic TRSs

Demographic TRSs (Aggarwal, 2016a; Ricci et al., 2015a) suggest items based on the demographic profile of the user. They generate different recommendations according to different demographic niches (Figure 5).

2.5 Hybrid TRSs

Hybrid TRSs use a combination of the abovementioned methods, exploiting the advantages of one technique to compensate the shortcomings of another, thereby improving the overall performance (Burke, 2002). So far different classes of hybrid TRSs are introduced (Burke, 2002). A weighted hybrid TRS combines several recommendation techniques to derive an aggregated recommendation, while a switching hybrid TRS system selects the most appropriate recommendation techniques depending on a series of criterion. The temporal dimension has been also considered as a valuable parameter to take into account. When applying a TRS in cascade, the first recommendation technique produces a coarse ranking of candidates, while the following ones should refine the recommendations given. Last, a meta-level hybrid TRS uses the model learned by one recommender system as an input for another.

Figure 5. Demographic TRS

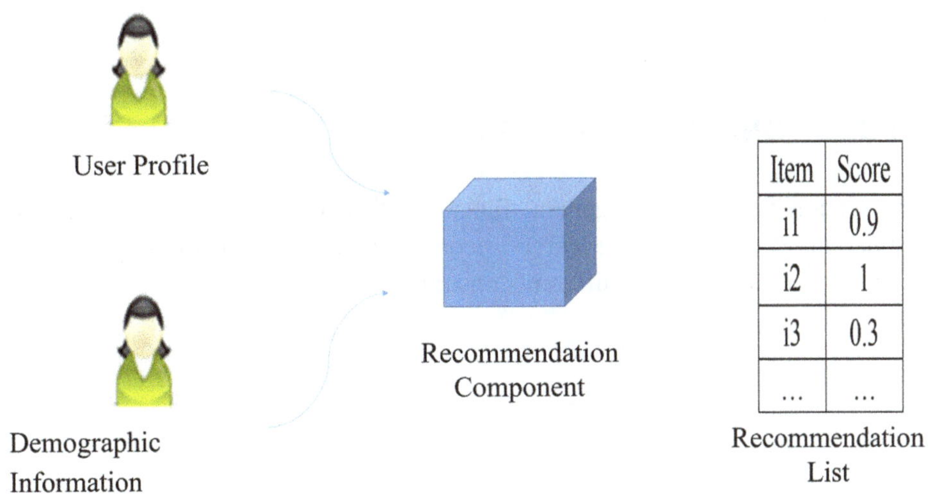

3. USER PROFILE REPRESENTATION

User profile plays a key role in the recommendation process since it represents the user preferences and interests. Therefore, every TRS needs to develop or maintain a user profile in order to identify the preferences of the user. Different representation of user profiles have been introduced as follows (Borràs et al., 2014).

- In the simplest form of user profile representation, a list of preferred keywords or items categories is associated to each user. However, this information is often too general and cannot provide accurate recommendations.
- In the more widespread representation approach, a vector with numerical ratings is associated to each user. Each rating indicates the degree of interest of the user with respect to each attributes of items. In this representation approach, the inclusion of other types of features in the user profile is facilitated.
- In the ontology-based representation method, the system uses semantic models and represents the domain knowledge in the form of an ontology. In other words, the keyword list contains classes' names in the ontology (i.e. concepts) and the ratings vector is also defined in the space of the concepts in the ontology.

Uncertainty-based representation methods are used to handle the credibility of the user profile information. In many situations, especially when the rating values are not given explicitly, rating values are uncertain and a confidence degree can be associated to each rating in the user profile. The confidence degree can be used as a weighting factor in the exploitation stage to infer from the user interaction with the domain items.

4. FUNCTIONALITIES OF TRSs

This section describes the general functionalities of different TRSs including destination, POIs list, route, itinerary recommendation, and the ones applied to a group of users according to different constraints.

4.1. Travel Destination Recommendation and Tourist Pack Construction

Some TRSs focus on the recommendation of a destination or an area of interest according to the user's preferences. They can also provide purchasing services and booking a trip costumers in order to find the best travel packages according to their preferences (Koceski and Petrevska, 2012; Lorenzi et al., 2011; Seidel et al., 2009).

4.2. Recommendation of a Ranked List of POIs

TRS may quantify the interest of the user in a particular POI and propose a list of the most interested POIs according to the user preferences when he decides to visit one specific destination or is already there (Borràs et al., 2011; Fenza et al., 2011; Montejo-Ráez et al., 2011; Ruiz-Montiel and Aldana-Montes, 2009; Ruotsalo et al., 2013; Sebastia et al., 2009).

4.3. Route Recommendation

Some TRSs suggest personalized route to the user. (Mocholi et al., 2012) collect contextual data related to the users' journeys and learn the sequences that emerge and that are associated to this contextual information. They apply semantic multi-criteria and ant colony algorithm to learn the user's routes and predict the most likely route a user is following based on his current location and context, as well as the spatial and semantic proximity to the predicted routes.

4.4. Trip or Itinerary Recommendation

Some TRSs not only suggest a list of POIs according to the user's preferences but also take a route through them into account (Borràs et al., 2012; Kurata and Hara, 2013; Luberg et al., 2011; Montejo-Ráez et al., 2011; Santiago et al., 2012; Vansteenwegen et al., 2011).

4.5. Recommendation to a Group of Users

In many situations, recommendations should be made to a group of users rather than to an individual. For instance, a group recommender system has been applied to recommend a particular item to a group of users rather than a single user (Masthoff, 2004, 2015).

4.6. Recommendation of Mandatory POIs or Mandatory POI Types

If a POI is considered as a mandatory point of interest by the system provider or the user, the user must see it when he closes to it. In other words, it should be at the top of his preference list. A mandatory POI type extends the concept of mandatory points of interest to the one of a category of mandatory POI. If one type of point of interest is determined as a mandatory point of interest type, the recommendation should contain at least one visit of this given POI type (Castillo et al., 2008; Vansteenwegen et al., 2011).

4.7. Recommendation of a Multiple-Days Trip

A TRS can plan a trip for multiple days. Indeed and most often each POI can be visited only once (Lee et al., 2009; Vansteenwegen et al., 2011).

4.8. Recommendation of POIs With Temporal Constraints

TRS might for example take into account the opening and closing hours of the POIs considered as suggested and applied in (Castillo et al., 2008; Vansteenwegen et al., 2011).

4.9. Recommendation of POIs With Semantic Constraints

When the user has a maximum amount of time or money to spend in the POIs selection and routing process, the TRS might constraints the selection of POIs that are visited by the users and routing between them (Castillo et al., 2008; Shiraishi et al., 2005).

5. RECOMMENDATION BASED ON LOCATION-BASED SOCIAL NETWORK

Location-based Social Networks (LBSNs) add additional social content when delivering services to the users. This leads to strengthen the connection between the social networking services and social networks. LBSNs (Figure 6) are defined as a type of social networking that use geographic services and capabilities such as geocoding and geotagging and enable additional social dynamics (Quercia et al., 2010).

Figure 6. Basic Concepts of LBSNs (Symeonidis et al., 2014)

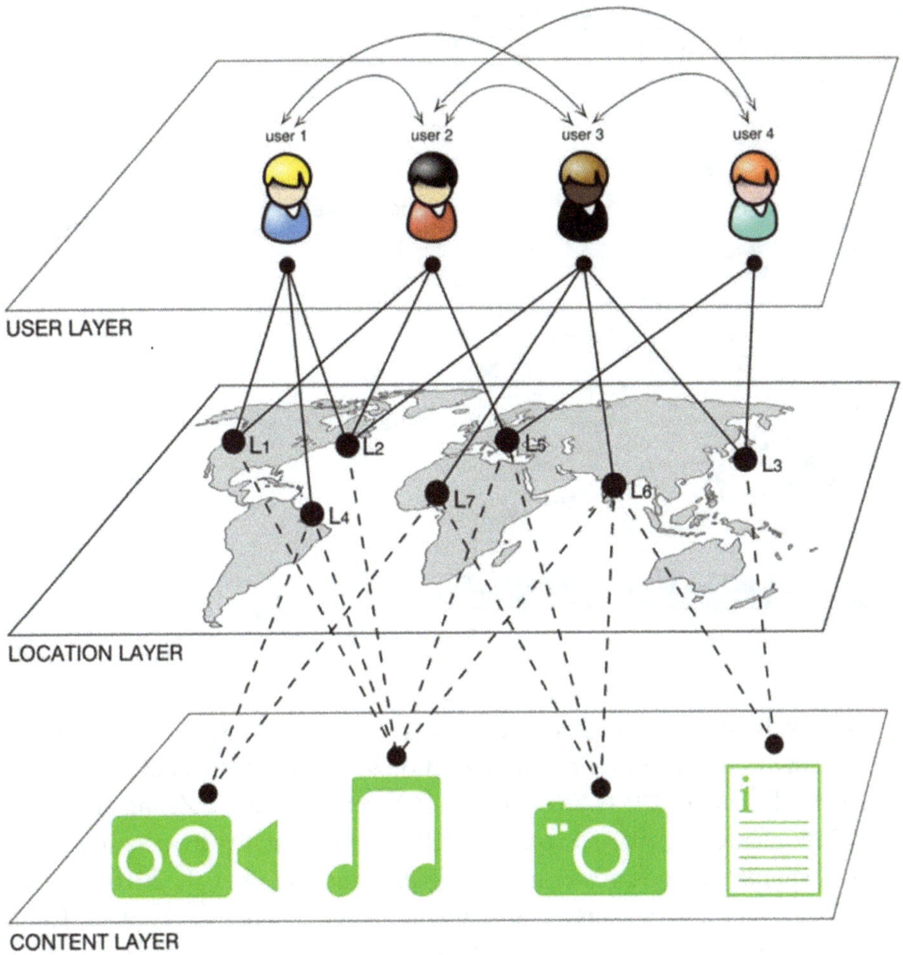

5.1. Social Factors in TRSs and LBSN

When related to LBSNs, several characteristics can affect POI visiting preferences or check-in behaviors (Figure 7).

5.1. 1. Geographical Influence in POI Recommendation

Geography is indeed an important factor that distinguishes the POI recommendation from common non geograpical item recommendation, as location plays a central role when recommending features. Several analysis on users' check-in data show that usually a user acting geographically in a given area is likely to prefers reaching

Figure 7. Influential factors in LBSN (Zhao et al., 2016a)

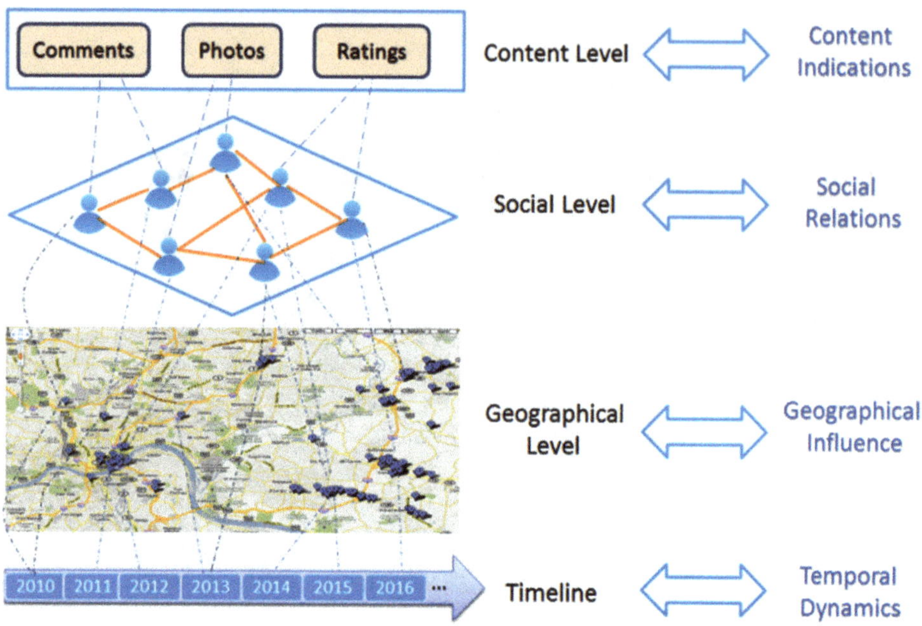

POIs nearby those where the user is currently located. Several studies study the role of geography to improve POI recommendation systems (Cheng et al., 2012; Lian et al., 2014; Liu et al., 2014; Ye et al., 2011; Yuan et al., 2013; Zhang and Chow, 2013, 2015; Zhao et al., 2013). In particular, three representative models apply some statistical approaches, i.e., power law distribution model (Ye et al., 2010; Ye et al., 2011; Yuan et al., 2013), Gaussian distribution model (Cheng et al., 2012; Cho et al., 2011; Zhao et al., 2013), and kernel density estimation model (Zhang and Chow, 2013) to capture the geographical influence in POI recommendation.

Basic POI geographical information (i.e., latitude and longitude coordinates) significantly affects users' check-in behaviors. For instance, people tend to visit POIs close to their homes or offices, and also may be interested in exploring the nearby places to their currently visited POIs when performing an itinerary.

5.1.2. Temporal Influence in POI Recommendation

Time is an important factor that greatly influences human activities, especially for POI recommendation and according to different modalities such as periodicity, consecutiveness, and non-uniformness. Users' check-in behaviors in LBSNs often exhibit some periodic patterns. For instance, users always check-in restaurants at noon and might have fun at night. Also users visit places around the office on weekdays and spend time in shopping malls on weekends (Cho et al., 2011; Gao et al., 2013; Yuan et al., 2013; Zhang and Chow, 2016).

Consecutiveness reflect check-in sequences, while successive check-ins are usually correlated (Cheng et al., 2013; Feng et al., 2015; He et al., 2016; Rendle et al., 2010; Zhang et al., 2014; Zhao et al., 2016b). For instance, users may have fun after diner in a restaurant. For instance, cinemas may be usually visited after restaurants as users would like to relax after dinner, and checking in stadiums first and then restaurants is better than the reverse way because it is not healthy to exercise right after a meal.

Often human movements exhibit sequential patterns. For instance, cinemas or bars may be usually visited after restaurants as users would like to relax after dinner, and checking in stadiums first and then restaurants is better than the reverse way because it is not healthy to exercise right after a meal. Therefore, the influence of sequential patterns is also important for users' check-in behaviors. The non-uniformness feature depicts a user's check-in preference variance at different hours of a day, or at different months of a year, or at different days of a week as well (Gao et al., 2013).

5.1.3. Content Indication

In LBSNs, users often generate additional semantic contents including tips and POI ratings as well as photographs which can be used to enhance POI recommendation (Gao et al., 2015; Hu and Ester, 2014; Lian et al., 2015; Yang et al., 2013; Yin et al., 2014).

5.2. Categorization of TRSs in LBSN by Theirs Objectives

TRSs as combined with LBSNs can be categorized according to their objectives according to some spatial constraints (the stand-alone locations and traveling routes), or semantic constraints related to the category of users (expert users, friends recommendation, and community discovery) and/or activities.

5.2.1. Stand-Alone and Sequential POI Recommendations

Existing location-based TRSs can be categorized into two groups, that are, stand-alone and sequential location recommender systems. Stand-alone location recommender systems provide the users with individual locations, such as POIs or destinations, which match their preferences (Yoon et al., 2010; Zheng and Xie, 2010). Instead, sequential location recommender systems suggest a series of locations, such as a travel route to the users based on their preferences and constraints (Wei et al., 2012; Yoon et al., 2012).

5.2.2. User-Based POI Recommendations

User-based recommendations combined with LBSN (Hung et al., 2009; Li et al., 2008; Xiao et al., 2010; Ying et al., 2011; Ying et al., 2010) take into account users' location histories in addition to the underlying social structure and user interaction patterns. LBSN provide a new way to deliver user recommendations by also considering users' location histories. Location histories are rich contextual information and are correlated significantly to social behaviors (Cranshaw et al., 2010) and user relationships within social networks (Goldenberg and Levy, 2009; Liben-Nowell et al., 2005; Scellato et al., 2011). User recommendations for LBSNs are categorized based on their objectives according to popular user discovery, friend recommendation, and community discovery.

5.2.3. Activity-Based Recommendations

An activity-based recommendation in a LBSN applies user-labeled tags, or semantic data related to the users' location histories to retrieve one or more activities that are appropriate for a query location (Noulas et al., 2013; Zheng et al., 2012).

5.2.4. Social Media-Based Recommendations

Social media recommendations in LBSN use location information to suggest photos, videos, or other web content to the user and improve the effectiveness and efficiency of recommendations (Daly and Geyer, 2011; Sarwat et al., 2013).

6. COMPUTATIONAL INTELLIGENCE TECHNIQUES APPLIED TO TRSs

This section review Computational Intelligence (CI) techniques applied in TRSs including statistics, fuzzy sets, artificial neural network, evolutionary computing, and swarm computing.

6.1. Statistics

A common way to recommend to a new user is employing the K-nearest neighbors approach (Larose, 2014) to calculate the average ratings of the k most similar users for each item and recommending the most rated items (Martinez et al., 2009; Noguera et al., 2012). Several researchers have applied K-means algorithms to group the users into different classes (Gavalas and Kenteris, 2011; Lucas et al., 2011). The seeds of the K-user clusters are initialized in some application-dependent ways. At each step of an iterative process, the users are sorted into the nearest user cluster and the cluster seeds are recalculated. The process is iterated until the users belong to the same user clusters in two consecutive iterations. SigTur/E-Destination (Moreno et al., 2013) applies a K-means algorithm to obtain initial tourist segments, and classify the users according to their demographic characteristics and their explicit ratings. It takes into account the demographic characteristics of users such as country of origin, travel group composition, travel budget and the accommodation type and applies aggregation operators like the Ordered Weighted Average (OWA) (Yager, 1988) and Logic Scoring of Preference (LSP) operator (Dujmović and Nagashima, 2006) to classify the users. Moreover, it applies Pearson correlation as similarity measure to classify users according to their ratings.

SPETA (García-Crespo et al., 2009) stores user preferences on several kinds of tourist activities in a vector. Similarly, it also stores the characteristics of each activity. It uses a Support Vector Machines (SVMs) to compute the distance between the user's preferences and the recommendable items, in order to propose the most appropriate items.

6.2 Fuzzy Sets

Fuzzy set theory (Zadeh, 1965) is applicable where the information is incomplete or imprecise. Fuzzy logic is used to manage uncertainty and analyze the vagueness of natural language. Some TRSs use the fuzzy sets and fuzzy reasoning to represent user preferences and calculate their fitness with the characteristics of a POI (García-Crespo et al., 2009; Lee et al., 2009), to calculate the membership degree of each user to different users groups (Lucas et al., 2011). VISIT (Meehan et al., 2013) applies fuzzy sets and fuzzy reasoning to represent contextual aspects of the journey such as weather conditions with a value given between 0 and 1, and delivers a more fine grained reasoning about the influence of contextual information on the recommendation of each cultural activity.

6.3. Artificial Neural Networks

Artificial Neural Network is a branch of artificial intelligence. It comprises of connected nodes (i.e., artificial neurons) that are arranged in layers, and a computational model for processing. VISIT (Meehan et al., 2013) takes into account the user context (e.g., location, time, weather, social media sentiment and user preferences) in the recommendation process. It uses an artificial neural network to assess the relevance of each context component for each user. (Bahramiana et al., 2017) proposes a hybrid interactive context-aware TRS thanks to the combination of a case based reasoning framework and an artificial neural network. It offers personalized tours to the user who have limited knowledge about his own preferences.

6.4. Evolutionary Computing

Evolutionary Computing is inspired by biological evolution and is a computational intelligence paradigm that focuses on algorithms following the concept of evolution (Zhang et al., 2011). Evolutionary Computing algorithms have several variants namely, genetic algorithm (GA), genetic programming (GP), evolutionary programming (EP), and evolutionary strategies (ES).

CT-Planner4 (Kurata and Hara, 2013) uses a genetic algorithm to recommend the plan to visit a city. At each iteration, it considers a population of different possible plans, and evaluates each plan according to their utility for the user. Then, it selects the best ones accordingly and mutates and recombines them via crossover to generate another population at the next iteration. The process is iterated in a certain number of iterations, to finally select the best plan.

6.5. Swarm Computing

The concept of Swarm Intelligence (SI) (Eberhart et al., 2001) originates from the swarms of social organisms and can be used for optimization. Ant Colony Optimization (ACO) (Gong et al., 2009) is a Swarm Intelligence technique in which the ants form a network of paths that connects their nests with the sources of food. The agent-based travel route recommender for Tainan (Lee et al., 2009) uses Ant Colony Optimization techniques and organizes a set of autonomous entities as the ants that cooperate through pheromone-mediated indirect and global communication. It finds a good solution to the travelling salesman problem (TSP) and plans a route that goes through different POIs around the city.

7. EVALUATING TRSs

This section reviewed some evaluation metrics that are commonly taken into account to decide which recommendation approach be selected.

7.1. Accuracy

Accuracy is the most discussed evaluation metrics in TRS literature. Three main classes of prediction accuracy are considered including accuracy of ratings predictions, accuracy of rankings prediction, and accuracy of utility-based ranking

7.1.1. Accuracy of Ratings Prediction

Let r_{uj} be the value of the (hidden) rating of user u for item j (entry $(u,j) \in E$), which is used in the test set. Moreover, let $\hat{r}_{uj}$ be the predicted rating of the entry (u,j) by the specific used training algorithm. The entry-specific error is given by

$$e_{uj} = \hat{r}_{uj} - r_{uj} \tag{1}$$

TRS uses some metric to evaluate accuracy of predicted ratings over the set E of entries in the ratings matrix on which the evaluation is performed. Mean Squared Error (MSE), Root Mean Squared Error (RMSE), and Mean Absolute Error (MAE) are computed as follows, respectively:

$$MSE = \frac{\sum_{(u,j)\in E} e_{uj}^2}{|E|} \tag{2}$$

$$RMSE = \sqrt{\frac{\sum_{(u,j)\in E} e_{uj}^2}{|E|}} \tag{3}$$

$$MAE = \frac{\sum_{(u,j)\in E} |e_{uj}|}{|E|} \tag{4}$$

As the RMSE sums up the squared errors, it is more influenced by large error values or outliers than MAE. Normalized RMSE (NMRSE) and Normalized MAE (NMAE) are versions of RMSE and MAE that is normalized by the ratings range (r_{max}, r_{min}).

$$NRMSE = \frac{RMSE}{r_{max} - r_{min}} \quad (5)$$

$$NMAE = \frac{MAE}{r_{max} - r_{min}} \quad (6$$

7.1.2. Accuracy of Rankings Prediction

TRS ranks items according to a given user to recommend the top-k items. Consider that the items ratings of the set I_u for user u are hidden by a hold-out or cross-validation strategy. In order to measure how well the ground-truth orderings of the ratings in I_u are related to the ordering predicted by the recommender system for the set I_u, Kendall rank correlation coefficient is computed. For each pair of items $j, k \in I_i$, the predicted ranking of these items is compared with their ground-truth ranking to compute the credit $C\left(j,k\right)$ as follows.

$$C\left(j,k\right) = \begin{cases} +1 \text{ if items } j \text{ and } K \text{ are in the same relative order in ground } - \text{ truth ranking and predicted ranking } \left(\text{concordant}\right) \cdot \\ -1 \text{ if items } j \text{ and } k \text{ are in a different relative order in ground } - \text{ truth ranking and predicted ranking } \left(\text{discordant}\right) \\ 0 \text{ if items } j \text{ and } k \text{ are tied in either the ground } - \text{ truth ranking or predicted ranking} \end{cases}$$

(7)

Then, the Kendall rank correlation coefficient for user u (τ_u) is computed as the average value of $C\left(j,k\right)$ over all the pairs of test items for the user.

$$\tau_u = \frac{\sum_{j<k} C\left(j,k\right)}{\left|I_u\right| \cdot \left(\left|I_u\right| - 1\right) / 2} \quad (8)$$

The Kendall rank correlation coefficient is also defined as follows.

$$\tau_u = \frac{\text{Number of concordant pairs} - \text{ Number of discordant pairs}}{\text{Number of pairs in } I_u} \quad (9)$$

A number of other measures, such as the normalized distance-based performance measure (NDPM), have been proposed to measure the accuracy of rankings prediction (Gunawardana and Shani, 2015).

7.1.3. Accuracy of Utility-Based Ranking

In the utility-based method, the rating is substituted with 0 or 1 value, depending on whether or not the user has visited a given item. Assume that the size of the recommended list is restricted to a maximum value of L. The overall R_score is as follows.

$$R\_score = \sum_{u=1}^{m} \sum_{j \in I_u, v_j \leq L} \frac{max\left\{r_{uj} - C_u, 0\right\}}{2^{\left(v_j - 1\right)/\alpha}} \tag{10}$$

where C_u is a break-even (neutral) rating value for user u, v_j is the rank of item j in the list of recommended items valued from 1 to n (the total number of items), and α is a half-life parameter.

Due to the exponential decay in the utility, the users are only interested in top-ranked items rather than lower-ranked items. In order to set the discount rate in a milder way, the discounted cumulative gain (DCG) is introduced.

$$DCG = \sum_{u=1}^{m} \sum_{j \in I_u, v_j \leq L} \frac{2^{rel_{uj}} - 1}{log_2\left(v_j + 1\right)} \tag{11}$$

where rel_{uj} is the ground-truth relevance of item j for user u, which is computed as a heuristic function of the ratings or hits. Then, the normalized discounted cumulative gain (NDCG) is defined as ratio of the discounted cumulative gain to its ideal value (IDCG).

$$NDCG = \frac{DCG}{IDCG} \tag{12}$$

7.2. Coverage

Due to the sparseness of the ratings matrices, a TRS may not be able to recommend a certain proportion of the items forever, or it may not be able to recommend to a certain proportion of the users forever. This measure is named as coverage (Aggarwal, 2016c; Ricci et al., 2015b). It is required to take into account a trade-off between accuracy and coverage in the evaluation process. Two types of coverage are considered including user-space coverage and item-space coverage (Aggarwal, 2016c).

User-space coverage measures the fraction of users for which at least k ratings may be predicted, where k is the expected size of the recommendation list. Also, it can be defined for a user as the minimum amount of user profile needed to make recommendations for that user (Aggarwal, 2016c). Similarly, item- space coverage or catalog coverage measures the fraction of items for which the ratings of at least k users can be predicted (Aggarwal, 2016c).

7.3. Confidence

Confidence in a recommendation process can be defined as the system's trust in its recommendations or predictions (Herlocker et al., 2000; Swearingen and Sinha, 2001). It can be measured as the probability that the predicted value is indeed true, or the interval around the predicted value where a predefined portion, e.g. 95%, of the true values lie. If the system confidence on a recommended item is low, the user may filter the recommended item or further research the item before decision making. The accuracy of two recommenders can be compared after their results were filtered by removing low confidence items.

7.4. Trust

While confidence refers to the system trust in its ratings (Aggarwal, 2016c; Ricci et al., 2015b), trust or system credibility is the user's trust in the system recommendation. Some ways exist to enhance trust in the system including recommending a few items that the user already knows and likes, and explaining the recommendations that the system provide.

Some methods exist in order to evaluate the user trust in an online test including asking users whether the system recommendations are reasonable or not, considering the number of recommendations that were followed (higher trust leads to more recommendations being used), and existence of repeated users. However, it is still unclear how to measure trust in an offline experiment due to the unavailability of an interaction between the system and a user.

7.5. Novelty

Novel recommendations are recommendations that users are not aware of, or that they have not seen before (Konstan et al., 2006). Therefore, popular items are less likely to be novel. In order to measure novelty through online experimentation, the system asks users whether they were familiar with a recommended item previously. In order to estimate novelty with offline methods, all ratings that were created after a certain time point (t_0) and some of the ratings occurring before t_0 are removed from the training data. Then, the system is trained with these ratings removed and uses these removed items for scoring purposes. For the item rated before t_0 and correctly recommended, the novelty evaluation score is penalized. On the other hand, for the item rated after t_0 and correctly recommended, the novelty evaluation score is rewarded.

7.6. Serendipity

Serendipity is a measure of the level of surprise or unexpectedness in successful recommendations (Aggarwal, 2016c; Kotkov et al., 2016). While, novelty only requires that the user was not aware of the recommendation earlier. Therefore, serendipity is a stronger condition than novelty. All serendipitous recommendations are novel, but the converse is not always true.

In order to measure serendipity through online methods, the recommender system asks the user the usefulness and obviousness of a recommendation. The fraction of recommendations that are both useful and non-obvious, is used as a measure of the serendipity. In order to measure serendipity through offline methods, the system uses a primitive recommender (such as a content-based recommender, that has a high tendency to recommend obvious items) to generate the information about the obviousness of a recommendation. The fraction of the recommended items in the top-k lists that are correct (i.e., high values of hidden ratings), and are also not recommended by the primitive recommender provides a measure of the serendipity. A number of metrics to evaluate serendipity are discussed in (Ge et al., 2010; Murakami et al., 2007).

7.7. Diversity

The diversity is determined by making a relation between the set of proposed items with a single recommended list. The vector-space representation of each item description is used to compute the content-centric similarity between each pairs of items. The average similarity between all pairs can be reported as the diversity. Lower average similarity value indicates greater diversity (Aggarwal, 2016c; Kunaver and Požrl, 2017).

7.8. Robustness and Stability

A TRS is said to be robust and stable if the recommendations are not significantly affected in the presence of attacks such as fake ratings or in significantly evolvement of the data patterns over time (Aggarwal, 2016c).

7.9. Scalability

A TRS is scalable if it can perform effectively and efficiently in the presence of large amounts of data. In recent years, scalability has become of great importance due to the increasing importance of the "big-data" paradigm. A variety of measures are used to determine the scalability of a TRS including training time, prediction time, and memory requirements. Training time is the overall time required to train a model in the training phase. Prediction time is the time used to determine the top recommendations for a particular user after the model has been trained. It determines the latency with which the user receives the responses. Memory requirements determine the space needed to hold the entire matrix in the main memory that is essential to design the algorithm to minimize it.

8. A COLLABORATIVE TOURISM RECOMMENDER SYSTEM IMPLEMENTATION

This section introduces a user-based collaborative recommender system applied to the Tourism domain. It recommends a list of venues (as items) and routes that pass through them. It uses the Foursquare LBSN available in the New York State, US including 4979 users, and 4428 venues (Figure 8).

Figure 8. Case study area

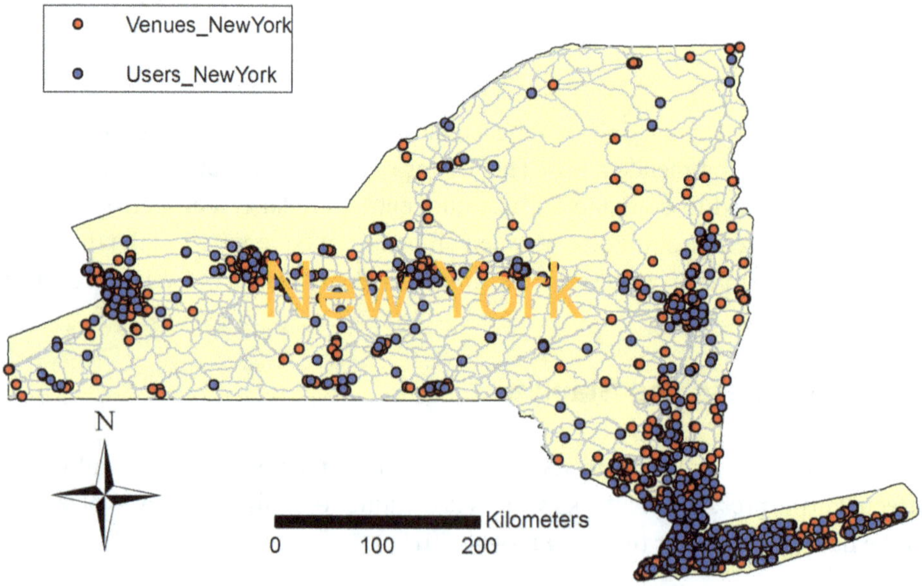

This experimental TRS relies on the ratings of a user as well as other users in the system. The available user-item ratings are used to predict ratings for new items. The rating of a user for a new item is likely to be similar to that of other similar users that rated other items in a similar way. For each user *u*, the system finds the neighbor users who have similar rating patterns. It evaluates the interest of the user *u* for an item i according to the interest of the similar user for item *i*.

Suppose *U* is the set of users in the system, and *I* is the set of items. Moreover, let us suppose that r_{ui} is the rating that user *u* assigned to a particular item *i*, *R* is the set of user-item ratings recorded in the system with a set of possible values for a rating (e.g. [0,1]). U_i, and I_u denote the subset of users that rated an item *i*, and the subset of items that were rated by a user *u*, respectively. Finally, I_{uv}, and U_{ij} represent the subset of items that were rated by two users u and v ($I_u \cap I_v$), and the set of users that have rated both items *i* and *j* ($U_i \cap U_j$), respectively.

In order to recommend the most *N* rated items to a particular user *u*, the proposed method uses the following steps: calculate *sim(u,v)* as the similarity between the user *u* and all other user *v*; find *k* most similar users (nearest neighbors) to user u; produce a rating prediction for the user u on each item j by taking the weighted average of all the ratings of the similar users on item *j*; and generate a top-N recommendation.

Similarity computation for the different users is a critical step. The proposed method calculates the similarity between the users u and v who have both rated the same item. We apply two different methods to compute similarity between users including Cosine-based similarity, and Pearson correlation measures.

- Cosine-based similarity measures can be measured by treating each user as a vector of items ratings and computing the cosine of the angle formed by the ratings vectors corresponding to the two users. The cosine-based similarity between user u and v is defined as

$$sim_{cosin}\left(u,v\right) = \cos\left(\vec{r}_u, \vec{r}_v\right) = \frac{\vec{r}_u . \vec{r}_v}{\left|\vec{r}_u\right|\left|\vec{r}_v\right|} \tag{13}$$

where "." denotes the dot-product of the two vectors.

- Pearson correlation between two users measures the extent to which two users linearly related each other. Pearson correlation between two users u and v is given by

$$sim_{Pearson}\left(u,v\right)=\frac{\sum_{i\in I_{uv}}\left(r_{ui}-\overline{r}_{u}\right)\left(r_{vi}-\overline{r}_{v}\right)}{\sqrt{\sum_{i\in I_{uv}}\left(r_{ui}-\overline{r}_{u}\right)^{2}}\sqrt{\sum_{i\in I_{uv}}\left(r_{vi}-\overline{r}_{v}\right)^{2}}} \tag{14}$$

where $i \in I_{uv}$ indicates the items that both the users *u* and $v \in I_{uv}$ have rated, and $\overline{r}$ is the average rating of the co-rated items of user *u*.

In order to predict user ratings on the items, the proposed approach find the k most similar users after computing the similarities, and aggregate their item ratings. It takes into account the fact that similar users can have different levels of similarities and weighs the contribution of each similar users based on its similarity to the given user. In addition, it normalizes the rating prediction to avoid the rating be outside the range of allowed values. Therefore, in order to predict user *u* ratings on the item *i*, a subset of nearest neighbors of the user u are chosen based on their similarity with him, and a weighted average aggregate of their ratings is used to predict his rating. The predicted rating for user *u* on a certain item *i* is calculated as

$$\hat{r}_{ui}=\frac{\sum_{u\in U_{i}} sim\left(u,v\right) r_{vi}}{\sum_{u\in U_{i}} sim\left(u,v\right)} \tag{15}$$

where $sim\left(u,v\right)$ is either cosine-based similarity or a Pearson coefficient measure between user *u* and *v*.

After computing the items ratings for the user, the proposed method recommends the most top-N rated items to the given user. Finally, it organizes the list of recommended POIs into an optimal route. Figure 9 shows the recommended tour to one of the users based on cosine similarity measure. It compares the results thanks to different similarity measures.

In order to compare the results according to the different similarity measures, the MSE, RMSE, and MAE are compared. Table 1 shows the evaluation metrics of the proposed recommendation method considering Cosine-based similarity, and Pearson coefficient measures. The results show that recommendations taking into account cosine-based similarity are more efficient than the Pearson coefficient measures in terms of MSE, RMSE, and MAE.

Figure 9. Recommended tour based on cosine similarity measure

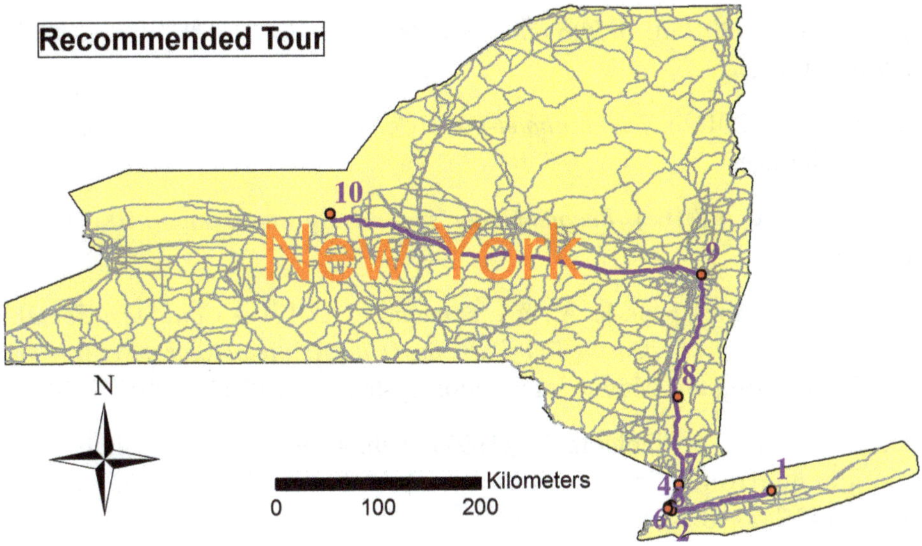

Table 1. Evaluation metrics of the recommendations considering different similarity measures

	MSE	RMSE	MAE
Recommendations considering cosine-based similarity	0.0016	0.0394	0.0019
Recommendations considering Pearson coefficient measure	0.0032	0.0564	0.0020

9. CONCLUSION

This chapter reviews current TRSs and approaches, how user profiles are taken into account, as well as the different categories of functionalities provided. The potential of TRSs is discussed when particularly combined with LBSNs. We also provide a review of the most common computational intelligence techniques applied in TRSs and provide some insights to the different evaluation metrics in TRSs. Finally, a case study illustrates a TRSs using a reference dataset extracted from Foursquare LBSN.

The main challenges in TRSs including improving recommendation accuracy, coverage, confidence and trust, diversity, novelty, serendipity, robustness and stability, scalability, and user satisfaction. Moreover, this chapter introduces a number of emerging topics in TRSs including context-aware TRSs, group TRSs, and solving the sparsity problem in TRSs, and the cold start problem in TRSs.

REFERENCES

Aggarwal, C. C. (2016a). *An Introduction to Recommender Systems, Recommender Systems*. Springer.

Aggarwal, C. C. (2016b). *Neighborhood-Based Collaborative Filtering, Recommender Systems*. Springer.

Aggarwal, C. C. (2016c). *Recommender systems*. Springer. doi:10.1007/978-3-319-29659-3

Bahramian, Z., Abbaspour, R. A., & Claramunt, C. (2017). *A Cold Start Context-Aware Recommender System for Tour Planning Using Artificial Neural Network and Case Based Reasoning*. Mobile Information Systems; doi:10.1155/2017/9364903

Balabanović, M., & Shoham, Y. (1997). Fab: Content-based, collaborative recommendation. *Communications of the ACM*, *40*(3), 66–72. doi:10.1145/245108.245124

Baltrunas, L. (2011). *Context-aware collaborative filtering recommender systems*. The Department of Computer Science, Free University of Bozen-Bolzano.

Berka, T., & Plößnig, M. 2004. Designing recommender systems for tourism. *ENTER 2004: 11th International Conference on Information Technology in Travel & Tourism*.

Borràs, J., de la Flor, J., Pérez, Y., Moreno, A., Valls, A., Isern, D., ... Anton-Clavé, S. (2011). *SigTur/E-Destination: a system for the management of complex tourist regions. In Information and Communication Technologies in Tourism 2011* (pp. 39–50). Springer. doi:10.1007/978-3-7091-0503-0_4

Borràs, J., Moreno, A., & Valls, A. (2014). Intelligent tourism recommender systems: A survey. *Expert Systems with Applications*, *41*(16), 7370–7389. doi:10.1016/j.eswa.2014.06.007

Borràs, J., Moreno, A., Valls, A., Ferré, M., Ciurana, E., Salvat, J., . . . Anton-Clavé, S. (2012). Uso de técnicas de inteligencia artificial para hacer recomendaciones enoturísticas personalizadas en la Provincia de Tarragona. Academic Press.

Burke, R. (2002). Hybrid recommender systems: Survey and experiments. *User Modeling and User-Adapted Interaction*, *12*(4), 331–370. doi:10.1023/A:1021240730564

Castillo, L., Armengol, E., Onaindía, E., Sebastiá, L., González-Boticario, J., Rodríguez, A., ... Borrajo, D. (2008). SAMAP: An user-oriented adaptive system for planning tourist visits. *Expert Systems with Applications*, *34*(2), 1318–1332. doi:10.1016/j.eswa.2006.12.029

Cheng, C., Yang, H., King, I., & Lyu, M. R. (2012). Fused Matrix Factorization with Geographical and Social Influence in Location-Based. Social Network, the Twenty-Sixth AAAI Conference on Artificial Intelligence, Toronto, Ontario, Canada, July 22 – 26.

Cheng, C., Yang, H., Lyu, M. R., & King, I. (2013). Where You Like to Go Next: Successive Point-of-Interest Recommendation. Academic Press.

Cho, E., Myers, S. A., & Leskovec, J. (2011). Friendship and mobility: user movement in location-based social networks. The 17th ACM SIGKDD International Conference on Knowledge Discovery and Data Mining, 1082-1090. doi:10.1145/2020408.2020579

Colombo-Mendoza, L. O., Valencia-García, R., Alor-Hernández, G., & Bellavista, P. (2017). *Special issue on context-aware mobile recommender systems*. Elsevier.

Cranshaw, J., Toch, E., Hong, J., Kittur, A., & Sadeh, N. (2010). Bridging the gap between physical location and online social networks. The 12th ACM International Conference on Ubiquitous Computing, 119-128. doi:10.1145/1864349.1864380

Daly, E. M., & Geyer, W. (2011). Effective event discovery: using location and social information for scoping event recommendations. The Fifth ACM Conference on Recommender Systems, 177-280, 277-280. doi:10.1145/2043932.2043982

Dujmović, J. J., & Nagashima, H. (2006). LSP method and its use for evaluation of Java IDEs. *International Journal of Approximate Reasoning*, *41*(1), 3–22. doi:10.1016/j.ijar.2005.06.006

Eberhart, R. C., Shi, Y., & Kennedy, J. (2001). *Swarm intelligence*. Elsevier.

Feng, S., Li, X., Zeng, Y., Cong, G., Chee, Y. M., & Yuan, Q. (2015). Personalized Ranking Metric Embedding for Next New POI Recommendation. The 24th International Conference on Artificial Intelligence, 2069-2075, Buenos Aires, Argentina, July 25 - 31, 2069-207.

Fenza, G., Fischetti, E., Furno, D., & Loia, V. (2011). *A hybrid context aware system for tourist guidance based on collaborative filtering*. IEEE International Conference on Fuzzy Systems. doi:10.1109/FUZZY.2011.6007604

Gao, H., Tang, J., Hu, X., & Liu, H. (2013). *Exploring temporal effects for location recommendation on location-based social networks*. The 7th ACM conference on Recommender systems. doi:10.1145/2507157.2507182

Gao, H., Tang, J., Hu, X., & Liu, H. (2015). Content-Aware Point of Interest Recommendation on Location-Based. *Social Network. the Twenty-Ninth AAAI Conference on Artificial Intelligence, Austin, Texas, January 25 - 30*, 1721–1727.

García-Crespo, A., Chamizo, J., Rivera, I., Mencke, M., Colomo-Palacios, R., & Gómez-Berbís, J. M. (2009). SPETA: Social pervasive e-Tourism advisor. *Telematics and Informatics*, *26*(3), 306–315. doi:10.1016/j.tele.2008.11.008

Gavalas, D., & Kenteris, M. (2011). A web-based pervasive recommendation system for mobile tourist guides. *Personal and Ubiquitous Computing*, *15*(7), 759–770. doi:10.1007/s00779-011-0389-x

Gavalas, D., Konstantopoulos, C., Mastakas, K., & Pantziou, G. (2014). Mobile recommender systems in tourism. *Journal of Network and Computer Applications*, *39*, 319–333. doi:10.1016/j.jnca.2013.04.006

Ge, M., Delgado-Battenfeld, C., & Jannach, D. (2010). *Beyond accuracy: evaluating recommender systems by coverage and serendipity*. The fourth ACM conference on Recommender systems, 257-260. doi:10.1145/1864708.1864761

Goldenberg, J., & Levy, M. (2009). *Distance is not dead: Social interaction and geographical distance in the internet era*. arXiv preprint arXiv:0906.3202

Gong, S. (2009). Joining case-based reasoning and item-based collaborative filtering in recommender systems. Second International Symposium on Electronic Commerce and Security (ISECS'09), 40-42. doi:10.1109/ISECS.2009.172

Gong, Y.-j., Xu, R.-t., Zhang, J., & Liu, O. (2009). *A clustering-based adaptive parameter control method for continuous ant colony optimization*. IEEE International Conference on Systems, Man and Cybernetics (SMC 2009), 1827-1832. doi:10.1109/ICSMC.2009.5346726

Gunawardana, A., & Shani, G. (2015). *Evaluating recommender systems, Recommender Systems Handbook*. Springer. doi:10.1007/978-1-4899-7637-6_8

He, J., Li, X., Liao, L., Song, D., & Cheung, W. K. (2016). Inferring a Personalized Next Point-of-Interest Recommendation Model with Latent Behavior Patterns. The 13th AAAI Conference on Artificial Intelligence, 137-143.

Herlocker, J. L., Konstan, J. A., & Riedl, J. (2000). *Explaining collaborative filtering recommendations*. The 2000 ACM conference on Computer supported cooperative work, 241-250.

Hinze, A., & Junmanee, S. (2005). *Travel recommendations in a mobile tourist information system*. Academic Press.

Horozov, T., Narasimhan, N., & Vasudevan, V. (2006). Using location for personalized POI recommendations in mobile environments. International symposium on Applications and the internet (SAINT 2006). doi:10.1109/SAINT.2006.55

Hu, B., & Ester, M. (2014). *Social topic modeling for point-of-interest recommendation in location-based social networks*. IEEE International Conference on Data Mining (ICDM), 845-850. doi:10.1109/ICDM.2014.124

Hung, C.-C., Chang, C.-W., & Peng, W.-C. (2009). *Mining trajectory profiles for discovering user communities*. The 2009 International Workshop on Location Based Social Networks, 1-8. doi:10.1145/1629890.1629892

Husain, W., & Dih, L. Y. (2012). A framework of a personalized location-based traveler recommendation system in mobile application. *International Journal of Multimedia and Ubiquitous Engineering*, *7*, 11–18.

Jadhav, A., & Sonar, R. (2009). An integrated rule-based and case-based reasoning approach for selection of the software packages. *Information Systems, Technology and Management*, 280-291.

Jian, C., Jian, Y., & Jin, H. (2005). *Automatic content-based recommendation in e-commerce*. The 2005 IEEE International Conference on e-Technology, e-Commerce and e-Service, 748-753.

Kabassi, K. (2010). Personalizing recommendations for tourists. *Telematics and Informatics*, *27*(1), 51–66. doi:10.1016/j.tele.2009.05.003

Koceski, S., & Petrevska, B. (2012). Empirical evidence of contribution to e-tourism by application of personalized tourism recommendation system. *Annals of the Alexandru Ioan Cuza University-Economics*, *59*(1), 363–374. doi:10.2478/v10316-012-0025-6

Konstan, J. A., McNee, S. M., Ziegler, C.-N., Torres, R., Kapoor, N., & Riedl, J. (2006). Lessons on applying automated recommender systems to information-seeking tasks. The 21st national conference on Artificial intelligence, Boston, Massachusetts, July 16 - 20, 1630-1633.

Koren, Y., & Bell, R. (2015). *Advances in collaborative filtering, Recommender systems handbook*. Springer. doi:10.1007/978-1-4899-7637-6_3

Kotkov, D., Wang, S., & Veijalainen, J. (2016). A survey of serendipity in recommender systems. *Knowledge-Based Systems*, *111*, 180–192. doi:10.1016/j.knosys.2016.08.014

Kunaver, M., & Požrl, T. (2017). Diversity in recommender systems–A survey. *Knowledge-Based Systems*, *123*, 154–162. doi:10.1016/j.knosys.2017.02.009

Kurata, Y., & Hara, T. (2013). *CT-planner4: Toward a more user-friendly interactive day-tour planner, Information and communication technologies in tourism 2014*. Springer.

Larose, D. T. (2014). *Discovering knowledge in data: an introduction to data mining*. John Wiley & Sons. doi:10.1002/9781118874059

Lee, C.-S., Chang, Y.-C., & Wang, M.-H. (2009). Ontological recommendation multi-agent for Tainan City travel. *Expert Systems with Applications*, *36*(3), 6740–6753. doi:10.1016/j.eswa.2008.08.016

Li, Q., Zheng, Y., Xie, X., Chen, Y., Liu, W., & Ma, W.-Y. (2008). *Mining user similarity based on location history*. The 21st national conference on Artificial intelligence, Boston, Massachusetts, July 16 - 20, 1630-1633. doi:10.1145/1463434.1463477

Lian, D., Ge, Y., Zhang, F., Yuan, N. J., Xie, X., Zhou, T., & Rui, Y. (2015). *Content-aware collaborative filtering for location recommendation based on human mobility data*. IEEE International Conference on Data Mining (ICDM). doi:10.1109/ICDM.2015.69

Lian, D., Zhao, C., Xie, X., Sun, G., Chen, E., & Rui, Y. (2014). GeoMF: joint geographical modeling and matrix factorization for point-of-interest recommendation. the 20th ACM SIGKDD international conference on Knowledge discovery and data mining, 831-840. doi:10.1145/2623330.2623638

Liben-Nowell, D., Novak, J., Kumar, R., Raghavan, P., & Tomkins, A. (2005). Geographic routing in social networks. *Proceedings of the National Academy of Sciences of the United States of America*, *102*(33), 11623–11628. doi:10.1073/pnas.0503018102 PMID:16081538

Liu, Y., Wei, W., Sun, A., & Miao, C. (2014). *Exploiting geographical neighborhood characteristics for location recommendation*. The 23rd ACM International Conference on Information and Knowledge Management, 739-748. doi:10.1145/2661829.2662002

Lops, P., De Gemmis, M., & Semeraro, G. (2011). *Content-based recommender systems: State of the art and trends, Recommender systems handbook*. Springer. doi:10.1007/978-0-387-85820-3_3

Lorenzi, F., Loh, S., & Abel, M. (2011). *PersonalTour: A recommender system for travel packages*. The 2011 IEEE/WIC/ACM International Conferences on Web Intelligence and Intelligent Agent Technology, 333-336.

Luberg, A., Tammet, T., & Järv, P. (2011). *Smart city: a rule-based tourist recommendation system, Information and Communication Technologies in Tourism 2011*. Springer. doi:10.1007/978-3-7091-0503-0_5

Lucas, J., da Silva Coelho, B., García, M., de Almeida Figueiredo, A., & Martins, C. (2011). Applying Recommender Methodologies in Tourism Sector. *Highlights in Practical Applications of Agents and Multiagent Systems*, 101-108.

Mahmood, T., & Ricci, F. (2009). *Improving recommender systems with adaptive conversational strategies*. The 20th ACM conference on Hypertext and hypermedia, 73-82. doi:10.1145/1557914.1557930

Martinez, L., Rodriguez, R. M., & Espinilla, M. (2009). *Reja: a georeferenced hybrid recommender system for restaurants*. IEEE/WIC/ACM International Joint Conferences on Web Intelligence and Intelligent Agent Technologies (WI-IAT'09), 187-190.

Masthoff, J. (2004). *Group modeling: Selecting a sequence of television items to suit a group of viewers, Personalized digital television*. Springer.

Masthoff, J. (2015). *Group recommender systems: aggregation, satisfaction and group attributes, Recommender Systems Handbook*. Springer. doi:10.1007/978-1-4899-7637-6_22

Meehan, K., Lunney, T., Curran, K., & McCaughey, A. (2013). *Context-aware intelligent recommendation system for tourism*. IEEE International Conference on Pervasive Computing and Communications Workshops (PERCOM Workshops), 328-331.

Mocholi, J. A., Jaen, J., Krynicki, K., Catala, A., Picón, A., & Cadenas, A. (2012). Learning semantically-annotated routes for context-aware recommendations on map navigation systems. *Applied Soft Computing*, *12*(9), 3088–3098. doi:10.1016/j.asoc.2012.05.010

Montejo-Ráez, A., Perea-Ortega, J. M., García-Cumbreras, M. Á., & Martínez-Santiago, F. (2011). Otiŭm: A web based planner for tourism and leisure. *Expert Systems with Applications*, *38*(8), 10085–10093. doi:10.1016/j.eswa.2011.02.005

Moreno, A., Sebastiá, L., & Vansteenwegen, P. (2015). *Recommender Systems in Tourism*. Academic Press.

Moreno, A., Valls, A., Isern, D., Marin, L., & Borràs, J. (2013). Sigtur/e-destination: Ontology-based personalized recommendation of tourism and leisure activities. *Engineering Applications of Artificial Intelligence, 26*(1), 633–651. doi:10.1016/j.engappai.2012.02.014

Murakami, T., Mori, K., & Orihara, R. (2007). *Metrics for evaluating the serendipity of recommendation lists.* Annual Conference of the Japanese Society for Artificial Intelligence, 40-46.

Ning, X., Desrosiers, C., & Karypis, G. (2015). *A comprehensive survey of neighborhood-based recommendation methods, Recommender systems handbook.* Springer. doi:10.1007/978-1-4899-7637-6_2

Noguera, J. M., Barranco, M. J., Segura, R. J., & Martínez, L. (2012). A mobile 3D-GIS hybrid recommender system for tourism. *Information Sciences, 215*, 37–52. doi:10.1016/j.ins.2012.05.010

Noulas, A., Mascolo, C., & Frias-Martinez, E. (2013). *Exploiting foursquare and cellular data to infer user activity in urban environments.* IEEE 14th International Conference on Data Management (MDM), 167-176. doi:10.1109/MDM.2013.27

Okabe, M., Yanagisawa, M., Yamazaki, H., Kobayashi, K., Yoshioka, A., & Yamaguchi, T. (2008). Organizational knowledge transfer of intelligence skill using ontologies and a rule-based system. *Practical Aspects of Knowledge Management*, 207-218.

Pazzani, M. J., & Billsus, D. (2007). *Content-based recommendation systems, The adaptive web.* Springer.

Quercia, D., Lathia, N., Calabrese, F., Di Lorenzo, G., & Crowcroft, J. (2010). *Recommending social events from mobile phone location data.* IEEE 10th International Conference on Data Mining (ICDM), 971-976. doi:10.1109/ICDM.2010.152

Rendle, S., Freudenthaler, C., & Schmidt-Thieme, L. (2010). *Factorizing personalized markov chains for next-basket recommendation.* The 19th international conference on World wide web, 811-820. doi:10.1145/1772690.1772773

Ricci, F., Rokach, L., & Shapira, B. (2015a). *Recommender systems: introduction and challenges, Recommender systems handbook.* Springer. doi:10.1007/978-1-4899-7637-6

Ricci, F., Rokach, L., Shapira, B., & Kantor, P. B. (2015b). *Recommender systems handbook*. Springer. doi:10.1007/978-1-4899-7637-6

Ruiz-Montiel, M., & Aldana-Montes, J. (2009). *Semantically enhanced recommender systems*. On the move to meaningful internet systems (OTM 2009 workshops), 604-609.

Ruotsalo, T., Haav, K., Stoyanov, A., Roche, S., Fani, E., Deliai, R., ... Hyvönen, E. (2013). SMARTMUSEUM: A mobile recommender system for the Web of Data. *Journal of Web Semantics*, *20*, 50–67. doi:10.1016/j.websem.2013.03.001

Santiago, F. M., López, F. A., Montejo-Ráez, A., & López, A. U. (2012). GeOasis: A knowledge-based geo-referenced tourist assistant. *Expert Systems with Applications*, *39*(14), 11737–11745. doi:10.1016/j.eswa.2012.04.080

Sarwat, M., Eldawy, A., Mokbel, M. F., & Riedl, J. (2013). *Plutus: leveraging location-based social networks to recommend potential customers to venues*. IEEE 14th International Conference on Mobile Data Management (MDM), 26-35.

Savage, N.S., Baranski, M., Chavez, N.E., & Höllerer, T. (2012). I'm feeling loco: A location based context aware recommendation system. *Advances in Location-Based Services*, 37-54.

Scellato, S., Noulas, A., Lambiotte, R., & Mascolo, C. (2011). Socio-spatial properties of online location-based social networks. *ICWSM*, *11*, 329–336.

Schafer, J. B., Frankowski, D., Herlocker, J., & Sen, S. (2007). *Collaborative filtering recommender systems, The adaptive web*. Springer.

Sebastia, L., Garcia, I., Onaindia, E., & Guzman, C. (2009). e-Tourism: A tourist recommendation and planning application. *International Journal of Artificial Intelligence Tools*, *18*(05), 717–738. doi:10.1142/S0218213009000378

Seidel, I., Gärtner, M., Pöttler, M., Berger, H., Michael, D., & Merkl, D. (2009). *Itchy feet: a 3D e-tourism environment. In Tourism Informatics: Visual Travel Recommender Systems, Social Communities, and User Interface Design* (pp. 209–242). Hershey, PA: IGI Global.

Shiraishi, T., Nagata, M., Shibata, N., Murata, Y., Yasumoto, K., & Ito, M. (2005). A personal navigation system with functions to compose tour schedules based on multiple conflicting criteria. *IPSJ Digital Courier*, *1*, 528–536. doi:10.2197/ipsjdc.1.528

Srisuwan, P., & Srivihok, A. (2008). Personalized trip information for e-tourism recommendation system based on Bayes theorem. *Research and Practical Issues of Enterprise Information Systems, II*, 1271–1275. doi:10.1007/978-0-387-76312-5_53

Su, X., & Khoshgoftaar, T. M. (2009). A survey of collaborative filtering techniques. *Advances in Artificial Intelligence, 2009*, 4. doi:10.1155/2009/421425

Swearingen, K., & Sinha, R. (2001). Beyond algorithms: An HCI perspective on recommender systems (6th ed.). ACM SIGIR 2001 Workshop on Recommender Systems, 1-11.

Symeonidis, P., Ntempos, D., & Manolopoulos, Y. (2014). *Recommender systems for location-based social networks*. Springer Science & Business Media. doi:10.1007/978-1-4939-0286-6

Vansteenwegen, P., Souffriau, W., Berghe, G. V., & Van Oudheusden, D. (2011). The city trip planner: An expert system for tourists. *Expert Systems with Applications, 38*(6), 6540–6546. doi:10.1016/j.eswa.2010.11.085

Wei, L.-Y., Zheng, Y., & Peng, W.-C. (2012). *Constructing popular routes from uncertain trajectories*. The 18th ACM SIGKDD international conference on Knowledge discovery and data mining, 195-203.

Xiao, X., Zheng, Y., Luo, Q., & Xie, X. (2010). *Finding similar users using category-based location history*. The 18th SIGSPATIAL International Conference on Advances in Geographic Information Systems, 442-445. doi:10.1145/1869790.1869857

Yager, R. R. (1988). On ordered weighted averaging aggregation operators in multicriteria decisionmaking. *IEEE Transactions on Systems, Man, and Cybernetics, 18*(1), 183–190. doi:10.1109/21.87068

Yang, D., Zhang, D., Yu, Z., & Wang, Z. (2013). *A sentiment-enhanced personalized location recommendation system*. The 24th ACM Conference on Hypertext and Social Media, 119-128.

Ye, M., Yin, P., & Lee, W.-C. (2010). *Location recommendation for location-based social networks*. The 18th SIGSPATIAL international conference on advances in geographic information systems.

Ye, M., Yin, P., Lee, W.-C., & Lee, D.-L. (2011). *Exploiting geographical influence for collaborative point-of-interest recommendation*. The 34th international ACM SIGIR conference on Research and development in Information Retrieval, 325-334. doi:10.1145/2009916.2009962

Yin, H., Cui, B., Sun, Y., Hu, Z., & Chen, L. (2014). Lcars: A spatial item recommender system. *ACM Transactions on Information Systems*, *32*(3), 11. doi:10.1145/2629461

Ying, J. J.-C., Lee, W.-C., Ye, M., Chen, C.-Y., & Tseng, V. S. (2011). *User association analysis of locales on location based social networks*. The 3rd ACM SIGSPATIAL International Workshop on Location-Based Social Networks, 69-76. doi:10.1145/2063212.2063214

Ying, J. J.-C., Lu, E. H.-C., Lee, W.-C., Weng, T.-C., & Tseng, V. S. (2010). *Mining user similarity from semantic trajectories*. The 2nd ACM SIGSPATIAL International Workshop on Location Based Social Networks, 19-26.

Yoon, H., Zheng, Y., Xie, X., & Woo, W. (2010). *Smart itinerary recommendation based on user-generated GPS trajectories*. Springer. doi:10.1007/978-3-642-16355-5_5

Yoon, H., Zheng, Y., Xie, X., & Woo, W. (2012). Social itinerary recommendation from user-generated digital trails. *Personal and Ubiquitous Computing*, *16*(5), 469–484. doi:10.1007/s00779-011-0419-8

Yuan, Q., Cong, G., Ma, Z., Sun, A., & Thalmann, N. M. (2013). *Time-aware point-of-interest recommendation*. The 36th international ACM SIGIR conference on Research and development in information retrieval, 363-372.

Zadeh, L. A. (1965). Fuzzy sets. *Information and Control*, *8*(3), 338–353. doi:10.1016/S0019-9958(65)90241-X

Zhang, J., Zhan, Z., Lin, Y., Chen, N., Gong, Y., Zhong, J., ... Shi, Y. (2011). Evolutionary computation meets machine learning: A survey. *IEEE Computational Intelligence Magazine*, *6*(4), 68–75. doi:10.1109/MCI.2011.942584

Zhang, J.-D., & Chow, C.-Y. (2013). iGSLR: Personalized geo-social location recommendation: a kernel density estimation approach. The 21st ACM SIGSPATIAL International Conference on Advances in Geographic Information Systems, 334-343.

Zhang, J.-D., & Chow, C.-Y. (2015). *Geosoca: Exploiting geographical, social and categorical correlations for point-of-interest recommendations. IEEE Transactions on Services Computing*. doi:10.1145/2766462.2767711

Zhang, J.-D., & Chow, C.-Y. (2016). TICRec: A probabilistic framework to utilize temporal influence correlations for time-aware location recommendations. The 22nd ACM SIGSPATIAL International Conference on Advances in Geographic Information Systems, 103-112 , *9*(4), 633–646. doi:10.1109/TSC.2015.2413783

Zhang, J.-D., Chow, C.-Y., & Li, Y. (2014). *Lore: Exploiting sequential influence for location recommendations*. The 38th International ACM SIGIR Conference on Research and Development in Information Retrieval, 443-452. doi:10.1145/2666310.2666400

Zhao, S., King, I., & Lyu, M. R. (2013). *Capturing geographical influence in POI recommendations*. International Conference on Neural Information Processing, 530-537.

Zhao, S., King, I., & Lyu, M. R. (2016a). *A survey of point-of-interest recommendation in location-based social networks*. arXiv preprint arXiv:1607.00647

Zhao, S., Zhao, T., Yang, H., Lyu, M. R., & King, I. (2016b). STELLAR: Spatial-Temporal Latent Ranking for Successive Point-of-Interest Recommendation. Academic Press 315-322.

Zheng, V. W., Zheng, Y., Xie, X., & Yang, Q. (2012). Towards mobile intelligence: Learning from GPS history data for collaborative recommendation. *Artificial Intelligence*, *184*, 17–37. doi:10.1016/j.artint.2012.02.002

Zheng, Y., & Xie, X. 2010. Learning Location Correlation from GPS trajectories. In *Proceedings of the International Conference on Mobile Data Management 2010 (MDM 2010)*. IEEE Press. doi:10.1109/MDM.2010.42

Zhu, X., Ye, H., & Gong, S. (2009). *A personalized recommendation system combining case-based reasoning and user-based collaborative filtering*. Control and Decision Conference (CCDC'09), 4026-4028. doi:10.1109/CCDC.2009.5192712

Chapter 11
A Survey of Recommender Systems and Geographical Recommendation Techniques

Khaled Soliman
Cairo University, Egypt

Mahmood Mahmood
Cairo University, Egypt

Ahmed El Azab
Cairo University, Egypt

Hesham Ahmed Hefny
Cairo University, Egypt

ABSTRACT

Advancement of location-acquisition technologies with fast development of mobile devices and wireless communication caused a revolution of information. It has been used in location-based social networks (LBSNs), has attracted millions of users to Facebook places, Gowalla, and Foursquare, is an important task to make location recommendations to users, and utilizes user preferences and other information that not only help users explore new places but also make LBSNs more attractive to users. This chapter discusses recommender systems (RS) and its application in different fields like LBSN, big data, and real life. It describes traditional recommendation approaches as well as modern approaches and explains smart community as one of powerful techniques to be used. It also introduces the state-of-art geographical techniques and presents a comparative study of recommendation techniques that can be served as a good guide and a roadmap for research and practice in this area. Finally, the authors discuss measurements and the limitations of RS.

DOI: 10.4018/978-1-5225-5088-4.ch011

INTRODUCTION

Nowadays, with the high availability of data, the wide use of social networks and the rapid growth of the internet caused a huge amount of data, it requires a complex process to extract useful information that can be presented to the user in order to help him manage the data properly leading to making correct decisions. Researches have been done that help in managing data to produce this useful information (Al-Otaibi & Ykhlef, 2012). Many users are interested in systems that recommends some products to them based on certain factors, and hence a system is needed in order to support users in selecting a product or any item taking into consideration that the user may have less knowledge about the domain, we call this system a recommender system (RS) (Anderson & Hiralall, 2009). (Ricci, Rokach & Shapira, 2011) define RS as an intelligent system that provides advice to the user about a specific item aiding him in the decision making process (see Figure. 1).

Such a system can gain information explicitly (e.g. Collect rating from user) or implicitly (monitoring user behavior, such as a book read or a song heard) (Bobadilla, Ortega, Hernando & Gutiérrez, 2013). With the advancement of location acquisition technologies and fast development of mobile devices and wireless communication, LBSNs have attracted millions of users such as Facebook places, Gowalla, and Foursquare. In LBSNs, users can share their experiences of visiting specific locations, also known as points-of-interests (POIs), for example museums, restaurants, and stores. These visits are also known as check-in activities that reflect the user's preferences on locations. In LBSNs, it is an important task to make location recommendation to users - it utilizes user preferences and other

Figure 1. The model of process for recommender system

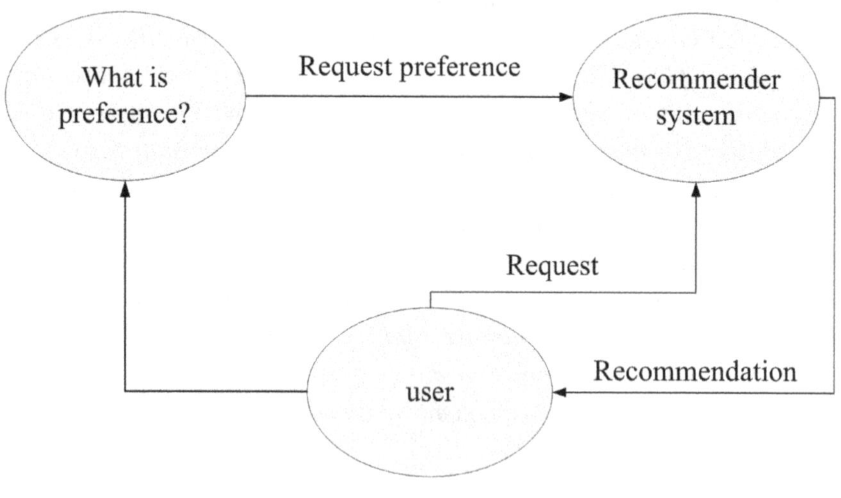

information (e.g., social friendships), which not only helps users explore new places but also makes LBSNs more attractive to users (Zhang & Chow, 2013). To gain a rich knowledge about users' interests, we can use available check-in data in LBSNs; it is beneficial to a wide range of applications such as location, activity, and friend's recommendations (Ye, Lee & Lee, 2011). In the LBSNs, the check-in data contains the following unique characteristics:

- **Frequency Data and Sparsity:** user-location matrix indicates the frequency of a user visiting a place. The density of the data set makes the POI recommendation task very tough.
- **Multi-Centers and Normal Distribution:** Where the check-in locations follow a Gaussian distribution at each center, because users tend to check in around several centers.
- **Inverse Distance Rule:** The probability of visiting a place is inversely proportional to the distance from its nearest center, although each user has a different personalized taste for POI. This implies that if a place is too far away from the location a user lives in, it is less likely that he would visit that place even if he likes it.

Friendship Influence: This implies that social influence exists; and may have an effect on the check-in activity (Cheng, Yang, King, & Lyu, 2012) (see Figure 2).

Generally, "social friends have similar behavior". This is known as a social influence theory (Karimi & Yektaei, 2015). The geographical information on locations plays a significant influence on users' check-in behavior known as geographical influence, which introduces new challenges in inferring users' preferences (Zhang & Chow, 2015; Zhang, Chow, & Li, 2014). There are several characteristics of LBSNs, which distinguish POI recommendation from traditional recommendation tasks:

Figure 2. Location-based social network (Zhang & Chow, 2015)

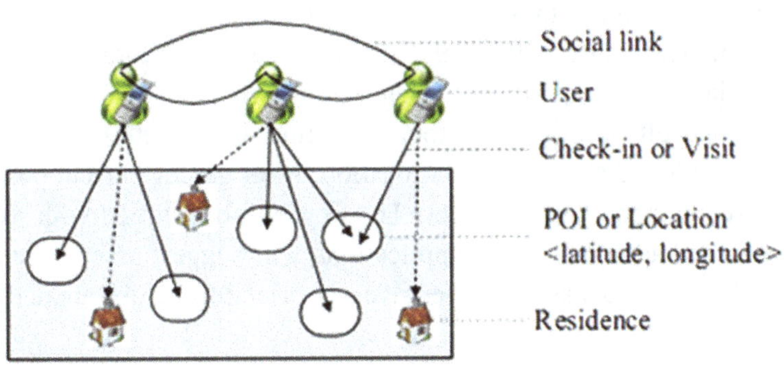

- **Tobler's First Law of Geography:** The law of geography states, "Everything is related to everything else, but nearby things are more related than distant things".
- **Geographical Influence:** The probability of a user visiting POI is inversely proportional to the geographic distance between them, due to geographical constraints and the cost of traveling large distances.
- **Regional Popularity:** Two POIs can have different popularities if they are located in different regions, although it is similar or the same semantic topics.
- **User Mobility:** That imposes huge challenges on POI recommendations, due to the dynamic movement, especially when travelling to new cities/regions.
- **Implicit User Feedback:** RS has to infer user preferences from implicit user feedback (e.g., check-in frequency) (Liu, Papadimitriou & Yao, 2015; Liu, Fu, Yao, & Xiong, 2013).

APPLIED RECOMMENDER SYSTEM IN DIFFERENT AREA

RS With LBSN

Nowadays, the evolution in broadband wireless networks and location sensing technologies led to the existence of smart mobile, tablets, etc. that allowed web access to users from anywhere via mobile device where they can benefit by getting access to location-based services. Moreover, they can share location-related information with friends to influence the community social knowledge by using location based social networks (LBSNs). It is clear that a user can make use of information from each user independently to influence recommendations. For instance, with each pair of places in the location we can compute the geographical distance between it. Moreover, we can calculate the similarity among users based on the social network, and then the best location can be recommended (Symeonidis, Ntempos & Manolopoulos, 2014).

We can use this for tourist those want visit places with the use of smart mobile. This may also be useful to those who want to visit nearby places. The purpose is to make an interactive system that helps the tourist to identify tourist location and recommend places they would like to visit. A smart mobile application which will show nearest as well as categorized tourist visiting places with ratings. Users are tracked using GPS and getting current location. After getting current position, the system will recommend nearby places. The user will be notified with weather of selected places. And some of the applications will suggest assets, according to weather conditions. And maybe another gives the brief information about the places.

RS With Big Data

Big Data refers to the huge amount of data which isn't easily stored, managed, and processed using the current methods, such as that created by social networks-containing all sort of real-world information such as time, human relations, space, etc. This increases the effort to analyze and store, making it difficult to obtain meaningful information for each person, so it important to create a recommendation system that reflects personal characteristics using Big Data with high accuracy (Lee, 2014 ; Han, Tian, Yoon & Lee, 2012).

RS With Data Mining

In RS, data mining is used to describe the collection of analysis techniques that are used to infer recommendation rules or build recommendation models from large data sets. RS incorporates techniques that build their recommendation using knowledge from user attribute and actions. It is often based on the development of user profile- these techniques include classification, clustering, association rules and the production of similarity graphs through different techniques (Panniello, Tuzhilin, Gorgoglione, Palmisano, & Pedone, 2009).

RS With Real Live World

Nowadays, there has been great progress in the field of RS making it a required application in different areas in real-life. For example, it is used to address some of the problems like drinking water, air pollution, climate change, agriculture.... Etc.. And a number of researches have been conducted to tackle the problem of drinking water quality monitoring and prediction using Machine Learning techniques proposing an intelligent system to tackle the problem (Mahmoud, El-Bendary, Mahmood, & Hassanien, 2013). On the other hand, some research presents a recommender system that deals with air pollution when it occurs with a sufficient concentration that would otherwise threaten human health and the environment. Moreover, research presented a recommender system that predicts the best cultivation according to climate changes (Mahmood, Al-Shammari, El-Bendary, Hassanien, & Hefny, 2013; Salam, Mahmood, Awad, Hazman, El Bendary, Hassanien,... & Saleh, 2014).

TRADITIONAL APPROACH TO RECOMMENDER SYSTEMS

Collaborative Filtering (CF)

It is considered one of the simplest yet popular approaches, it is used as a wide range implementation for RS applications, recommending an item for users that have the same preference, the similarity is calculated based on user's rating history – hence, it can be renamed as "people-to-people correlation " (Ricci, Rokach & Shapira, 2011). The important task for CF algorithms is predicting the rating of a specific user for specific items that have not been rated yet based on similarity between user and observed item (See Figure 3.) (Sharma & Gera, 2013).

There are two types of CF algorithms:

- **Memory Based:** It is simpler, using a matrix to determine the prediction. In order to select a user or item for an active user, it uses a similarity measure and then calculates predicted from neighbors rating, which can be classified

Figure 3. Process of CF technique

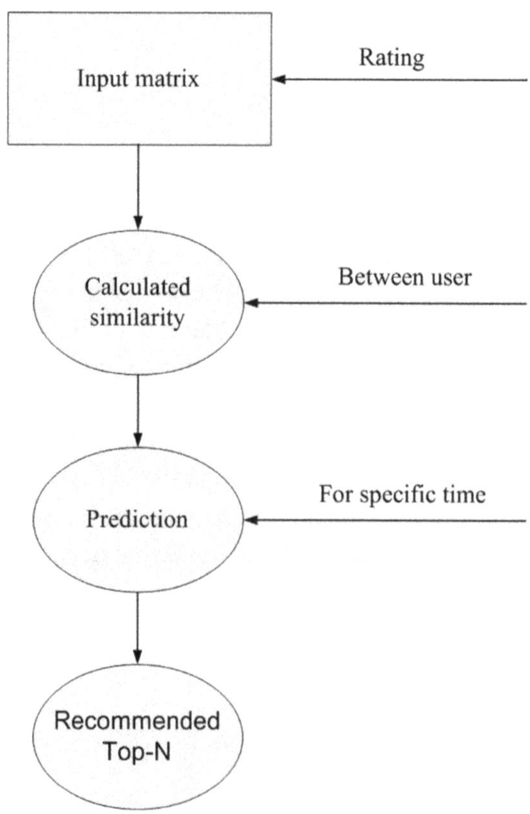

based on the process of getting a neighbor into user-based algorithms and item-based algorithms.

- **Model Based:** To predict the rating by making a model to represent users' behavior. It uses rating matrix data to estimate the parameter of the model-their multiple approaches used in machine learning such as neural networks, clustering, factor analysis, Bayes networks, and graph (Mani, 2014).

Content-Based

This approach analyses document of items the user has previously rated and then builds a profile for user interest based on that rating. Matching is done between user profile attributes and item attributes that concern the recommendation process. The result represents the level of interest of the user to the item (See Figure 4.). To gain user feedback there are two techniques: explicit and implicit (Lops, De Gemmis & Semeraro, 2011).

Explicit: A technique, it's used when the system requires from the user to evaluate an item and this indicates the user's interest in that item. There are three approaches to get this:

- **Like \ Dislike:** By binary rating scale-divided to "relevant" or "not relevant"
- **Rating:** Such as "hot", "cold", "lukewarm" this has possibility rating for web page.
- **Text Comments:** Collected and presented all comments about one item to the user to help in the decision making process.

Figure 4. Process of content based technique

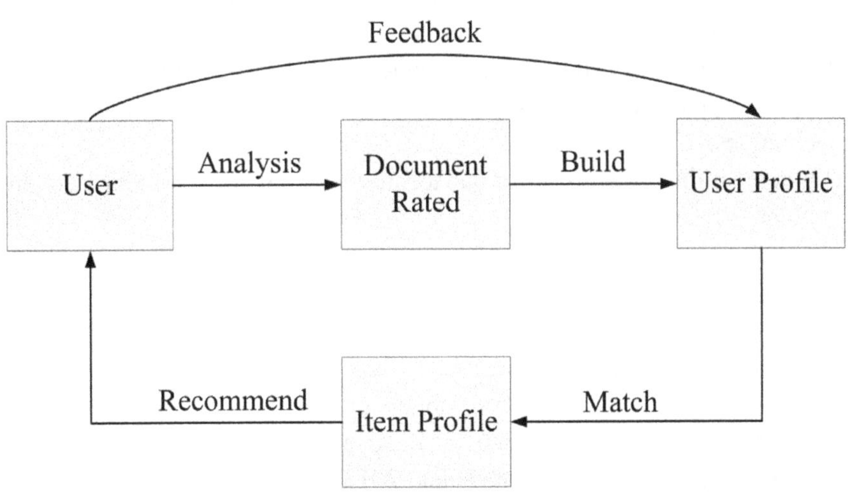

Implicit: which does not request an evaluation of user, but rather monitors and analyses users' behavior and activities (Lops, De Gemmis & Semeraro, 2011).

There are various machine learning techniques used for content-based recommendation such as Bayesian classifiers, neural network, decision tree that are different from information retrieval approaches. One of the most popular techniques used to provide recommendation are nearest neighbor algorithms, which are based on textual information stored in memory. Content in text is usually described with a keyword, the importance of a word is determined with some weighting measure and one of the best measures is "Term Frequency / Inverse Document Frequency" (TF-IDF) (Sharma & Mann, 2013).

Hybrids

This approach combines between different approaches in order to achieve the best performance or provides quality of recommendation. Also solves problems like cold-start problem overcoming some disadvantages with other approaches.

There is a different way of this:

1. **Weighted Hybrid Recommender:** Used all recommendation techniques that are available in the system and calculated score of item recommendation from the results.
2. **Switching Hybrid Recommendation:** Switching between techniques and use some measure to do that.
3. **Mixed:** Applied simultaneously for a large number of recommendations.
4. **Feature Combination:** Use content- based techniques to improve dataset and use collaborative as an additional feature.
5. **Cascade:** One technique used to produce a rough ranking of candidates and other regions the recommendation.
6. **Feature Augmentation:** One technique used to classify an item or calculate the rating and then that information is used in the processing of the other techniques.
7. **Model:** Simply used results of one techniques' input for another (Al-Otaibi & Ykhlef, 2012).

Demographic

In this approach, demographic data are used like (age, gender, location, etc.) as input for recommendation, this approach combines groups that contain similarity users based on similarity demographic data, so the user with similar data will rate items similarly (See Figure.5.) (Safoury & Salah, 2013).

Figure 5. Process of demographic technique

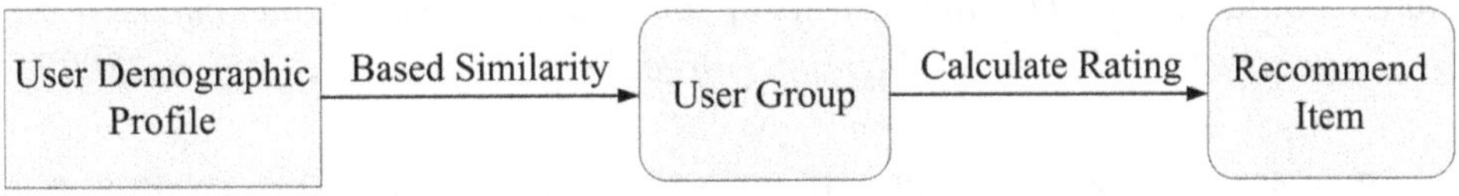

Knowledge

This approach recommends items based on domain knowledge about items that meet user preference and needs, similarity function have important rules for problem description by estimating how much user needs to match the solution of the problem - that similarity score can be used as the utility of recommendation (See Figure.6.) (Ricci, Rokach & Shapira, 2011).

MODERN APPROACHES FOR RS

Context-Aware Approaches

Context is the information about the environment of a user a bit of details about the situation he is currently in that may play an important role in recommendations more than ratings of items (Asanov, 2011). In the recommendation process, most traditional techniques recommended without taking into consideration the circumstance and other contextual information that may be changing the result of a recommendation. Some research has shown that context information must be added in RS because it improves the performance of RS, and this field describes the way, by using a multidimensional approach to recommendation, where the contextual information is added so that the rating the function R is defined as:

R: Users x Items x Context →Ratings

Figure 6. Process of knowledge, technique

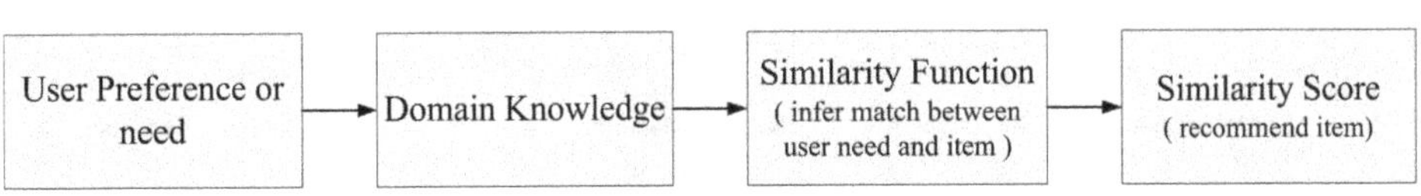

Where Users and Items are the sets of users and items, Ratings is a discrete and finite set of ratings, and Context is a set of contextual attributes K that can have an elaborate structure (Panniello, Tuzhilin, Gorgoglione, Palmisano & Pedone, 2009).

There are many ways to represent contextual information and its relation to a user and an item, one of which is contextual graph. It uses a graph algorithm and helps improve prediction. User, item and contextual object are represented as a node of the graph while another edge between nodes represents the ratings of items and similarity between users (Asanov, 2011).

Their three forms take on context-aware recommendation process as shown in Figure7. That is based on contextual user preference elicitation and estimation:

- **Contextual Pre-Filtering:** In this paradigm, the contextual information used for selecting the relevant set of data records that can be predicted by 2D RS.
- **Contextual Post-Filtering:** In this paradigm, we do not start with contextual information, but rather predict by traditional 2D RS after which the result is contextualized using contextual information for each user.

Figure 7. Three forms take on context-aware recommendation process (Adomavicius & Tuzhilin, 2015)

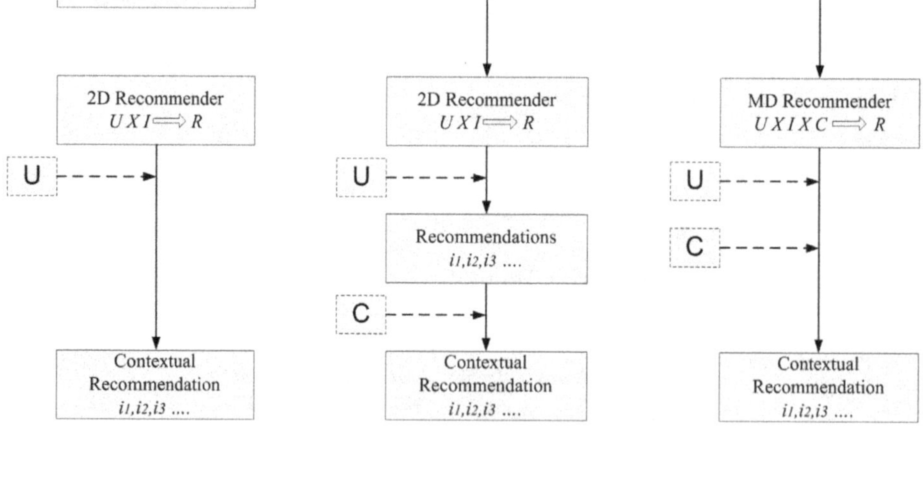

- **Contextual Modelling:** This paradigm uses contextual information directly in the modelling technique as part of rating estimation; it also called contextualization of recommendation function.

Often in RS, combining between approaches, introduces a significant performance improvement than when relying on individual approaches, the three paradigms for context-aware recommender systems can be introduced as a combination approach (Adomavicius & Tuzhilin, 2015).

Cross-Domain Based Approach

Computing similarity between two users over their appreciation of an item and building accurate neighborhood is important processes in RS, but similar appreciation in individual domain do not imply same similarity in another domain. In traditional RS based on CF, comparison is made between users without splitting an item in different domains, but in cross-domain systems, the similarity of user dependent domain is computed. First, according to domains, the engine creates local neighborhoods for each user. Then, computed similarity values and finite set of nearest-neighbors are sent for overall similarity computation. RS determines the overall similarity, creates overall neighborhoods and makes predictions and recommendations (Asanov, 2011).

Peer-to-Peer Approaches

Peer-to-Peer (P2P) network is modelled as a bi-partite graph, which is a special case of the traditional CF matrix, representing the ranking of an item by a user as a link in the graph. This is used for sharing content e.g., files by the huge number of users that make them an extremely valuable resource for tasks related to information retrieval (Shavitt, Weinsberg & Weinsberg, 2010). Several RS have been proposed in (P2P) systems that can be operated in different environments such as unstructured networks, super-peer based networks, Distributed Hash Tables that are dependent on decentralization idea- that is; no need for information analysis of user in a centralized database, the data about item or user's preferences is spread on the network. These approaches are useful for privacy of data for the user and work to solve the scalability problem (Pussep, Kaune, Flick & Steinmetz, 2009).

Group Recommendations Approaches

According to different types of relations in real-life, users may have friendship in the form of financial and sports relations, these relations create groups. They could be sharing their comments and opinions, which is considered a very important feature of social networks. The major goal of RS is to recommend an item/s for individual users, so different suggestions are sent to different users. However, in circumstances where the item was selected not only for the user to use, but also for group of user's e.g. DVD could be watched by a group of family or friends. These groups can vary from stable groups to ad hoc groups requiring recommendations only occasionally, so some recent work has identified the problem "good for" a group of users. Group recommendation approach is based on integrated group profile generations; the system is also designed to work in different domain, such as: music, tourism, TV program, webpage (Baltrunas, Makcinskas & Ricci, 2010).

There are four common recommendation techniques:

1. **Groups With More Users:** This technique recommends groups with a high number of users; it is non-personalized and recommends the same groups to all users.
2. **Collaborative Filtering:** In this technique, put user's similar memberships into similar categories, and then in each category user is recommended to the two groups with the highest number of users. In order to put users into different categories it uses Hierarchical clustering as one of the personalized techniques.
3. **Recommendation Based on Users' Friends:** In this technique, recommendation is given to the user about the two groups, which have the highest popularity between a user's friend, it is based on the concept of graphs and is also a personalized method.
4. **Recommendation Based on Association Rules:** By using association rules and considering user memberships, it found two more closely related groups and recommends to user a group he has not joined yet that is closely related to a group he is currently in (Baltrunas, Makcinskas & Ricci, 2010).

Most algorithms of group recommendation predict the individual preference and then aggregate this amongst group members. In order to aggregate this preference, there are several strategies:

- **Utilitarian Strategy:** This not only uses ranking information, but also takes account of utility values of other alternative expressions for an expected happiness.
- **Approval Voting Strategy:** This asks users to vote as many items as they wish to promote the election of moderate alternatives, which are not strongly disliked.

- **Least Misery Strategy:** Which considers for each item the rating with the minimum value and the one with the highest minimum value is recommended to the group (Song, Hu, Liu, Shi & Tian, 2013).

J. Bobadilla, F. Ortega, A. Hernando, A. Gutierrez provides classification recommendation to groups in CF RS, this consists of four basic levels: similarity metric, establishing the neighborhood, prediction phase, determination of recommended items. In the first case, data unification is performed in the prediction phase of the CF process: n individual predictions of n users of the group are combined in one prediction of the group (predictions aggregation). The second case acts on the sets of neighbors of the group users, by unifying them in one neighborhood for the whole group. In the third case, the recommendations obtained for each individual user of the group are merged into one recommendation for the group. The fourth case uses a similarity metric that acts directly on the set of ratings of the group of users. This solution is the only one that directly provides a set of neighbors for the group of users (Bobadilla, Ortega, Hernando & Gutiérrez, 2013).

SMART COMMUNITY

It can be defined as a group of connected of (social) objects that deliver smart services to possibly all members, this object, it interacts with each other over ubiquitous networks, the members of a smart community are objects that can be

Figure 8. "Integration of the social, physical and cyber worlds that makes it possible to build smart communities"

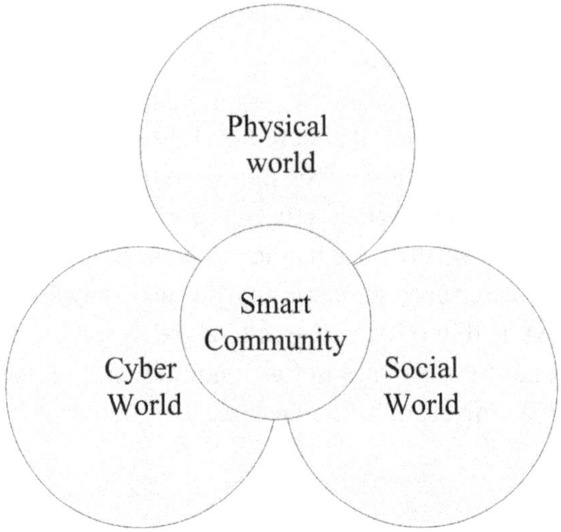

human individuals, as well as physical things such as a pen, a door. It is also possible might be included some other living things generally these objects have implicit links among them (see figure 8). For example, they may Strive together to achieve the common goal

Smart communities are time-evolving the scale change over time, one smart community may have only several members, while another may have a very huge number of members, smart communities should have good scalability. Some smart communities may function in (local) environments, it is not a necessary connection to the internet, the lifecycle of smart communities it depends on the application supported, it could possibly be very long in some cases while it's short in another (Xia, Asabere, Ahmed, Li & Kong, 2013).

Community is one of important feature derived from social network and widely defined a group of nodes and its nodes belonging to the same community tend to have the same properties, Graph clustering approaches, hierarchical clustering and modularity-based methods that a way of method to identify communities (Fatemi & Tokarchuk, 2013).

There are two approaches called strictly divided communities and non-strictly divided communities to form communities. The approach of strictly divided communities means that a member (user) can only belong to one community and cannot simultaneously belong to multiple communities On the contrary, in the other approach the member can belong to multiple communities (Wen, Liu, Zhang, Xiong & Cao, 2014).

The number of subject's community network has been growing rapidly with the growth of social networking websites and it helps in social media analysis to understand more about users' collective behavior Sahebi and Cohen tries to give the solution for the cold start problem by use different dimensions of social network to extract latent communities and use it to solve this problem (Sahebi & Cohen, 2010).

Cross-Lingual Approaches

This approach is to directly translate the source and find the related in another language, it breaks the language barrier and gives opportunities to look for items, papers, information, book in different languages that lets the users receive recommendations for the items that described in different languages for user do not understand or speak it. Moreover, some research used semantic analysis to make a language-independent representation of text, and in other research to avoid synonyms problem and build more accurate relations between tags in folksonomies used online WordNet lexical ontology with EuroWordNet multilingual lexical resource (Asanov, 2011; Yang, Chen & Wu, 2008).

Semantic Based Approaches

Usually user and item are described in a textual form in RS and the rest of the web, when using tags and keywords without any semantic meaning often its effect on recommendation accuracy, so understanding text and it structuring is very important part recommendation. Traditional approaches of text mining show descriptions that can be understood by a user, but not a computer or RS, it is based on lexical and syntactic analysis. Therefore, that was a reason of creating new text mining techniques that were based on semantic analysis. RS with such techniques are called semantic based recommender systems. The performance of semantic recommender systems is based on knowledge base usually defined as a concept diagram (like taxonomy) or ontology (Wanaskar, Vij & Mukhopadhyay, 2013).

GEOGRAPHICAL RECOMMENDATION TECHNIQUES

1. **Multi-Center Gaussian Model (MGM):** This is based on the observations that users tend to check-in around several centers, where the check-in probability follows a Gaussian distribution at each center, (Liu, Liu, Aberer & Miao, 2013) MGM improves the accuracy of location recommendations. However, the improvement is considerably limited, because it models the geographical influence as a universal distribution for all users and considers the distance between a location and a center instead of between every pair of locations visited by the same user. It uses a fixed distance to define a district. When check-ins in a district are more than a threshold, the mean of all check-ins is the center. It utilizes a greedy method to find the district and requires no overlapping between two districts (Zhao, King & Lyu, 2013).
2. **Power Law Distribution (PD):** By modelling the distance between every pair of locations visited by the same user as a power-law distribution for all users, PD further enhances the performance of recommending locations, but it still inherits the limitation of the universal distance distribution for all users (Zhang, Chow & Li, 2015).
3. **Nonnegative Matrix Factorization (NMF):** As one of the most powerful collaborative filtering techniques, matrix factorization models are superior to classic nearest-neighbor models for producing personalized recommendations NMF is still inferior to, PD and MGM, since it ignores the geographical influence of users and locations on users' check-in behavior (Zhang, Chow & Li, 2015).

Also, there are some techniques that concern just one measure, which recommends based only on the distance and its named Nearest location like Distance-Based, User Home, This technique tend to recommend to user to visit locations that are near to the locations or user home so that its most intuitive and simple geographic-aware recommender techniques(Nunes & Marinho, 2013). Table1illustratessummarizing of papers on state-of-art geographic recommendation techniques since 2011.

MEASUREMENTS AND EVALUATIONS OF RECOMMENDATION SYSTEM

To do a good evaluation, measurement that depends on the user tasks, types of analysis and datasets being used in the system. However, there is no single good evaluation method that has been used generally due to the complex nature of recommendation problems:

Table 1. Depicted Summarize Surveys of Papers on State-of-art Geographic Recommendation Techniques

Name	MGM	PD	NMF	Objective	Pros	Cons	Compare with: (techniques)	Year
Exploiting Geographical Influence for Collaborative Point-of-Interest Recommendation (Ye, Yin, Lee & Lee, 2011)		✓		- Employ a powerlaw probabilistic model to capture the geographical influence among POIs	-Fusing geographical influence user preference, and social influence to devise a check-in probability prediction model for a given user to visit POI	- One dimensional geographic distance influence - Non personalized geographical influence	-User based -Friend based -GI based recommendation -Random walk with restart -User preference/social influence based recommendation - User preference / geographical influence based recommendation	2011
Fused Matrix Factorization with Geographical and Social Influence in Location-Based Social Networks (Cheng, Yang, King, & Lyu, 2012)	✓			FMF with MGM it fused matrix factorization framework with the Multi-center Gaussian Model (FMFMGM).	-Modelling the probability of a user's check-in at a location as a (MGM) - Consider the social influence	- One dimensional geographic distance influence - Non personalized geographical influence	- Multi-center Gaussian Model (MGM) - PMF - PMF with Social Regularization (PMFSR) -Probabilistic Factor Models (PFM)	2012
A Gaussian Kernel Approach for Location Recommendations (Nunes & Marinho, 2013)				New geographic-aware recommender system based on Gaussian kernels	Infer the region most likely to be visited by the user	- Don't take into consider other aspects of the LBSN domain, such as the temporal and social context.	- Nearest location -Linear Regression -Multi-Center Gaussian Model	2013

continued on following page

Table 1. Continued

Name	MGM	PD	NMF	Objective	Pros	Cons	Compare with: (techniques)	Year
Capturing Geographical Influence In POI Recommendations (Liu, Liu, Aberer & Miao, 2013)	✓			-Two models, Gaussian mixture model (GMM) and genetic algorithm based Gaussian mixture model (GA-GMM)	-Capture geographical influence and to find activity centers more accurately and eliminate outliers	- One dimensional geographic distance influence - Non personalized geographical influence - Don't take in considering the influence of user preference and social relationship	- Gaussian model (GM) - Multi-center Gaussian model (MGM)	2013
Personalized Point-of-Interest Recommendation by Mining Users' Preference Transition (Liu, Liu, Aberer & Miao, 2013)				Propose to cluster similar users for accurate POI recommendations	- Taking into account both category information and temporal effects - Predict the possible categories of a user's next check-in location for each user cluster	- Just take in consider the geographical influence	- BasicMF - GeoCF - MGMMF - Markov - ML	2013
Learning Geographical Preferences for Point-of-Interest Recommendation (Liu, Fu, Yao & Xiong, 2013)				A General framework to learn geographical preferences for POI recommendation in LBSNs	- Takes all these factors, which influence the user check-in decision process into consideration -Model is flexible	- One dimensional geographic distance influence - Non personalized geographical influence	-Singular value decomposition (SVD), -Probabilistic matrix factorization (PMF), -Nonnegative matrix factorization (NMF), -Bayesian nonnegative matrix factorization (BNMF)	2013
iGSLR: Personalized Geo-Social Location Recommendation - A Kernel Density Estimation Approach (Zhang & Chow, 2013)	✓	✓		**Unified Geo-social recommendation framework fusing user preference, social influence, geographical influence of users, and the personalized geographical influence of locations**	**- Estimate the probability of a user checking in at a new location more accurately - Considering the geographical influence of both users and locations when recommending locations - Test the performance for cold-start users with only a few check-in records -Also performance in the problem of data sparsity**	**- One dimensional geographic distance influence**	**- User-based CF - Social CF - Social & Geographical CF**	**2013**
GeoMF: Joint Geographical Modeling and Matrix Factorization for Point-of-Interest Recommendation (Lian, Zhao, Xie, Sun, Chen & Rui, 2014)			✓	**Exploit weighted matrix factorization to conduct POI recommendation**	**- The modeling of the spatial clustering phenomenon is easily incorporated into matrix factorization**	**- Don't take into consideration social network information or show how to extend to it.**	**- UCF - MF-01 - MF-Freq - B-NMF - WMF-B**	**2014**

continued on following page

Table 1. Continued

Name	MGM	PD	NMF	Objective	Pros	Cons	Compare with: (techniques)	Year
LORE: Exploiting Sequential Influence for Location Recommendations (Zhang, Chow, & Li, 2014)				**Propose a new location recommendation approach with sequential influence based on an additive Markov chain**	**- Enhance the quality of location recommendations**	**- Don't take temporal influence into account to capture the change of user preferences - Don't show how to recommend a trip of POIs**	**- Sequential category: FMC, AMC - Geographical-social category: IGSLR, GS2D - Sequential + geographical-social category: FMC+GS2D**	**2014**
iGeoRec: A Personalized and Efficient Geographical Location Recommendation Framework (Zhang, Chow & Li, 2015).	✓	✓	✓	**Propose a probabilistic approach to predict the probability of the user visiting any new location based on his personalized distance distribution**	**- Personalize the geographical influence on a user's check-in behavior**	**-One dimensional geographic distance influence -Don't explain how to incorporate the category information of locations in our personalized geographical location recommendation framework - And how to take temporal influence into account to capture the change of users' preferences**	**-Nonnegative Matrix Factorization (NMF) -Multi-center Gaussian Model (MGM) -Power-law Distribution (PD)**	**2015**
A General Geographical Probabilistic Factor Model for Point of Interest Recommendation (Liu, Papadimitriou & Yao, 2015)			✓	**A framework for geographical probabilistic factor modeling**	**-It can take various factors into consideration strategically**	**-One dimensional geographic distance influence**	**-Probabilistic Matrix Factorization (PMF) -Bayesian Non-negative Factorization (BNMF) -Poisson Factor Model (PoiFM) -Fused Poisson factor model (FuPoiFM) -Geo-BNMF**	**2015**
Core: Exploiting the personalized influence of two-dimensional geographic coordinates for location recommendations (Zhang & Chow, 2015)	✓	✓		**Framework, it infers users' preferences by exploiting the personalized two-dimensional geographical influence for each user**	**- Improve system efficiency and scalability - Enhance the user preference model and obtain better quality of location recommendations**		**- Geographical category: MGM, PD, IGSLR, Exact - Social category: SCF - Fusion category: Prod, Sum**	**2015**

1. Accuracy of recommendation systems:
    a. **Mean Absolute Error (MAE)*:*** Some related measures include the Mean Squared Error, Root Mean Squared Error, and Normalized Mean Absolute Error. It measures the average absolute deviation between a prediction rating and the user's true rating

$$\text{MAE} = \sum_{i=1}^{N} \left(pi - ri \right) / N$$

Where N is the total number of predictions pi is the predicted value and ri is the true value

b. **Precision and Recall:** It's considered popular metrics for evaluating information retrieval systems.

$$\text{Precision (P)} = \frac{Nrs}{Ns} \text{Recall (R)} = \frac{Nrs}{Ns}$$

c. "Where *Nrs* is number of accurate predictions. *Ns* is the number of predictions *Nr* is the number of possible accurate predictions." Some ways have been taken to combine precisions and recalls, MAP (Mean Average Precision) one of them

$$\text{MAP (F)} = \frac{2PR}{P + R}$$

d. **Classification Accuracy Metrics:** It measures the frequency with which a recommender system makes correct or incorrect decisions about whether an item is good

e. **Rank Accuracy Metrics:** It measures the ability of a recommendation algorithm to produce recommended ordering of items that matches how the users would have ordered the same items.

f. **Prediction-Rating Correlation:** Two variables are correlated if the variance in one variable can be explained by the variance in the second. Three of the most well-known correlation.

2. **Confidence:** There are two conflicting dimensions often faced researchers in deciding how to interpret the recommendation results first is the strength of the recommendation: how much does the recommender system think this user will like this item, second is the confidence of the: how sure is the recommender system that its recommendation is accurate.
3. **Novelty and Serendipity:** The general consideration is that a system may want to try to estimate the probabilities that a user will be familiar with an item. The dimension to measure the non-obviousness of the recommendation is novelty. A serendipitous recommendation helps the user find a surprisingly interesting item he might not have otherwise discovered.
4. **Coverage of an Evaluation System:** Coverage must be measured in combination with accuracy, so recommenders are not tempted to raise coverage by making

bogus predictions for every item this is the domain of items in the system over which the system can form predictions or make recommendations.

5. **User Evaluation:** Important aspect of the evaluations by answers the question of how to directly evaluate user reactions to a recommender system (Chen, Owusu & Zhou (2013).

Limitation of RS

Cold Start Problem

Is one major problem with RS, it happens when entering a new user or new item, with little information about this user or item, which leads to the system trying to infer the user recommendation. There are three types of this problem: new user, new item and new system problem. It is difficult to recommend with this type of problem because of the lack of information about new user or rating about the new item. This problem is a sub-problem of the coverage problem because it measures the system coverage given a specific set (user and item).

Scalability

With the fast growth of information through the internet, scalability refers to the ability to handle a huge amount of information. RS is having an explosion of information and it keeps handling this with continuously growing demand. Some algorithms of RS have challenges, which increase with the growing number of item and user.

Data Sparsity

Is typically a problem with RS, the main reason for this problem is that most items don't get rated by most users and the availability of the rating is usually sparse, this problem affects RS quality, also CF suffers because it is based on rating matrix. Another problem caused by this problem is the cold start problem.

Over Specialization Problem

RS recommends an item to a user based on what is already known or defined in their profile so the user becomes restricted with this information and it prevents from discovering new items and options that are available.

Privacy

User preference or information is known for RS regarding items that need recommendation; however, it is important that no third party uses RS to obtain information or preferences about the user that he wishes to remain private. Thus, the accuracy of a recommendation should also be evaluated according to how private it is.

Trust

Users may be likely to trust their recommendation system that could be good at recommending a few items that are liked by and familiar to a user, but this is not much of a gain to the user about a recommendation process. The system should provide a more reasonable recommendation, which would also lead to an increase in system trust (Mani, 2014; Shani & Gunawardana, 2011; Zhang, Chow & Li, 2015).

FUTURE WORK

The wide use of social networks and the rapid growth of the internet caused a huge amount of data, and we use RS to extract useful information that can be presented to the user in order to help him manage the data properly leading to making correct decisions. Nowadays, there has been great progress in the field of RS making it a required application in different areas in real-life, LBSNs have attracted millions of users such as Facebook places, Gowalla, and Foursquare, and they can share their experiences of visiting specific locations. RS it is divided to traditional approaches and modern approaches. Traditional approaches it is considered simplest and popular approaches, it is used as a wide range implementation for RS applications, but most traditional techniques recommended without taking into consideration the circumstance and other information that may be changing the result of a recommendation. In addition, modern approaches it considers complex and powerful approaches that can be present smart service for all users.

We face many limitations in the RS process like cold start problem, data sparsity… etc. And there are also some limitation that appears in a geographical recommender system such as some techniques don't take into consider personalized geographical influence or aspects of the LBSN domain such as the temporal and social context. So that we try to build a novel personalize geographical approach avoid those problems based on smart communities and demographic techniques in social networks, which can be used to solve those problems with huge and complex structure data, according to provide of smart services to all members.

REFERENCES

Adomavicius, G., & Tuzhilin, A. (2015). Context-aware recommender systems. In Recommender systems handbook (pp. 191-226). Springer US. doi:10.1007/978-1-4899-7637-6_6

Al-Otaibi, S. T., & Ykhlef, M. (2012). A survey of job recommender systems. *International Journal of Physical Sciences, 7*(29), 5127–5142. doi:10.5897/IJPS12.482

Anderson, C., & Hiralall, M. (2009). *Recommender systems for e-shops*. Business Mathematics and Informatics paper.

Asanov, D. (2011). *Algorithms and methods in recommender systems*. Berlin, Germany: Berlin Institute of Technology.

Baltrunas, L., Makcinskas, T., & Ricci, F. (2010, September). Group recommendations with rank aggregation and collaborative filtering. In *Proceedings of the fourth ACM conference on Recommender systems* (pp. 119-126). ACM. doi:10.1145/1864708.1864733

Bhaidkar, Y., Bhagwat, P., Bhalere, P., Gujar, R., & Walunj, S. (2015). *Tourist place recommendation system*. Academic Press.

Bobadilla, J., Ortega, F., Hernando, A., & Gutiérrez, A. (2013). Recommender systems survey. *Knowledge-Based Systems, 46*, 109–132. doi:10.1016/j.knosys.2013.03.012

Chen, S., Owusu, S., & Zhou, L. (2013, September). Social Network Based Recommendation Systems: A Short Survey. In *Social Computing (SocialCom), 2013 International Conference on* (pp. 882-885). IEEE. doi:10.1109/SocialCom.2013.134

Cheng, C., Yang, H., King, I., & Lyu, M. R. (2012, July). *Fused Matrix Factorization with Geographical and Social Influence in Location-Based Social Networks* (Vol. 12). AAAI.

Fatemi, M., & Tokarchuk, L. (2013, September). A Community Based Social Recommender System for Individuals & Groups. In *Social Computing (SocialCom), 2013 International Conference on* (pp. 351-356). IEEE. doi:10.1109/SocialCom.2013.55

Han, X., Tian, L., Yoon, M., & Lee, M. (2012, November). A big data model supporting information recommendation in social networks. In *Cloud and Green Computing (CGC), 2012 Second International Conference on* (pp. 810-813). IEEE. doi:10.1109/CGC.2012.125

Karimi, B., & Yektaei, M. H. (2015). Location Recommendation based on location-based social networks for Entertainment services. *Advances in Computer Science: an International Journal*, *4*(1), 55–64.

Lee, M. (2014). A Recommendation System based on Big Data Technology. *International Journal (Toronto, Ont.)*, *3*(5).

Lian, D., Zhao, C., Xie, X., Sun, G., Chen, E., & Rui, Y. (2014, August). GeoMF: joint geographical modeling and matrix factorization for point-of-interest recommendation. In *Proceedings of the 20th ACM SIGKDD international conference on Knowledge discovery and data mining* (pp. 831-840). ACM. doi:10.1145/2623330.2623638

Liu, B., Fu, Y., Yao, Z., & Xiong, H. (2013, August). Learning geographical preferences for point-of-interest recommendation. In *Proceedings of the 19th ACM SIGKDD international conference on Knowledge discovery and data mining* (pp. 1043-1051). ACM. doi:10.1145/2487575.2487673

Liu, B., Xiong, H., Papadimitriou, S., Fu, Y., & Yao, Z. (2015). A general geographical probabilistic factor model for point of interest recommendation. *IEEE Transactions on Knowledge and Data Engineering*, *27*(5), 1167–1179. doi:10.1109/TKDE.2014.2362525

Liu, X., Liu, Y., Aberer, K., & Miao, C. (2013, October). Personalized point-of-interest recommendation by mining users' preference transition. In *Proceedings of the 22nd ACM international conference on Information & Knowledge Management* (pp. 733-738). ACM. doi:10.1145/2505515.2505639

Liu, Y., Wei, W., Sun, A., & Miao, C. (2014, November). Exploiting geographical neighborhood characteristics for location recommendation. In *Proceedings of the 23rd ACM International Conference on Conference on Information and Knowledge Management* (pp. 739-748). ACM. doi:10.1145/2661829.2662002

Lops, P., De Gemmis, M., & Semeraro, G. (2011). Content-based recommender systems: State of the art and trends. In Recommender systems handbook (pp. 73-105). Springer US.

Mahmood, M. A., Al-Shammari, E. T., El-Bendary, N., Hassanien, A. E., & Hefny, H. A. (2013, September). Recommender system for ground-level Ozone predictions in Kuwait. In *Computer Science and Information Systems (FedCSIS), 2013 Federated Conference on* (pp. 107-110). IEEE.

Mahmoud, S., El-Bendary, N., Mahmood, M. A., & Hassanien, A. E. (2013, December). An intelligent recommender system for drinking water quality. In *Hybrid Intelligent Systems (HIS), 2013 13th International Conference on* (pp. 285-290). IEEE. doi:10.1109/HIS.2013.6920498

Mani, M. (2014, September). Challenges & Limitation in Recommender system. *International Journal of Latest Trends in Engineering and Technology*, *4*(3), 138–142.

Nunes, I., & Marinho, L. (2013). A gaussian kernel approach for location recommendations. In Proceedings of KDMiLe-Symposium on Knowledge Discovery, Mining and Learning (Vol. 1060). Academic Press.

Panniello, U., Tuzhilin, A., Gorgoglione, M., Palmisano, C., & Pedone, A. (2009, October). Experimental comparison of pre-vs. post-filtering approaches in context-aware recommender systems. In *Proceedings of the third ACM conference on Recommender systems* (pp. 265-268). ACM. doi:10.1145/1639714.1639764

Pussep, K., Kaune, S., Flick, J., & Steinmetz, R. (2009, March). A peer-to-peer recommender system with privacy constraints. In *Complex, Intelligent and Software Intensive Systems, 2009. CISIS'09. International Conference on* (pp. 409-414). IEEE. doi:10.1109/CISIS.2009.32

Ricci, F., Rokach, L., & Shapira, B. (2011). Introduction to recommender systems handbook. In Recommender systems handbook (pp. 1-35). springer US. doi:10.1007/978-0-387-85820-3_1

Safoury, L., & Salah, A. (2013). Exploiting user demographic attributes for solving cold-start problem in recommender systems. *Lecture Notes on Software Engineering*, *1*(3), 303–307. doi:10.7763/LNSE.2013.V1.66

Sahebi, S., & Cohen, W. W. (2010). Community-based recommendations: a solution to the cold start problem. *Proceedings of WOODSTOCK'97*.

Salam, M. A., Mahmood, M. A., Awad, Y. M., Hazman, M., El Bendary, N., Hassanien, A. E., & Saleh, S. M. (2014). Climate recommender system for wheat cultivation in North Egyptian Sinai Peninsula. In *Proceedings of the Fifth International Conference on Innovations in Bio-Inspired Computing and Applications IBICA 2014* (pp. 121-130). Springer International Publishing. doi:10.1007/978-3-319-08156-4_13

Shani, G., & Gunawardana, A. (2011). Evaluating recommendation systems. *Recommender systems handbook*, 257-297.

Sharma, L., & Gera, A. (2013). A survey of recommendation system: Research challenges. *International Journal of Engineering Trends and Technology*, *4*(5), 1989–1992.

Sharma, M., & Mann, S. (2013). A survey of recommender systems: Approaches and limitations. *International Journal of Innovations in Engineering and Technology*, *2*(2), 8–14.

Shavitt, Y., Weinsberg, E., & Weinsberg, U. (2010, October). Building recommendation systems using peer-to-peer shared content. In *Proceedings of the 19th ACM international conference on Information and knowledge management* (pp. 1457-1460). ACM. doi:10.1145/1871437.1871646

Song, Y., Hu, Z., Liu, H., Shi, Y., & Tian, H. (2013, September). A novel group recommendation algorithm with collaborative filtering. In *Social Computing (SocialCom), 2013 International Conference on* (pp. 901-904). IEEE. doi:10.1109/SocialCom.2013.138

Symeonidis, P., Ntempos, D., & Manolopoulos, Y. (2014). *Recommender systems for location-based social networks*. Springer Science & Business Media. doi:10.1007/978-1-4939-0286-6

Wanaskar, U., Vij, S., & Mukhopadhyay, D. (2013). *A hybrid web recommendation system based on the improved association rule mining algorithm*. arXiv preprint arXiv:1311.7204

Wen, Y., Liu, Y., Zhang, Z. J., Xiong, F., & Cao, W. (2014). Compare two community-based personalized information recommendation algorithms. *Physica A*, *398*, 199–209. doi:10.1016/j.physa.2013.12.037

Xia, F., Asabere, N. Y., Ahmed, A. M., Li, J., & Kong, X. (2013). Mobile multimedia recommendation in smart communities: A survey. *IEEE Access: Practical Innovations, Open Solutions*, *1*, 606–624. doi:10.1109/ACCESS.2013.2281156

Yang, C. Z., Chen, I. X., & Wu, P. J. (2008). Cross-lingual news group recommendation using cluster-based cross-training. *Computational Linguistic and Chinese Language Processing*, *13*(1), 41-60.

Ye, M., Yin, P., Lee, W. C., & Lee, D. L. (2011, July). Exploiting geographical influence for collaborative point-of-interest recommendation. In *Proceedings of the 34th international ACM SIGIR conference on Research and development in Information Retrieval* (pp. 325-334). ACM. doi:10.1145/2009916.2009962

Zhang, J. D., & Chow, C. Y. (2013, November). iGSLR: personalized geo-social location recommendation: a kernel density estimation approach. In *Proceedings of the 21st ACM SIGSPATIAL International Conference on Advances in Geographic Information Systems* (pp. 334-343). ACM. doi:10.1145/2525314.2525339

Zhang, J. D., & Chow, C. Y. (2015). CoRe: Exploiting the personalized influence of two-dimensional geographic coordinates for location recommendations. *Information Sciences*, *293*, 163–181. doi:10.1016/j.ins.2014.09.014

Zhang, J. D., Chow, C. Y., & Li, Y. (2014, November). Lore: Exploiting sequential influence for location recommendations. In *Proceedings of the 22nd ACM SIGSPATIAL International Conference on Advances in Geographic Information Systems* (pp. 103-112). ACM. doi:10.1145/2666310.2666400

Zhang, J. D., Chow, C. Y., & Li, Y. (2015). iGeoRec: A personalized and efficient geographical location recommendation framework. *IEEE Transactions on Services Computing*, *8*(5), 701–714. doi:10.1109/TSC.2014.2328341

Zhao, S., King, I., & Lyu, M. R. (2013, November). Capturing geographical influence in POI recommendations. In *International Conference on Neural Information Processing* (pp. 530-537). Springer. doi:10.1007/978-3-642-42042-9_66

Chapter 12
GIS-Based Decision Support System for Village Level:
A Case Study in Andhra Pradesh

Subbu Lakshmi Esakki Pandian
VIT University, India

Kiran Yarrakula
VIT University, India

Probal Chaudhury
Bhabha Atomic Research Center, India

ABSTRACT

Decision support system (DSS) plays a vital role especially in rural areas to develop rural sector for sustainable development and socio-economic uplifting of the country. To make appropriate decisions and to develop village economy, decision support system is useful for the mandal revenue officer, collector, Surpanch, and different administrators. It deals both spatial and non-spatial data at village level and comprises various ancillary information including mandal maps and village-wise information of Anantapur and Kadapa districts of Andhra Pradesh such as number of houses, male and female, SC/ST/OBC/general (or) OC population, literate and illiterate population, total working and non-working population. Datasets are collected from district collector office, mandal revenue officer (MRO) and inserted in GIS database. This chapter makes an attempt to build features of various decisions at Anantapur and Kadapa districts by integrating various layers of information at village level.

DOI: 10.4018/978-1-5225-5088-4.ch012

IMPORTANCE OF DSS AT VILLAGE LEVEL

Geographical information system (GIS) has been developing at a remarkable pace over few decades and plays a vital role in development of nations in the 21st Century. Decentralized planning in India is still in the stage of experimentation. Now it's time for planners and decision makers to discuss the importance of GIS technology and its applications in rural development (Reddy et al, 2011). So there is a need for spatial DSS and its demand is likely to grow. In India 27% of the population resides in urban areas and 73% of population settled in rural area, a proper structured planning procedure is essential such that the infrastructure facilities and development activities are available at both rural and urban area. Since majority of population settled in rural areas are provided with the least infrastructure facilities, creates a regional imbalance in development, that cause migration of people from rural to urban areas. Selwal, 2008 addressed that existing systems make decision making for village developments manually based on information recorded on papers. Such manual process consume time and inefficient to take decision regarding further developments. Previously, lack of proper land records and inefficient judiciary makes concerning system vulnerable to high loss and it demands a system that keeps the accurate record of lands and makes it available in time (Patil 2014; Padma et al, 2015; Vijay Krishna, 2003). Hence decision-makers or administrators require a well organized tool like GIS will help to obtain the updated layout of the region (Kodge and Hiremath, 2012). Adinarayana et al., 2008 proposed a geographical information and communication technology (Geo-ICT) based information system for rural development to assist the user in analsying the rural information for rural development. Patil, 2014 used GIS platform using Gram++ software to assist administrators for village level planning. Asadi et al, 2012 used remote sensing and GIS techniques along with VB. NET for micro level planning and sustainable development of mandals that would be helpful for government officers, planners, geographers and a person who does not have prior knowledge on the software can avail village information at one place. DSS play a huge role in improving school facilities which is very useful for village Sarpanch and parents, district development officers for decision making at appropriate level (Hareesh and Valli Kumari, 2013). Govindaraj et al, 2013 addressed that information system based on spatial data at panchayath level will be useful for proper planning, monitoring, investigation and decision making to provide better public utility services. GIS and remote sensing act as an important tool has greatly reduced the problem by computerization of the spatially correlated data for handling, retrieving, analysis of spatial data and storage (Sitender et al, 2012; Penning de Vries et al, 1991). Sitender et al, 2012 developed information system to examine the status of infrastructure facilities using HTML environment. Kodge and Hiremath, 2012 used open source geospatial tool such as QGIS and web based information system using SQL to extract village level information system. To determine the impacts on future land use patterns, web based information system also known as village information system (VIS) allow planners and citizens to act efficiently and quickly create and examine the scenarios alternative development (Santosh et al, 2009). Some of the service organization such as NIC and ICT application play a vital role in facilitating grass root delivery of GIS based DSS for planning at micro levels and providing integrated spatial and non spatial data services for planning (Yadav and Singh, 2009). Rayed addressed SDSS converts existing datasets into new datasets by applying DSS models, expert system, and artificial intelligence, such data combination provide a model output that provide solutions with spatial relevance. In such a place where spatial and visual representation is important, GIS data play a key role (Rayed, 2010). The present book chapter highlights the power of GIS technology to understand and evaluate spatial data by creating graphic displays using attribute information stored in the GIS database such that the clients can view the information query. It is concluded that people living in rural area urgently needs a GIS based e-governance system in order to help the government in planning, implementation and monitoring of various projects for development of rural areas at a faster rate to develop the state more technologically advanced. The modern technology of remote sensing includes both aerial and satellite based systems, provide easy access to data on a repetitive basis with speed and spatial data can be analyzed together with the help of GIS, generating various options such as modeling, thereby the whole planning process can be optimized (Jhawar et al, 2012). The high resolution RESOURCESAT-II (LISS-IV Mx) sensors are used for rural and urban area detection and identification in the study area. Remote sensing approaches have been widely used due to its cost-effectiveness, extensibility and temporal frequency. Some of the important features of DSS such as it provide an accurate spatial view of district such as district, mandals, village level as well as road and rail network, drainage network, education and health related data, demographic data on desktop in a GIS environment. DSS assists them to find out the optimum locations for health centers, schools depending on certain parameters include total number of population, number of health centers and its possible location and in case of education number of primary schools, and percentage of population literate, vacant places, teachers recruited in a village etc.

STUDY AREA

The study area is located in the state of Andhra Pradesh covering parts of Anantapur and Kadapa districts. The area confines a total of 282935.09 ha and identified that 7 out of 16 Mandals of Anantapur district and 9 out of 16 Mandals of Kadapa district are covered in the study area. These are namely Lingala, Nallamada, Kadiri, Gandlapenta, Nambulapulikunta, Mudigubba, Talapula, Chakrayapet, Vempalle, Pendlimarri, Veerapunayunipalle, Thondur, Simhadripuram, Vemula, Pulivendula, and Bukkapatnam covering latitude 14° 0' 0" N to 14° 40' 0" N and Longitude 78° 0' 0"E to 78° 30' 0"E of the study area. The study area contains 163 numbers of villages. There are 236 Panchayath and 776 numbers of sub villages in total panchayath, out of which 136 number of villages and 27 number of sub villages in total Panchayath are covered in the study area. The total population of all mandals in the study area was 386682.Various ancillary information including mandal maps and village-wise information of 16 mandals were collected from the respective district collector office and respective mandal revenue officers (MRO). The village-wise data include the total population, number of houses, male and female, SC/ST/OBC population, literate and illiterate population; total working and non working population (male and female) are collected and inserted in GIS database. Figure 1 shows geographical location of the study area.

METHODOLOGY

Decision support systems (DSS) as `interactive computer programs that utilize analytical methods, such as decision analysis, optimization algorithms, program scheduling routines for model developing to help decision makers, analyze their impacts, and interpret and select appropriate options for implementation' (Adelman et al, 1992). The DSS at village level is done for Anantapur and Kadapa district covering all the mandals with villages. For analysis and interpretation, the following data is essential such as satellite data, toposheet and ground data where the coordinates of the location obtained using field survey or GPS and data from articles, internet, and textbooks (Subbu Lakshmi and Yarrakula, 2017). The attribute data are collected from the district collector offices and subsequently attached with the individual villages. In GIS, data is stored in layer format using original scale of the feature. Before entering the data into GIS format, datum information and projection system must be defined. Then various thematic layers are incorporated along with their entities of attributes into the GIS platform like census data, habitats, important places, schools, colleges, industries, villages etc. All features are digitized based on satellite imageries. However, wherever features are not visible/clearly marked, SOI toposheets are used for digitization. The data obtained from field survey is entered into a GIS database and stored in such a manner that it could be retrieved as per need. Using ARC GIS tools, statistical analysis is performed based on the gathered data such as Male/Female population, SC/ST/OBC/General population, literacy population. Based on the maps and associated feature data, queries are

Figure 1. Geographical location of the study area

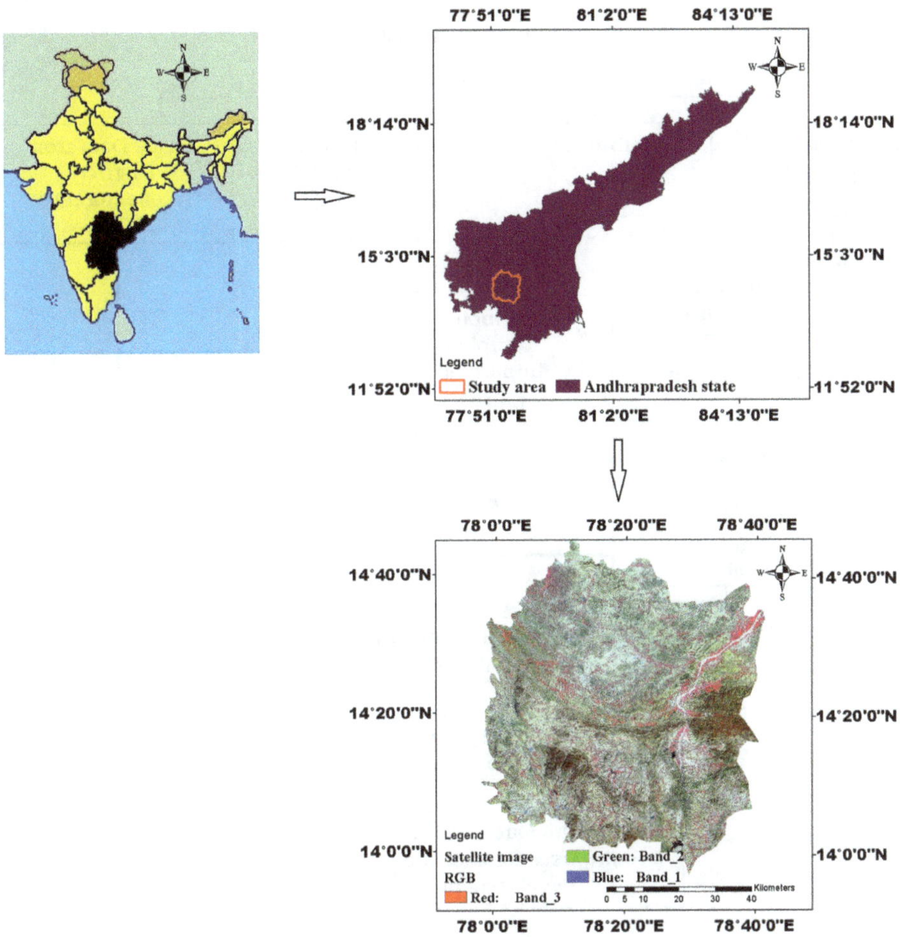

generated and documented based on the user's requirements to find the desired results in graphical as well as tabular format. One of the significant tools is to generate queries based on the attribute data is GIS. Generation of queries can be done in two ways: i) spatial queries, and ii) attribute queries using ARC GIS software. In the present study, the queries are generated based on attribute data. For example," Population Density (persons/sq. m) > 200 AND Number of House Holds >100)". Apart from the queries, statistical, raster and vector analysis can also be performed based on the maps and data stored in the features. Maps and geographic features show decision related information and relationship that will be very much useful to solve important problems. Figure 2 shows detailed methodology for DSS for village level. Table 1 shows Attributes of features selected for the study area. Table 2 shows detail information of census data.

Figure 2. Flow chart for DSS at village level

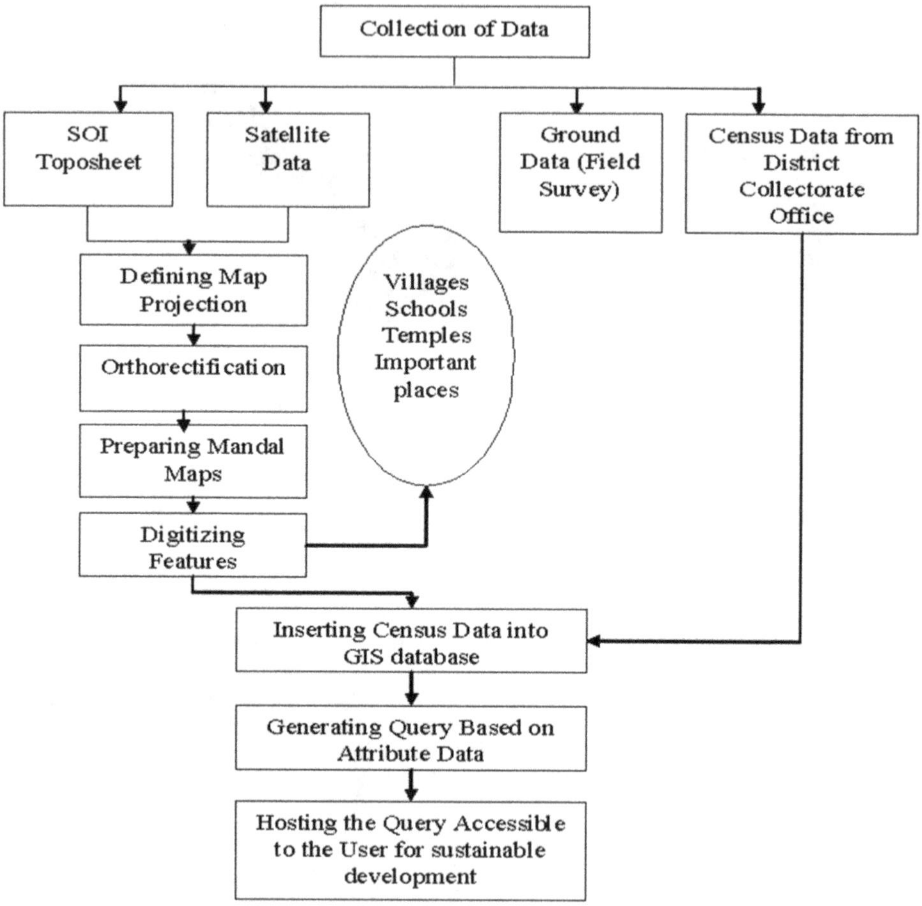

RESULTS

Administrative Mandals

Administrative mandals provide lot of facilities to people such as training facilities for administrative and technical personnel for regulatory and development of the villages by providing certain relaxation in the norm for backward and tribal areas. Figure 3 shows all the administrative mandals (7 out of 16 Mandals of Anantapur district and 9 out of 16 Mandals of Kadapa district) in the study area namely Lingala, Nallamada, Kadiri, Gandlapenta, Nambulapulikunta, Mudigubba, Talupula, Chakrayapet, Vempalle, Pendlimarri, Veerapunayunipalle, Thondur, Simhadripuram, Vemula, Pulivendula, and Bukkapatnam. The information about any mandal is

Table 1. Attributes of features selected for the study area

Feature Name	Attributes								
	1	2	3	4	5	6	7	8	9
Places of Interest	Temples	Object Id	Longitude	Latitude	Shape	Mandal	Village	Location	
	Schools	Object Id	Longitude	Latitude	Shape	Mandal	Village	Location	
	Important Places	Object Id	Longitude	Latitude	Shape	Mandal	Village	Location	
Village	Object Id	Longitude	Latitude	Shape	Mandal	Village	No_HH	Tru	Tot_P
	Tot_M	Tot_ F	P_Sc	M_Sc	F_Sc	P_St	F_St	M_St	P_Lit
	P_OBC	M_ OBC	F_ OBC	P_General/OC	M__General/ OC	F__General/OC	M_Lit	F_Ill	Lit_Rate
	F_Lit	M_Lit	P_Ill	F_Ill	M_Ill	P_06	Tot_Work_P	Tot_Work_M	Tot_Work_F
	Main Work_P	Main Work_M	Main Work_F	Main _Cl_P	Main _Cl_M	Main _Cl_F	Main _Al_P	Main _Al_M	Main _Al_F
	Main_HH_P	Main_HH_M	Main_HH_F	Main_OT_P	Main_OT_M	Main_OT_F	Marg Work_P	Marg Work_M	Marg Work_F
	Marg_Cl_P	Marg_Cl_M	Marg_Cl_F	Main_Al_P	Main_Al_M	Main_Al_F	Main_HH_P	Main_HH_M	Main_HH_F
	Main_OT_P	Main_OT_M	Main_OT_F	Marg Work_3_6_9_P	Marg Work _3_6_9_M	Marg Work _3_6_9_F	Marg_ CL_3_6_9_P	Marg_ CL_3_6_9_M	Marg_ CL_3_6_9_F
	Marg_ AL_3_6_9_P	Marg_AL_3_6_9_M	Marg_ AL_3_6_9_F	Marg_ HH_3_6_9_P	Marg_ HH_3_6_9_M	Marg_ HH_3_6_9_F	Marg_ OT_3_6_9_P	Marg_ OT_3_6_9_M	Marg_ OT_3_6_9_F
	Marg Work_0_3_P	Marg Work _0_3_M	Marg Work _0_3_F	Marg_ CL_0_3_P	Marg_ CL_0_3_M	Marg_CL_0_3_F	Marg_ AL_0_3_P	Marg_ AL_0_3_P	Marg_AL_0_3_P
	Marg_HH_0_3_P	Marg_HH_0_3_P	Marg_HH_0_3_P	Marg_ OT_0_3_P	Marg_ OT_0_3_P	Marg_OT_0_3_P	Non_ Work_P	Non_Work_M	Non_Work_F
	Panchayat Name								
Mandals	Fid	Shape	Shape Area	Shape length	Population	Total Male	Total_sc	Mandal name	Total Female
	Total_St	Total_Lit	Total_Illit						

Table 2. Detailed information of census data

Village name	Total Male	Total Female	Total population	Literacy rate	
Veerapunayunipalle	11986	1190	24497	TOT_LIT	TOT_IILT
Simhadripuram	3436	3505	6941	1354	1034
Lingala	13351	1283	26189	1571	1047
Mudigubba	41854	0	22525	0	0
Bukkapatnam	0	0	0	0	0
Nallamada	11587	1149	23080	1240	1067
Kadiri	51603	0	35944	0	0
Gandlapenta	13191	1299	26183	1358	1260
Nambulapulikunta	11885	1177	23663	1192	1173
Chakrayapet	16062	1519	31258	1583	1542
Talupula	37512	0	42392	0	0
Pendlimarri	6235	6111	12346	6624	5722
Pulivendula	2357	2334	13178	2824	1867
Vemula	27538	2634	13178	3018	2370
Thondur	13688	1402	23964	1531	1240
Vempalle	31618	3018	61344	0	2626

obtained by clicking on the mandal map or by locating the attribute table of mandals. The queries and reports of mandals are also generated based on the data (attributes) attached with mandal. Mandal attributes contains information such as male, female, total population, SC-ST/OBC population, and literacy rates.

Thematic Map of Villages/Urban Areas and Important Places of Administrative Mandals

Villages play a crucial role for proper planning, execution and maintenance of infrastructure facilities for sustainable development and to enhance the standard of life in rural areas to avoid migration of village population to urban areas. Figure 4 shows all geo-referenced villages, location of Schools, Temples, and Important places in the study area. The information about any village is obtained by clicking on the village map or by locating the attribute table of villages. Altogether 163 numbers of villages are digitized. The attribute data are also collected from the district collector offices and subsequently attached with the individual villages. The queries and report of villages are generated based on the data (attributes) attached

Figure 3. Administrative mandals of the study area

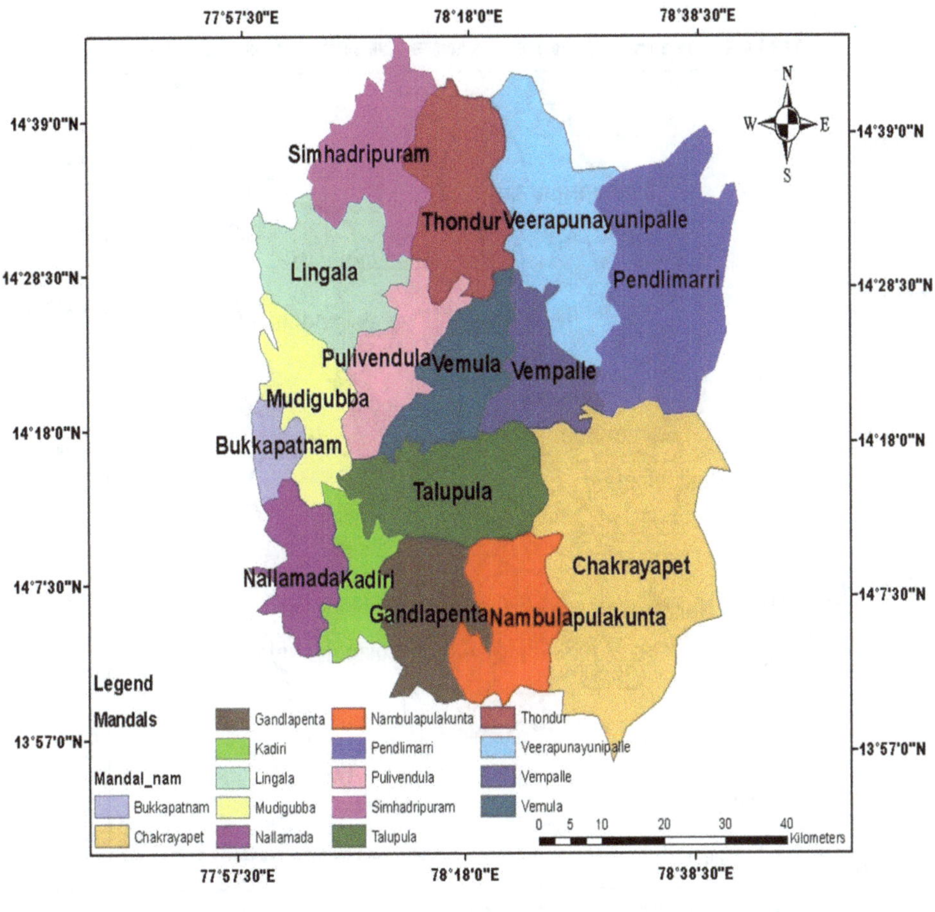

with villages. Village attributes are same as administrative mandals as given below: male, female, total population, SC-ST/OBC population, children population, literacy rates, working and non working population and others.

Important Places in the Study Area

Walkability score calculated based on the locations of schools, colleges, bus stops, and other essential amenities and to know the elementary facilities in schools, colleges for better education which are helpful in the development of the particular region. Important places such as hotels, parks, religious places and other tourism sites play a major role for sustainable growth of economy. Hospital, police stations, fire stations is very helpful in taking decisions at emergency time. Figure 4 depicts

Figure 4. All villages, location of Schools, Temples, and Important places in the study area

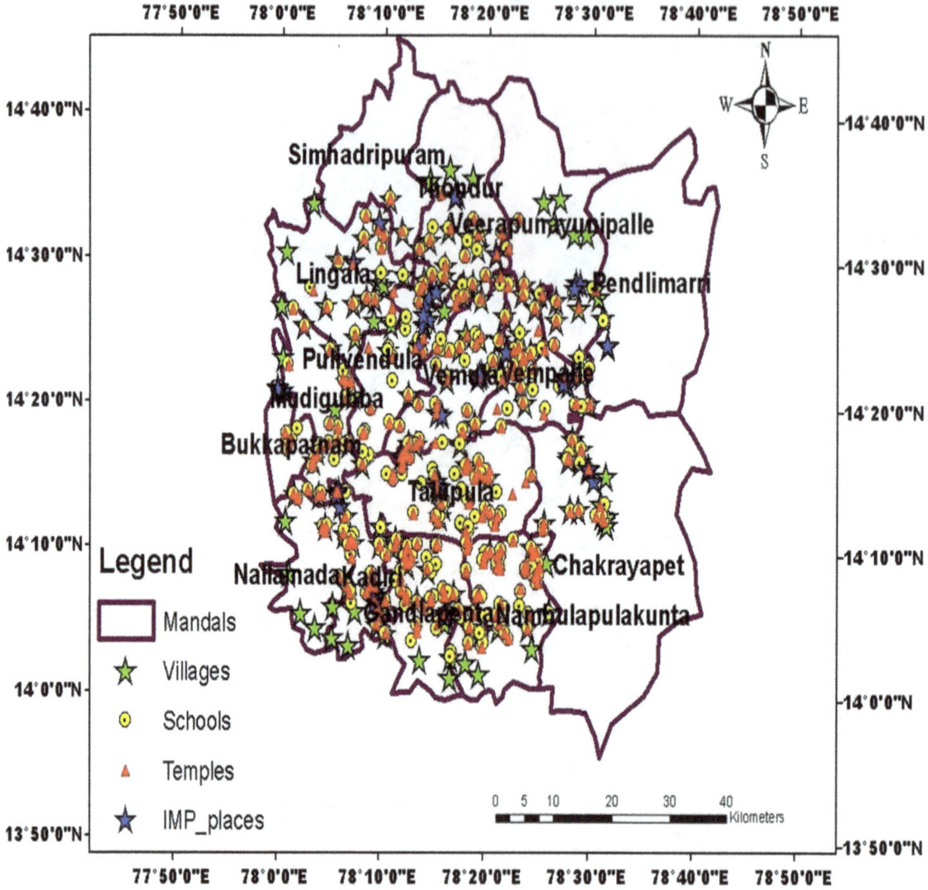

the locations of places of interest in the study area. The lat-long of these locations are placed on the map. Other than these, GPS survey is also carried out at different important locations including schools, colleges, hospitals, police stations, temples, masjid and other religious places and industries. The places of interest are digitized as point features having lat-long and attributes of choice. The information regarding the places of interest are been obtained by clicking the graphics feature.

Distribution of Village and Mandal Wise Population Density Map

Growth of population plays a huge role for the socio-economic development of a country. For the past few years, population growth decreases in rural areas due to lack of facilities and unemployment and the movement away from agriculture, but more into modern industries lead to migration from rural to urban areas (Subbu Lakshmi and Yarrakula, 2016). Population estimation of villages on mandal wise can be done easily on mandal wise using GIS database. Figure 5 shows the village wise total population density (number of people per sq. kilometers) of sixteen mandals. Figure 5 illustrates that the concentration of population is relatively higher industrial area and in villages that have townships. It is noted that Bukkapatnam do not contain any demographic data since the data is not available from the head quarters. As expected, the population density in Simhadripuram, Pendlimarri and Chakrayapet mandals are higher ranging between 18270 to over 46740 people per sq km. On the contrary, population density is much lower in Vempalle and Nallamada mandal ranging from 3243 to 4322 persons per sq. km. Overall the population density ranges between 3243 to 46740 population densities per sq. km.

Figure 5. Total population density of sixteen mandals

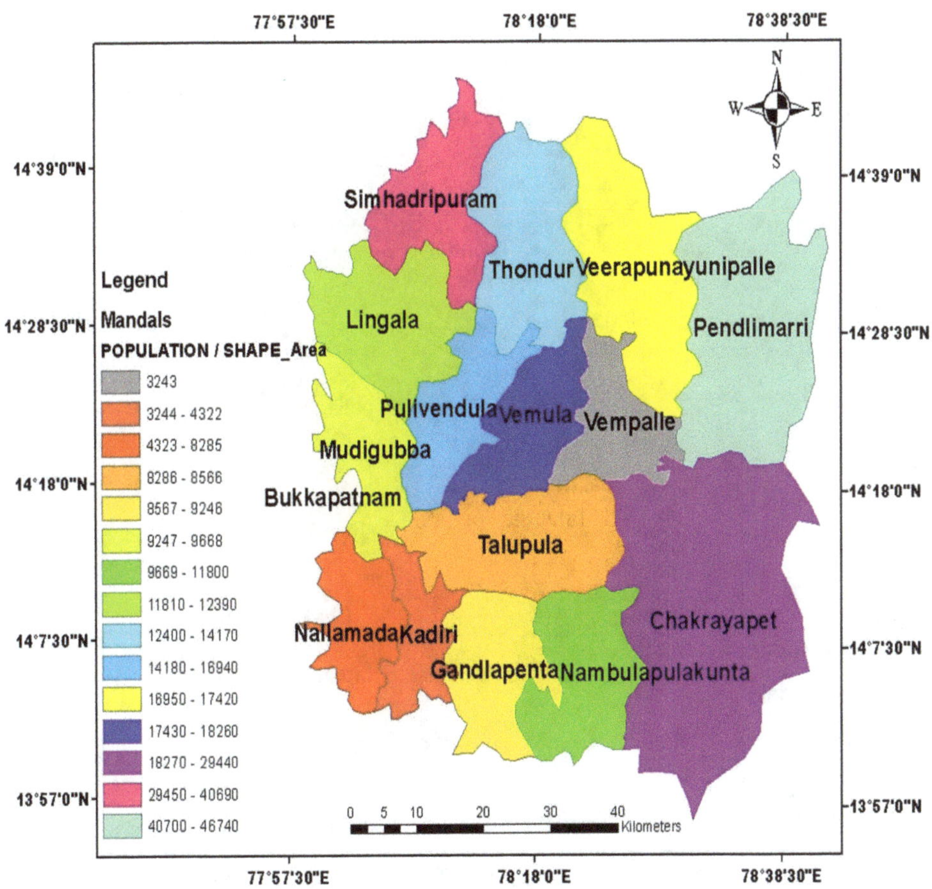

Statistical Map of Literacy Rate Village Wise

Literacy is an indispensable means for socio-economic participation contributing to poverty reduction and human development. From the literacy data using GIS database, distribution of literacy rate is plotted mandal wise as shown in Figure 6. It provides statistical information about the data associated with each set of objects and shows higher literacy around Vemula, Pulivendula, and Simhadripuram mandals. Literacy rate of Nallamada, Nambulapulikunta, Gandlapenta, Veerapunayunipalle mandals having less than 5% literacy rate and the literacy rate in Lingala, Thondur and Chakrayapet mandals having more than 5% literacy rate, while the literacy rate in Simhadripuram, Pulivendula, Vemula blocks having literacy rate of around 5-16%. Pendlimarri Mandal is having higher literacy rate of around 16-24%. It may be noted

Figure 6. Thematic map showing literacy rate (%) of all mandals

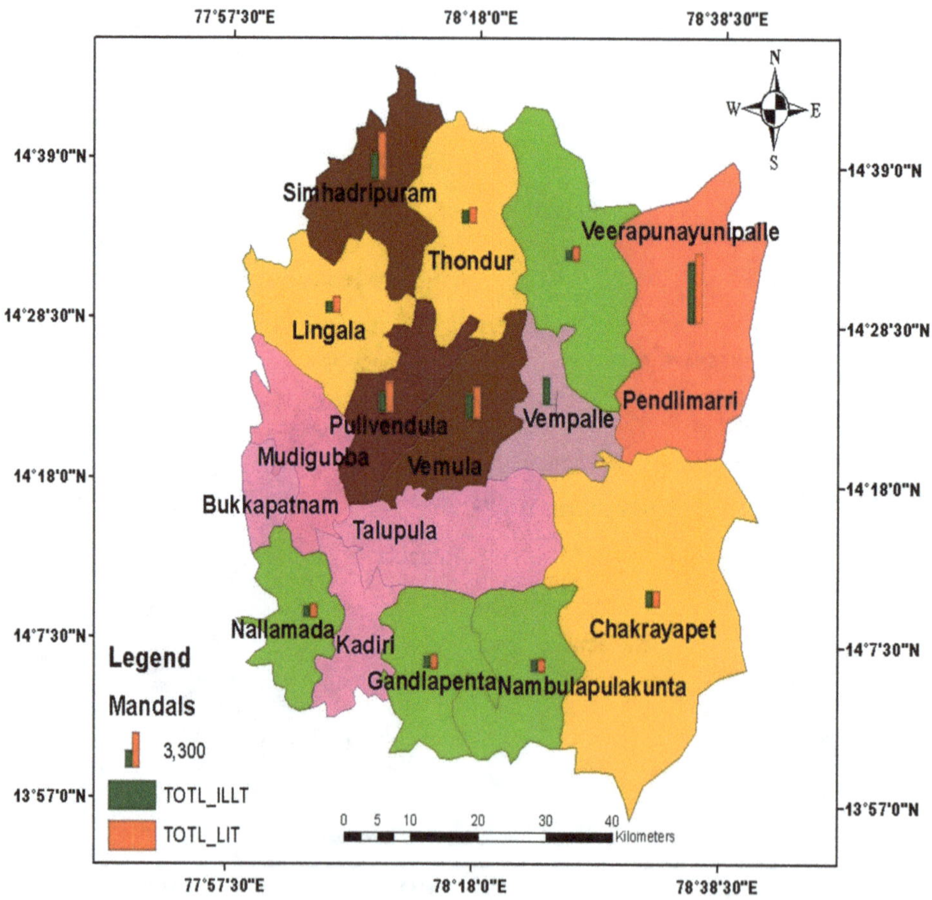

that Magenta color villages in the Mudigubba, Bukkapatnam, Talupula, Kadiri, and mandals do not contain any demographic data since they are not available from the district collector office.

Village Wise Thematic Map of Female to Male Population Ratio

Female to Male (F/M) ratio is an important social benchmark in India. Based on the data obtained from the Collectorate offices, the ratio is plotted thematically as shown in Figure 7. A red bar indicates total male population and green bar indicates total female population. The bars are scaled in proportion to the data values represented for each area. It is clear from the plot that male population is much higher as compared to female population. The mandals like Kadiri, Talupula, Mudigubba, Vemula, and Vempalle are having high male population whereas the mandals like Simhadripuram, Pendlimarri, and pulivendula are having equal male to female population ratio.

Figure 7. Thematic map of female to male population ratio

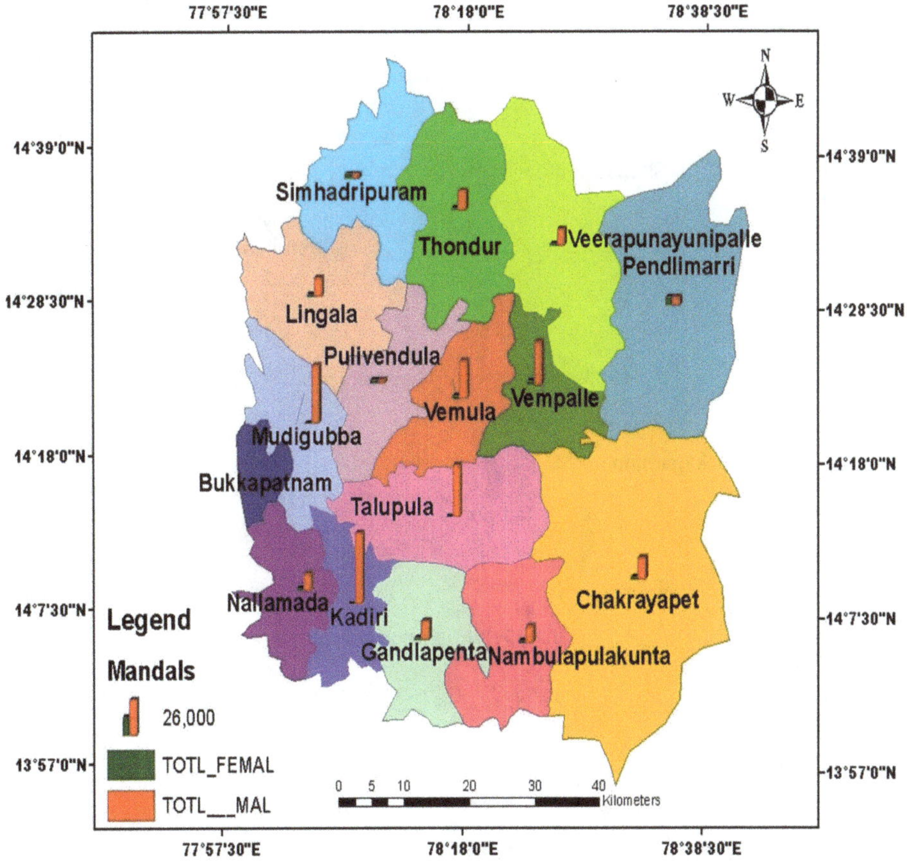

Mandal Wise SC – ST Population Thematic Map

Scheduled Castes and Scheduled Tribes population constitutes nearly one quarter of the total population. Concerted efforts have been planned to uplift social and economic status of schedule castes and Scheduled Tribes. The basic requirement of these population is the availability of data on marital status, education, age, sex, economy, fertility are being collected from 1951 onwards in each census and these information are provided to administrators and planners to evolve suitable plans and welfare measures for the development of SC-ST population. Figure 8 depicts the thematic map of SC – ST population in 16 mandals. A blue bar indicates ST population and green bar signifies SC population. The bars are scaled in proportion to the data values represented for each area. It is clear from the plot that SC population is much higher as compared to ST population in the study area. The mandals like Thondur, Veerapunayunipalle, Chakrayapet, Vemula, Vempalle, Talupula,

Figure 8. Thematic representation of SC and ST population in the study area

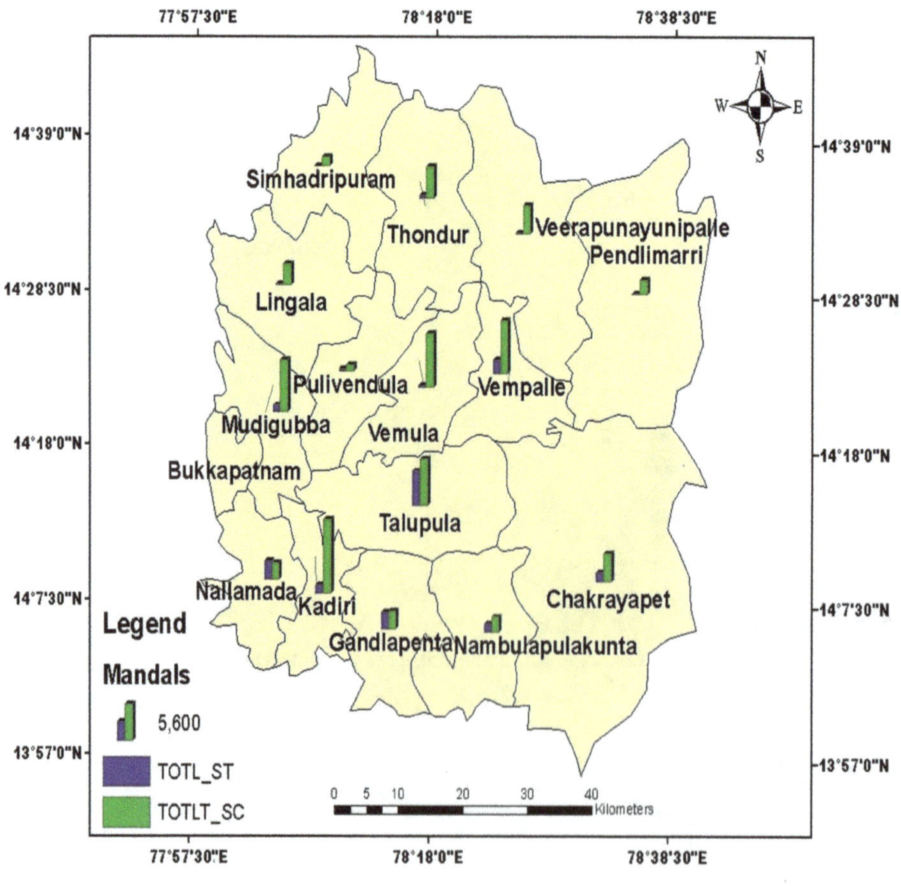

Mudigubba, Kadiri are having high SC population and mandals like Simhadripuram, Pendlimarri, Veerapunayunipalle, Lingala, Vemula, Thondur, Pulivendula having low ST population.

Generation of Query Based on GIS Database

One of the significant tools is to generate queries based on the attribute data. Generation of queries can be done in two ways: i) spatial queries, and ii) attribute queries using GIS database.

Village Level Query About Population Literate and Number of House Holds

Query: Number of House Holds (persons/sq. km) > 100 AND Population Literate > 200)

Using query based on GIS database we can able to identify the number of literate within the given area. Hence DSS will be very helpful in taking decisions for the village development through the knowledge of literates within the provided extent. The above query depicts to locate the villages where total population literate per sq km are more than 100 AND in the same villages number of House Holds are more than 100. Out of 163 villages, 150 villages satisfy the above query. Once the query is generated, information regarding the villages is obtained just by click on the village. More complex query is also be generated. Likewise, many queries are generated based on the requirements of the user and the attributed data. Figure 9 shows village level query on population literacy and number of households.

Village Level Query About Population Density and Number of ST Population

Query: Population Density (persons/sq. km) > 20 AND ST Population < 100)

Using DSS total number of population density and addition to that total number of ST population in a village can be determined easily so as to uplift their social and economic status by implementing various plans and measures for the village development and economic growth by analyzing population density because the country's economic growth were uplifted through the advance of human knowledge. The above query depicts to locate the villages, where population density are less than 20 and in the same villages number of ST population are less than 100. However, in this case, ST Population is of about 5.85% and population density is of about 4.91%. Out of 163 villages 114 villages satisfies the query. Figure 10 shows village level query on population density and no. of ST population.

Figure 9. Village level query on population literate and no. of house holds

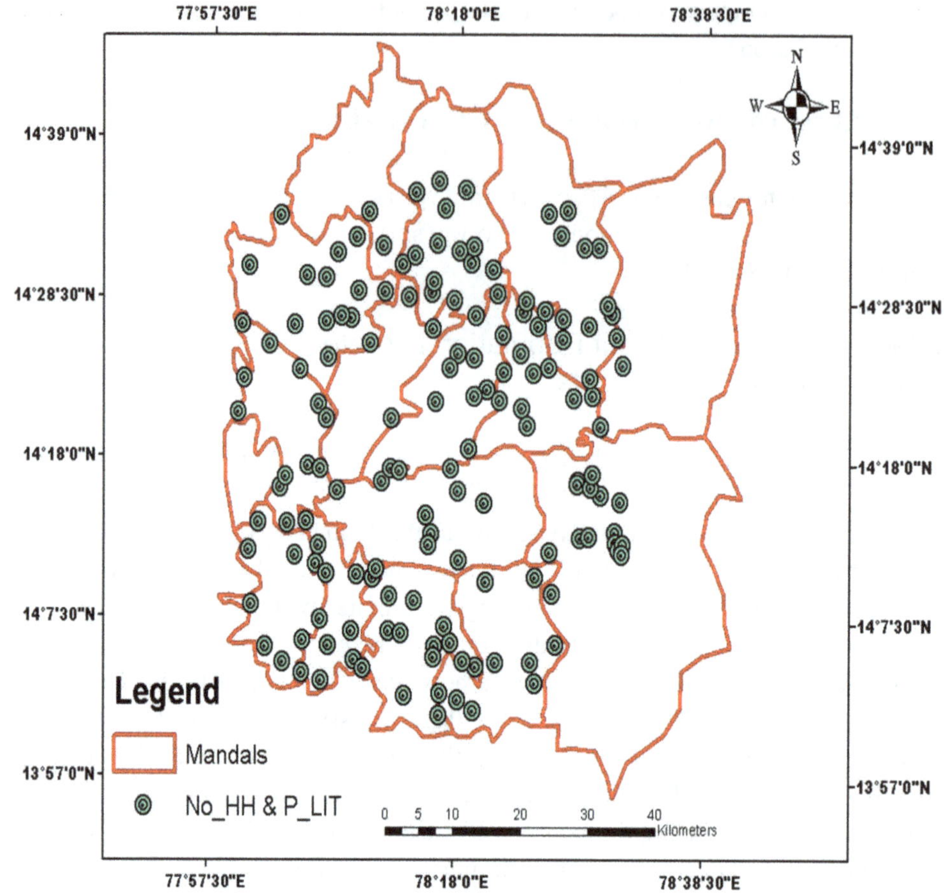

Village Level Query on Population Illiterate and Non-Working Population

Query: Population Illiterate > 100 AND Non-working Population Rate > 200

Using query based on GIS database we can able to identify the number of illiterate and Non-Working Population within the given area. The reasons for illiterate are due to poverty and physically disabled persons such as blindness, mental retardation etc. Government has reduced its expenditure on education from 13 percent in 1999 to 10 percent in 2010. Using census data, those information's are provided to administrators and planners to evolve suitable plans and welfare measures and providing facilities to the disabled person, providing proper education

Figure 10. Village level query on Population Density and no. of ST population

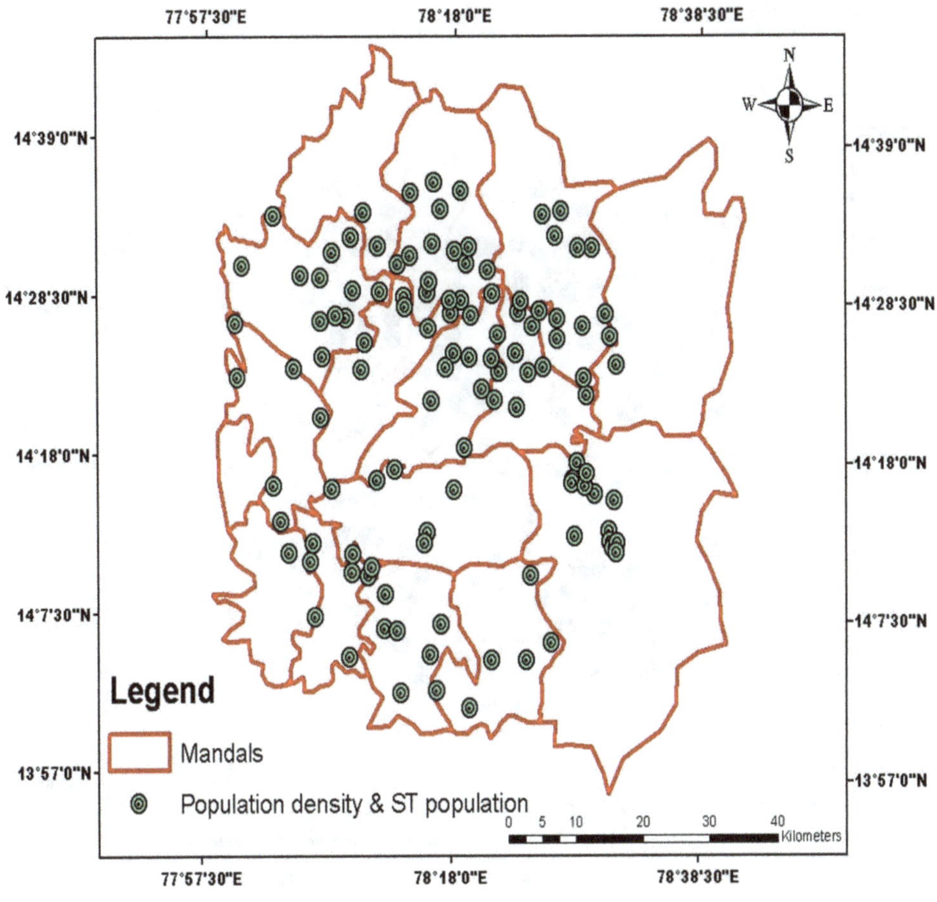

for the development of individuals enabling them to understand their political, social environment better and respond to it, thus DSS play a key role in solving problems and helpful in taking decisions based on GIS database. This query is generated to identify the villages where number of population illiterate is more than 100 and non-working population rate is greater than 200. The query is significant from social studies point of view. Out of 163 villages, 148 villages satisfy the above query. Once the query is generated, information regarding the villages are to be obtained just by clicking on the village. Figure 11 shows village level query on population illiterate and non-working population.

Figure 11. Village level queries on Population Illiterate and non-working population

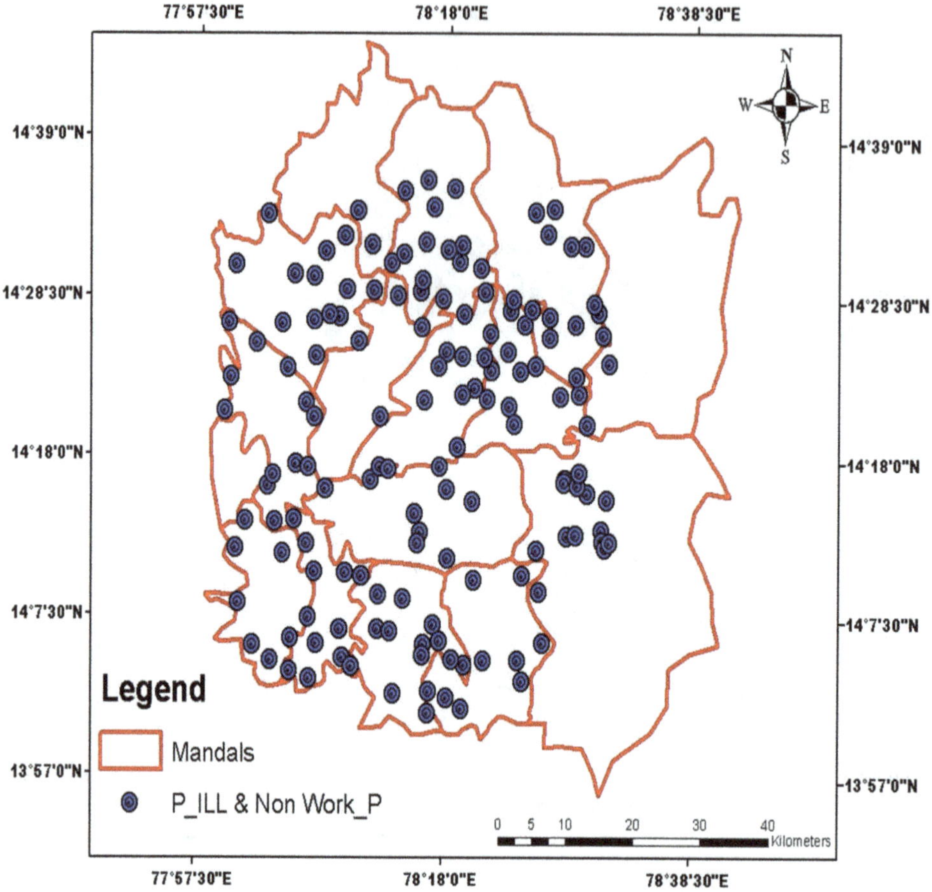

Village Level Combined Query of Multiple Attributes

Query: Population Density > 20 AND Number of House Holds > 100 AND Population Literate > 200

Using query based on GIS database we can able to identify total number of populations and among them total number of people are educated, and number households in a village, can be determined easily. So such information's will be very helpful in planning, implementing and taking certain decisions at emergency time and so as to improve the development of economy. The present query combines two previous queries and produce results in Figure 12. This query is generated to identify those villages where population density is more than 20 and number of households are more than 100 and population Literate are more than 200. It may be noted that out of 163 villages, 150 villages satisfied the above query. The Arc

Figure 12. Village level queries on combination query

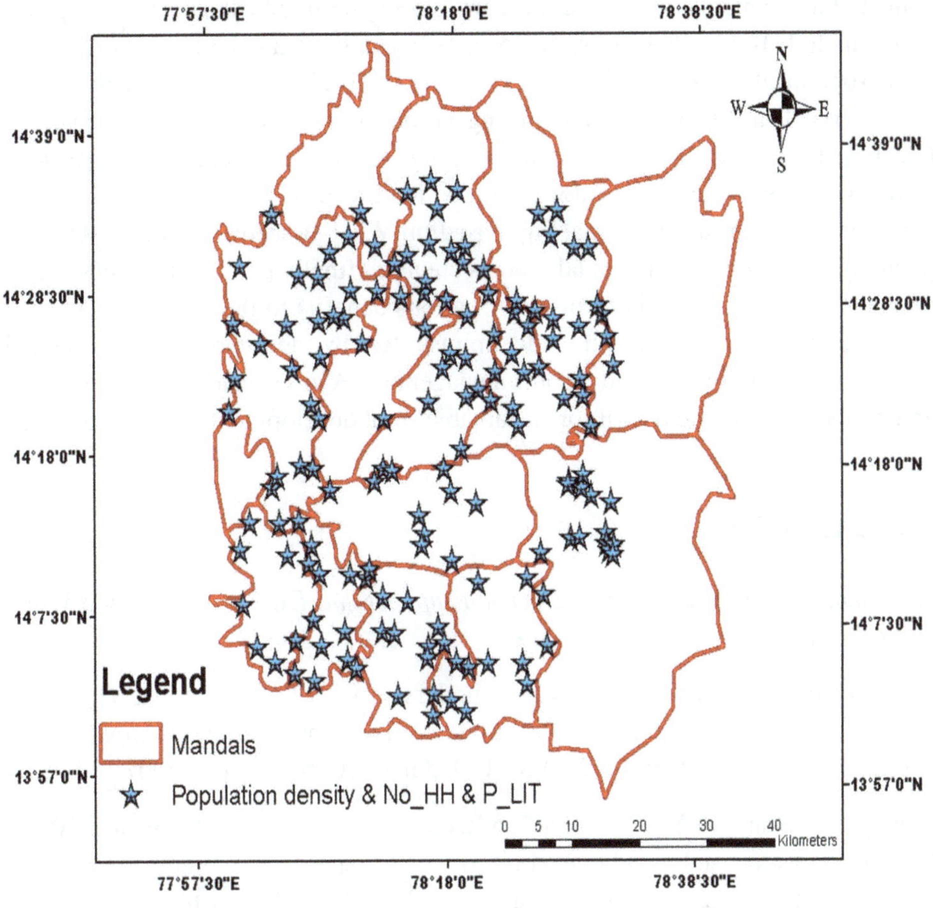

Info software has various conjuction operators such as AND, OR, NOR, XOR and others. Hence, complex combination of queries can also be built depending on the user requirement.

CONCLUSION

As the village maps are very old as they have undergone various stages of manual settlement results in degradation of quality. Therefore, updated information of village level maps plays a vital role to take good decisions for the administrator. So, an attempt has made to inbuilt features of various decisions at Anantapur and Kadapa districts by integrating various layers of information both spatial and attribute at village level for sustainable development and monitoring using remote sensing and

GIS techniques. DSS match practical and social expectations would be of great value to the planners and emergency team. DSS system for village level using GIS model include field survey to obtain location coordinates and collecting the census data to obtain attribute data, digitization of various GIS layers, generation of various queries based on attribute on data offering various possibilities for decision making. GIS based decision support System for village level has several advantages such as timely and easy access to information on land records will make faster, easier, and more effective decision making, based on What-If scenario studies strategic planning can be done, conceptual models creation, finally patterns presentation by thematic mapping. GIS platforms offer various benefits to the stake holders such as planners, policy makers and land administrators by the improved, efficient and effective methodology of land records management. A village based DSS using GIS and remote sensing are useful for sustainable rural development of the territories.

REFERENCES

Adelman, L. (1992). *Evaluating Decision support and Expert system*. John Wiley and sons Inc.

Adinarayana, J. S., Tewari, A. G., & Sundarsan, D. (2008). Gramya Vikas – A Distributed Collaboration Model for Rural Development Planning. *Computers and Electronics in Agriculture*, *62*(2), 128–140. doi:10.1016/j.compag.2007.12.008

Asadi, S. S., Raju, M. V., Vasantha Rao Reddy, B. V. T., & Anji Reddy, M. (2012). Creation of web based mandal level information system using remote sensing and GIS and visual basic program - A model study. *IACSIT International Journal of Engineering and Technology*, *3*(6), 361–372.

Hareesh, J., & Kumari, V. V. (2015). Decision Tree Based Information System for Improving School Facilities. *International Journal of Science and Research*, *4*(10), 2013–2016.

Kodge B.G., Hiremath P.S. (2012). Web based Geo-Spatial and Village Level Information Extraction System using FOSS. *Journal of Advances in Information Technology, 3*(4), 222-227. doi:10.4304/jait.3.4.222-227

Padma, G.V., Ramireddy, P.V., Tata Babu, C., Ramana Murty, M.V., & Prasada Rao, G. (2015). A Geospatial frame work for mapping of approximate cadastral sub divisions in Joniganuru village, santhipuram mandal, chittoor district, Andhra Pradesh - An Initiative. *International Journal of Engineering Sciences & Research Technology, 4*(7).

Patil, Y. M. (2014). GIS Enabled Tool for Effective Land Records Management. *International Journal on Recent and Innovation Trends in Computing and Communication*, *2*(6), 1627–1631.

Ravindran, A., & Jaishankar, S. (2006). *GIS Based Information System for Village Level planning*. Academic Press.

Rayed, C. A. (2012). Using GIS for Modeling a Spatial DSS for Industrial Pollution in Egypt. *American Journal of Geographic Information System*, *1*(4), 100–104. doi:10.5923/j.ajgis.20120104.01

Reddy, G.P., Reddy, M.N., Sontakki, B.S., & Kumar, K .V. (2011). A Framework of Participatory Geo-Spatial Information System for Micro Level Planning – A Case Study in Aquaculture. *Ind. Jn. of Agri. Econ, 66*(3).

Santosh, K., Lakshmi, T. V., & Reddy, M. A. (2009). *Development Of remote Sensing and GIS Based Information System For Village Level Planning - A Case Study Of Addanki Mandal, Prakasam District, Andhra Pradesh, India.* American Geophysical Union, Spring Meeting.

Sitender, S., & Kumar, R. (2012). Village Information System –A Case Study of Muklan Village, Hisar, Haryana (INDIA). *International Journal of Research in Social Sciences*, *2*(2).

Subbu Lakshmi, E., & Yarrakula, K. (2017). Comparative analysis of digital elevation models: A case study around Madduleru River. *Indian Journal of Geo Marine Sciences*, *46*(7), 1339–1351.

Subbu Lakshmi, E., & Yarrakula, K. (2016). Monitoring Land use Land Cover Changes using Remote Sensing and GIS Techniques: A Case Study around Papagni River, Andhra Pradesh, India. *Indian Journal of Ecology*, *43*(2), 383–387.

Vijay Krishna, A. (2003). *Records Management Guide* (vol. 2). Academic Press.

Vries, F. W. T. P., Teng, P., & Metselaar, K. (1991). Systems approaches for agricultural development. *Proceedings of the International Symposium on Systems Approaches for Agricultural Development*, 2–6.

Yadav, L. R., & Singh, R. S. (2009). GIS Applications in Statistical System and Micro Level Planning Strengthening. *10th ESRI India User Conference, Geography in Action*.

Compilation of References

Aanensen, D. M., Huntley, D. M., Feil, E. J., & Spratt, B. G. (2009). EpiCollect: Linking smartphones to web applications for epidemiology, ecology and community data collection. *PLoS One*, *4*(9), e6968. doi:10.1371/journal.pone.0006968 PMID:19756138

Abdel-Fadeel, M.M., Saad, S.K., & Omran, S. (2013). Opportunities and challenges of using GIS in sustainable tourism development: The case of Egypt. *Conference of Egypt-Thailand Relations Chiang Mai Rajabbat University*.

Abomeh, O.S., & Nuga, O.B. (2013). Utilization of GIS technology for tourism management in Victoria Island Lagos. *European Scientific Journal, 9*(3).

Adelman, L. (1992). *Evaluating Decision support and Expert system.* John Wiley and sons Inc.

Adinarayana, J. S., Tewari, A. G., & Sundarsan, D. (2008). Gramya Vikas – A Distributed Collaboration Model for Rural Development Planning. *Computers and Electronics in Agriculture*, *62*(2), 128–140. doi:10.1016/j.compag.2007.12.008

Adomavicius, G., & Tuzhilin, A. (2015). Context-aware recommender systems. In Recommender systems handbook (pp. 191-226). Springer US. doi:10.1007/978-1-4899-7637-6_6

Aggarwal, C. C. (2016a). *An Introduction to Recommender Systems, Recommender Systems.* Springer.

Aggarwal, C. C. (2016b). *Neighborhood-Based Collaborative Filtering, Recommender Systems.* Springer.

Aggarwal, C. C. (2016c). *Recommender systems.* Springer. doi:10.1007/978-3-319-29659-3

Agrawal, S., & Gupta, R. D. (2014). Development and comparison of open source based Web GIS Frameworks on WAMP and Apache Tomcat Web Servers. *The International Archives of the Photogrammetry, Remote Sensing and Spatial Information Sciences*, *40*(4), 1.

Aji, A., Wang, F., Vo, H., Lee, R., Liu, Q., Zhang, X., & Saltz, J. (2013, August). Hadoop GIS. *Proceedings of the VLDB Endowment International Conference on Very Large Data Bases*, *6*(11), 1009–1020. doi:10.14778/2536222.2536227 PMID:24187650

Akça, H. & Esengün, K. (2003). Coğrafi Bilgi Sistemlerinin Tarım Ekonomisi Alanında Kullanım Olanakları. *TKK Ekin Dergisi*, 25.

Akukwe Thecla, I. (n.d.). Determinants of Flooding in Port Harcourt Metropolis. *Journal of the Humanities and Social Sciences*, *19*(11), 62–72.

Albayrak, A. (2012). *Alternatif turizm*. Ankara: Detay Yayıncılık.

Albuquerque, H. C. D. C., Martins, F. M. C. P. F., Raposo, R. M. D. A., Cardoso, L. M. T. G., Beça Pereira, P. M. D. S., & Dias, P. (2016). Construction of a web-based geographical information system–the case of "Ria de Aveiro" region. *Anatolia*, *27*(1), 71–81. doi:10.1080/13032917.2015.1083210

Ali, T. A. T., Saeed, R. A., & Fageeri, S. O. (2017). Web-based GIS Business Hotels Tourism Sites in Khartoum, Sudan. In *Communication, Control, Computing and Electronics Engineering (ICCCCEE), 2017 International Conference on* (pp. 1-5). IEEE.

Al-Otaibi, S. T., & Ykhlef, M. (2012). A survey of job recommender systems. *International Journal of Physical Sciences*, *7*(29), 5127–5142. doi:10.5897/IJPS12.482

AlQeed, M. A., Bazazo, I. K., Hasoneh, A. I., & Alqaid, B. A. (2010). Using Geographic Information System to Visualize Travel Patterns and Market Potentials of Petra City in Jordan. *International Journal of Marketing Studies*, *2*(2), 144–159.

Ames, D. P., Horsburgh, J. S., Cao, Y., Kadlec, J., Whiteaker, T., & Valentine, D. (2012). HydroDesktop: Web services-based software for hydrologic data discovery, download, visualization, and analysis. *Environmental Modelling & Software*, *37*, 146–156. doi:10.1016/j.envsoft.2012.03.013

Anand, S., & Sen, A. (1997). Concepts or Human Development and Poverty: A Multidimensional Perspective. *United Nations Development Programme, Poverty and human development: Human development papers*, 1-20.

Anderson, C., & Hiralall, M. (2009). *Recommender systems for e-shops*. Business Mathematics and Informatics paper.

Anderson, V. M. (2004). *Developing Integrated Object-Oriented Conception of Geomarketing as a Tool for Promotion of Regional Sustainable Development: The Case Study of Ukraine*. Research Paper, 2004-7.

Annual Report 2009–10. (n.d.). Ministry of Tourism, Government of India.

Anumba, C., Dainty, A. R. J., Ison, S. G., & Sergeant, A. (2005). The application of GIS to construction labour market planning. *Construction Innovation*, *5*(4), 219–230. doi:10.1108/14714170510815267

Applebaum, W. (1954). Marketing Geography. *American Geography: Inventory and Prospect*, 245-251.

Applebaum, W. (1965). Can store location research be a science? *Economic Geography*, *41*(3), 234–237. doi:10.2307/141900

Applebaum, W. (1966). Guidelines for a Store-Location Strategy Study. *Journal of Marketing*, *30*(4), 42–45. doi:10.2307/1249498

Arnold, J. G., Srinivasan, R., Muttiah, R. S., & Williams, J. R. (1998). Large area hydrologic modeling and assessment part I: Model development. *JAWRA Journal of the American Water Resources Association*, *34*(1), 73–89. doi:10.1111/j.1752-1688.1998.tb05961.x

Arundel, J., Winter, S., Gui, G., & Keatley, M. (2016). A web-based application for beekeepers to visualise patterns of growth in floral resources using MODIS data. *Environmental Modelling & Software*, *83*, 116–125. doi:10.1016/j.envsoft.2016.05.010

Asadi, S. S., Raju, M. V., Vasantha Rao Reddy, B. V. T., & Anji Reddy, M. (2012). Creation of web based mandal level information system using remote sensing and GIS and visual basic program - A model study. *IACSIT International Journal of Engineering and Technology*, *3*(6), 361–372.

Asanov, D. (2011). *Algorithms and methods in recommender systems*. Berlin, Germany: Berlin Institute of Technology.

Ashley, C., Roe, D., & Goodwin, H. (2001). *Pro-poor Tourism Strategies: Making Tourism Work for the Poor. A Review of Experience*. London: Overseas Development Institute.

Athanasis, N., Karagiannis, F., Palaiologou, P., Vasilakos, C., & Kalabokidis, K. (2015). AEGIS App: Wildfire Information Management for Windows Phone Devices. *Procedia Computer Science*, *56*, 544–549. doi:10.1016/j.procs.2015.07.249

Australian Department of the Environment and Energy (2008). *National Wilderness Inventory*. Department of the Environment and Energy.

Avdimiotis, S., Mavrodontis, T., Dermetzopoulos, A.S., & Riavoglou, K. (2006). GIS applications as a tool for tourism planning and education: A case study of Chalkidiki. *Turizam: znanstveno-stručničasopis, 54*(4), 405-413.

Avdimiotis, S., & Christou, E. (2006). GIS Applications in Tourism Planning: A Tool for Sustainable Development Involving Local Communities. *23rd Euro Chrie Conference*.

Bahaire, T., & Elliott-White, M. (1999). The application of geographical information systems (GIS) in sustainable tourism planning: A review. *Journal of Sustainable Tourism*, *7*(2), 159–174. doi:10.1080/09669589908667333

Bahramian, Z., Abbaspour, R. A., & Claramunt, C. (2017). *A Cold Start Context-Aware Recommender System for Tour Planning Using Artificial Neural Network and Case Based Reasoning*. Mobile Information Systems; doi:10.1155/2017/9364903

Balabanović, M., & Shoham, Y. (1997). Fab: Content-based, collaborative recommendation. *Communications of the ACM*, *40*(3), 66–72. doi:10.1145/245108.245124

Baltrunas, L. (2011). *Context-aware collaborative filtering recommender systems*. The Department of Computer Science, Free University of Bozen-Bolzano.

Baltrunas, L., Makcinskas, T., & Ricci, F. (2010, September). Group recommendations with rank aggregation and collaborative filtering. In *Proceedings of the fourth ACM conference on Recommender systems* (pp. 119-126). ACM. doi:10.1145/1864708.1864733

Barik, R. K., Das, P. K., & Lenka, R. K. (2016). Development and implementation of SOA based SDI model for tourism information infrastructure management web services. In *Cloud System and Big Data Engineering (Confluence), 2016 6th International Conference* (pp. 748-753). IEEE. doi:10.1109/CONFLUENCE.2016.7723633

Barik, R. K., Samaddar, A. B., & Gupta, R. D. (2009). Investigations into the Efficacy of Open Source GIS Software. *Map World Forum, on Spatial Technology for Sustainable Planet Earth.*

Barik, R. K., & Samaddar, A. B. (2014a). Service Oriented Architecture Based SDI Model for Education Sector in India. *Proceedings of the International Conference on Frontiers of Intelligent Computing: Theory and Applications (FICTA)*, 555-562. doi:10.1007/978-3-319-02931-3_63

Barik, R. K., & Samaddar, A. B. (2014b). Service Oriented Architecture Based SDI Model for Mineral Resources Management in India. *Universal Journal of Geosciences (Prague)*, *2*, 1–6.

Barik, R. K., Samaddar, A. B., & Samaddar, S. G. (2011). Service Oriented Architecture based SDI Model for Geographical Indication Web Services. *International Journal of Computers and Applications*, *25*, 42–49. doi:10.5120/3017-4077

Baviskar, M. R. (2017). Designing and Developing a GIS Database for Tourism in Aurangabad: A Review. *International Journal of Computers and Applications*, *165*(5).

Bazazo, I. K., & Hasoneh, A. I. (2010). Using Geographic Information System to visualize travel patterns and market potentials of Petra City in Jordan. *International Journal of Marketing Studies*, *2*(2), 144.

Belussi, Catania, & Migliorini. (2012). Approximate queries for spatial data. In *Advanced Query Processing: Vol. 1. Issues and Trends*. Springer Science & Business Media.

Ben Brahim, M., Drira, W., Filali, F., & Hamdi, N. (2016). Spatial data extension for Cassandra NoSQL database. *Journal of Big Data*, *3*(1), 11. doi:10.1186/s40537-016-0045-4

Benkirane & Kettani. (n.d.). Retrieved from www.aui.ma/personal/~D.Kettani/courses/gis/GDB-benkirane.ppt

Bensghir, T. K., & Akay A. (2006). Bir Kamu Politika Aracı Olarak Coğrafi Bilgi Sistemleri (CBS): Türkiye'de Belediyelerin CBS Uygulamalarının Değerlendirilmesi. *Çağdaş Yerel Yönetimler, 15*(1), 31-46.

Berka, T., & Plößnig, M. 2004. Designing recommender systems for tourism. *ENTER 2004: 11th International Conference on Information Technology in Travel & Tourism.*

Berry, B. J. L. (1967). *Geography of Market Centers and Retail Distribution*. Englewood Cliffs, NJ: Prentice Hall.

Berry, J. K. (1987). Fundamental operations in computer-assisted map analysis. *International Journal of GIS*, *1*, 119–136.

Berry, J. K. (1991). GIS in island resource planning: A case study in map analysis. *Geographical Information Systems*, *2*, 285–295.

Bhaidkar, Y., Bhagwat, P., Bhalere, P., Gujar, R., & Walunj, S. (2015). *Tourist place recommendation system*. Academic Press.

Bhattacharya, G., & Haldar, S. K. (2014). Trend, Differential and Determinants of Deprivation of Reproductive and Child Health in the Districts of West Bengal, India. *Journal of Health Management*, *16*(1), 93–112. doi:10.1177/0972063413518686

Birking, M., Clarke, G., Clarke, M., & Wilson, A. (1996). *Intelligent GIS: Location Decisions and Strategic Planning*. Cambridge, UK: GeoInformation International.

Black, S., Powers, G., & Roche, M. (1994). A Gis-based approach to community analysis for targeted marketing. *Economic Development Review (Schiller Park, Ill.)*, *12*(2), 69–73.

Bobadilla, J., Ortega, F., Hernando, A., & Gutiérrez, A. (2013). Recommender systems survey. *Knowledge-Based Systems*, *46*, 109–132. doi:10.1016/j.knosys.2013.03.012

Bobov. (n.d.). *Spatial data Visualization*. Retrieved from https://portal.opengeospatial.org/files/?artifact_id=73214

Boonchieng, E., Boonchieng, W., Senaratana, W., & Singkaew, J. (2014). Development of mHealth for public health information collection, with GIS, using private cloud: A case study of Saraphi district, Chiang Mai, Thailand. *Computer Science and Engineering Conference (ICSEC), 2014 International*, 350-353. doi:10.1109/ICSEC.2014.6978221

Boori, M. S., Voženílek, V., & Choudhary, K. (2015). Land use/cover disturbance due to tourism in Jeseniky Mountain, Czech Republic: A remote sensing and GIS based approach. *The Egyptian Journal of Remote Sensing and Space Sciences*, *18*(1), 17–26. doi:10.1016/j.ejrs.2014.12.002

Borràs, J., Moreno, A., Valls, A., Ferré, M., Ciurana, E., Salvat, J., . . . Anton-Clavé, S. (2012). Uso de técnicas de inteligencia artificial para hacer recomendaciones enoturísticas personalizadas en la Provincia de Tarragona. Academic Press.

Borràs, J., de la Flor, J., Pérez, Y., Moreno, A., Valls, A., Isern, D., ... Anton-Clavé, S. (2011). *SigTur/E-Destination: a system for the management of complex tourist regions. In Information and Communication Technologies in Tourism 2011* (pp. 39–50). Springer. doi:10.1007/978-3-7091-0503-0_4

Borràs, J., Moreno, A., & Valls, A. (2014). Intelligent tourism recommender systems: A survey. *Expert Systems with Applications*, *41*(16), 7370–7389. doi:10.1016/j.eswa.2014.06.007

Bowler, B. (2014). How Spiritual Tourism Might Change the World. *The Huffinton Post*, *2*, 13.

Boyd, S., & Butler, R. (1996). Seeing the forest through the trees: Using GIS to identify potential ecotourism sites in Northern Ontario. In L. C. H. Harrison & W. J. Husbands (Eds.), *Practicing Responsible Tourism: International case studies in tourism planning, policy and development* (pp. 380–403). New York: John Wiley and Sons.

Brinkhoff, T. (2008). Supporting mobile GIS applications by spatial web services. *The International Archives of the Photogrammetry, Remote Sensing and Spatial Information Sciences*, *37*, 865–870.

Brody, R. (1999). Geographic Information Systems. *Journal of Business & Finance Librarianship*, *5*(1), 3–18. doi:10.1300/J109v05n01_02

Brovelli, M. A., Minghini, M., Moreno-Sanchez, R., & Oliveira, R. (2016). Free and open source software for spatial applications (FOSS4G) to support Future Earth. *International Journal of Digital Earth*, 1–19.

Brovelli, M. A., Minghini, M., & Zamboni, G. (2014). Public participation GIS: A FOSS architecture enabling field-data collection. *International Journal of Digital Earth*, *8*(5), 345–363. doi:10.1080/17538947.2014.887150

Brovelli, M. A., Minghini, M., & Zamboni, G. (2016). Public participation in GIS via mobile applications'. *ISPRS Journal of Photogrammetry and Remote Sensing*, *114*, 306–315. doi:10.1016/j.isprsjprs.2015.04.002

Buhalis, D., & Amaranggana, A. (2014). Smart tourism destinations. *Proceedings of the international conference: Information and communication technologies in tourism 2014*, 553-564.

Buhalis, D., & Law, R. (2008). Progress in information technology and tourism management: 20 years on and 10 years after the internet - The state of eTourism research. *Tourism Management*, *29*(4), 609–623. doi:10.1016/j.tourman.2008.01.005

Bunruamkaew, K., & Murayama, Y. (2012). Land use and natural resources planning for sustainable ecotourism using GIS in Surat Thani. *Thailand Sustainability*, *4*(3), 412–429. doi:10.3390/su4030412

Burke, R. (2002). Hybrid recommender systems: Survey and experiments. *User Modeling and User-Adapted Interaction*, *12*(4), 331–370. doi:10.1023/A:1021240730564

Burrough, P. (1986). *Principles of geographical information systems for land resources assessment*. Oxford, UK: Clarendon Press.

Buyya, R., Yeo, C. S., Venugopal, S., Broberg, J., & Brandic, I. (2009). Cloud computing and emerging IT platforms: Vision, hype, and reality for delivering computing as the 5th utility. *Future Generation Computer Systems*, *25*(6), 599–616. doi:10.1016/j.future.2008.12.001

Cassandra. (2017). *Contrasting Hadoop & Apache Cassandra.* Retrieved April 12, 2017 from https://www.datastax.com/nosql-databases/nosql-cassandra-and-hadoop

Castillo, L., Armengol, E., Onaindía, E., Sebastiá, L., González-Boticario, J., Rodríguez, A., ... Borrajo, D. (2008). SAMAP: An user-oriented adaptive system for planning tourist visits. *Expert Systems with Applications*, *34*(2), 1318–1332. doi:10.1016/j.eswa.2006.12.029

Catalog of Müze Asist. (n.d.). Retrieved July 15, 2017, from http://www.muzeasist.com

Cattell, R. B. (1966). The scree test for the number of factors. *Multivariate Behavioral Research*, *1*(2), 245–276. doi:10.1207/s15327906mbr0102_10 PMID:26828106

Çeken, H. (2003). *Küreselleşme, Yabancı Sermaye ve Türkiye Turizmi*. İstanbul: Değişim Yayınları.

Çeken, H., Karadağ, L., & Dalgın, T. (2007). Kırsal Kalkınmada Yeni Bir Yaklaşım Kırsal Turizm ve Türkiye'ye Yönelik Teorik Bir Çalışma. *Artvin Çoruh Üniversitesi Orman Fakültesi Dergisi*, *8*(1), 1–14.

Çelik, Z. (2005). *Planlı Dönemde Türkiye'deki Kırsal Kalkınma Politika ve Uygulamaları Üzerine Bir Değerlendirme, Planlama, TMMOB*. Ankara: Şehir Plancıları Odası Yayını.

Chang, Z. (2003). *Design and Implementation of distributed Web Services in Mobile GIS Application* (Master's Thesis). International Institute for Geo-information Science and Earth Observation, The Netherlands.

Chang, G., & Caneday, L. (2011). Web-based GIS in tourism information search: Perceptions, tasks, and trip attributes. *Tourism Management*, *32*(6), 1435–1437. doi:10.1016/j.tourman.2011.01.006

Chaudhuri, S., & Ray, N. (2015). Application of Web-Based Geographical Information System (GIS) In Tourism Development. In Encyclopedia of Information Science and Technology (3rd ed.; pp. 7026-7036). IGI Global.

Chen, H., & Xiao, K. (2011). The design and implementation of the geological data acquisition system based on mobile GIS. *Geoinformatics, 2011 19th IEEE International Conference* on, 1-6.

Chen, S., Owusu, S., & Zhou, L. (2013, September). Social Network Based Recommendation Systems: A Short Survey. In *Social Computing (SocialCom), 2013 International Conference on* (pp. 882-885). IEEE. doi:10.1109/SocialCom.2013.134

Cheng, C., Yang, H., King, I., & Lyu, M. R. (2012). Fused Matrix Factorization with Geographical and Social Influence in Location-Based. Social Network, the Twenty-Sixth AAAI Conference on Artificial Intelligence, Toronto, Ontario, Canada, July 22 – 26.

Cheng, C., Yang, H., Lyu, M. R., & King, I. (2013). Where You Like to Go Next: Successive Point-of-Interest Recommendation. Academic Press.

Cheng, C., Yang, H., King, I., & Lyu, M. R. (2012, July). *Fused Matrix Factorization with Geographical and Social Influence in Location-Based Social Networks* (Vol. 12). AAAI.

Chen, R. J. (2007). Geographic information systems (GIS) applications in retail tourism and teaching curriculum. *Journal of Retailing and Consumer Services*, *14*(4), 289–295. doi:10.1016/j.jretconser.2006.07.004

Chen, Z., Chen, N., Yang, C., & Di, L. (2012). Cloud computing enabled web processing service for earth observation data processing. *IEEE Journal of Selected Topics in Applied Earth Observations and Remote Sensing*, *5*(6), 1637–1649. doi:10.1109/JSTARS.2012.2205372

Chhetri, P., & Arrowsmith, C. (2008). GIS-based modelling of recreational potential of nature-based tourist destinations. *Tourism Geographies*, *10*(2), 233–257. doi:10.1080/14616680802000089

Chittick, W. (1992). Notes on Ibn al-`Arabi's influence in the Subcontinent. *The Muslim World*, *82*(1), 218–241.

Cho, E., Myers, S. A., & Leskovec, J. (2011). Friendship and mobility: user movement in location-based social networks. The 17th ACM SIGKDD International Conference on Knowledge Discovery and Data Mining, 1082-1090. doi:10.1145/2020408.2020579

Chunling, Y., Man, Y., & Jianwei, N. (2009). A Light Mobile GIS Framework Based on Mobile SVG. *Computer Science and Information Engineering, 2009 WRI World Congress on*, *7*, 616–619.

Clark, J. (1996). *Kalkınmanın Demokratikleşmesi. (Çev.) Serpil Ural.* Ankara: Türkiye Çevre Vakfı Yayını.

Coleman, D. J., Rajabifard, A., & Kolodziej, K. W. (2016). Expanding the SDI environment: Comparing current spatial data infrastructure with emerging indoor location-based services. *International Journal of Digital Earth*, *9*(6), 629–647. doi:10.1080/17538947.2015.1119207

Colombo-Mendoza, L. O., Valencia-García, R., Alor-Hernández, G., & Bellavista, P. (2017). *Special issue on context-aware mobile recommender systems*. Elsevier.

Conner, L. G., Ames, D. P., & Gill, R. A. (2013). HydroServer Lite as an open source solution for archiving and sharing environmental data for independent university labs. *Ecological Informatics*, *18*, 171–177. doi:10.1016/j.ecoinf.2013.08.006

Converse, P. D. (1949). New Laws of Retail Gravitation. *Journal of Marketing*, *14*(3), 379–384. doi:10.2307/1248191

Cook, W. J., Collins, S., Flynn, M. K., Guttman, M., Cohen, W., & Budiansky, S. (1994). 25 Breakthroughs That Are Changing the Way We Live and Work. *U.S. News & World Report*, *2*(May), 46–60.

Coşar, G. Ö., & Engindeniz, S. (2011). Tarım Arazilerinin Değerlemesinde Coğrafi Bilgi Sistemlerinden Yararlanma Olanakları. *Ege Üniversitesi Ziraat Fakültesi Dergisi*, *48*(3), 283–290.

Çoşgun, S. (2015). *Coğrafi bilgi sistemleri kullanılarak ekolojik faktörlerin bazı asli ağaç türlerinin konumsal dağılımına etkilerinin incelenmesi* (Unpublished master thesis). University of Süleyman Demirel, Isparta, Turkey.

Coyle, A. (2011). Interior library GIS. *Library Hi Tech*, *29*(3), 529–549. doi:10.1108/07378831111174468

Cranshaw, J., Toch, E., Hong, J., Kittur, A., & Sadeh, N. (2010). Bridging the gap between physical location and online social networks. The 12th ACM International Conference on Ubiquitous Computing, 119-128. doi:10.1145/1864349.1864380

da Cruz Albuquerque, H. C., Cardoso Pedrosa Martins, F. M., & Tomás Galiza Cardoso, L. M. (2014). *Geographical information systems, tourism*. Springer International Publishing.

Daly, E. M., & Geyer, W. (2011). Effective event discovery: using location and social information for scoping event recommendations. The Fifth ACM Conference on Recommender Systems, 177-280, 277-280. doi:10.1145/2043932.2043982

DataStax Apache Cassandra Documentation. (2016). Retrieved from http://www.odbms.org/wp-content/uploads/2013/11/cassandra10.pdf

de Abreu, F. (2014). Development of a Mobile Mapping Solution for Spatial Data Collection Using Open-Source Technologies. *Procedia Technology*, *16*, 481–490. doi:10.1016/j.protcy.2014.10.115

Delipetrev, B., Jonoski, A., & Solomatine, D. P. (2014). Development of a web application for water resources based on open source software. *Computers & Geosciences*, *62*, 35–42. doi:10.1016/j.cageo.2013.09.012

Directorate of Istanbul Historical Areas. (n.d.). *World Heritage Areas*. Retrieved from http://www.alanbaskanligi.gov.tr/unesco_dunya_miras_alanlar%C4%B1.html

Dittrich, Arnulfo, & Ruiz. (2012). Efficient Big Data Processing in Hadoop MapReduce. *Proceedings of the VLDB Endowment*, *5*(12).

Dogru, A. O., Ulugtekin, N. N., & Alkoy, S. (2007, August). GIS applications on epidemiology with cartographic perspective in Turkey. *23rd International Cartographic Conference*.

Dong, C., Liu, F., Wang, H., & Chen, F. (2010). Application research of mobile GIS in forestry informatization. *Computer Science and Education (ICCSE), 2010 5th IEEE International Conference on*, 351-355.

Donnelly, F. P. (2010). Evaluating open source GIS for libraries. *Library Hi Tech*, *28*(1), 131–151. doi:10.1108/07378831011026742

DPT Ulusal Kalkınma Stratejisi. (2006). Retrieved April 5, 2017, from www.ekonomi.name

Dubey, N. K. (2015). Efficient Approach To Find Nearest Location Using Geohashing On Hadoop And Pig. *International Journal Of Engineering Research-Online*, *3*(3), 771–777.

Dujmović, J. J., & Nagashima, H. (2006). LSP method and its use for evaluation of Java IDEs. *International Journal of Approximate Reasoning*, *41*(1), 3–22. doi:10.1016/j.ijar.2005.06.006

Dunteman, G. H. (1989). *Principal components analysis (No. 69)*. Sage. doi:10.4135/9781412985475

Dye, A. S., & Shaw, S. L. (2007). A GIS-based spatial decision support system for tourists of Great Smoky Mountains National Park. *Journal of Retailing and Consumer Services*, *14*(4), 269–278. doi:10.1016/j.jretconser.2006.07.005

Eberhart, R. C., Shi, Y., & Kennedy, J. (2001). *Swarm intelligence*. Elsevier.

Egresi, I. O., & Kara, F. (2016). Residents' attitudes to tourists visiting their mosques: A case study from Istanbul, Turkey. *Journal of Tourism and Cultural Change*, 1–21. doi:10.1080/14766825.2016.1192182

Elliott, R. (2014). Geographic information systems (GIS) and libraries: Concepts, services and resources. *Library Hi Tech News*, *31*(8), 8–11. doi:10.1108/LHTN-07-2014-0054

Elliott-White, M. P., & Finn, M. (1997). Growing in Sophistication: The Application of Geographical Information Systems in Post-Modern Tourism Marketing. *Journal of Travel & Tourism Marketing*, *7*(1), 65–84. doi:10.1300/J073v07n01_05

Entidade Regional de Turismo da Região de Lisboa. (2015). *Plano Estratégico para o Turismo na Região de Lisboa 2015-2019*. Available online: http://www.ertlisboa.pt/pt/institucional/plano-estrategico-para-o-turismo-na-regiao-de-lisboa-2015-2019/

Erşen, G., Yüncü, H. R., Metin, T. C., & Çabuk, A. (2016). Gastronomi Turizminde Yeni Açılımlar: Karaburun Zeytin Rotalarının Coğrafi Bilgi Sistemleri İle Analizi. Balıkesir Gastronomi Turizmi Kongresi.

Evangelidis, K., Konstantinos, N., Stathis, M., & Constantine, P. (2014). Spatial services in the Cloud. *Computers & Geosciences*, *63*, 116–122. doi:10.1016/j.cageo.2013.10.007

Fadanhusi, J. T. (2011). Application of Geographical Information System (GIS) Technology to Tourism Management in Ile-Ife, Osun State, Nigeria. *Pacific Journal of Science and Technology*, *12*(2), 274–283.

Fajuvigbe, O., Balogun, V. F., & Obembe, O. M. (2007). Web-Based Geographic Information System (GIS) for Tourism in Oyo State, Nigeria. *Information Technology Journal*, *6*(5), 613–622. doi:10.3923/itj.2007.613.622

Farsari, Y. (2003, April). GIS-based support for sustainable tourism planning and policy making. In *Proceedings of the XII International Leisure and Tourism symposium* (pp. 3-4). Academic Press.

Farsari, Y., & Prastacos, P. (2004). GIS Applications in the planning and management of tourism. In A Companion to Tourism. Blackwell Publishing. doi:10.1002/9780470752272.ch47

Farsari, Y., Butler, R., & Prastacos, P. (2007). Sustainable tourism policy for Mediterranean destinations: Issues and interrelationships. *International Journal of Tourism Policy*, *1*(1), 58–78. doi:10.1504/IJTP.2007.013898

Fatemi, M., & Tokarchuk, L. (2013, September). A Community Based Social Recommender System for Individuals & Groups. In *Social Computing (SocialCom), 2013 International Conference on* (pp. 351-356). IEEE. doi:10.1109/SocialCom.2013.55

Feng, S., Li, X., Zeng, Y., Cong, G., Chee, Y. M., & Yuan, Q. (2015). Personalized Ranking Metric Embedding for Next New POI Recommendation. The 24th International Conference on Artificial Intelligence, 2069-2075, Buenos Aires, Argentina, July 25 - 31, 2069-207.

Fenza, G., Fischetti, E., Furno, D., & Loia, V. (2011). *A hybrid context aware system for tourist guidance based on collaborative filtering*. IEEE International Conference on Fuzzy Systems. doi:10.1109/FUZZY.2011.6007604

Fidan, H. (2009). Marketing Information System (MIS) and USAGE of Geographic Information System (GIS) in Marketing. *Journal of Yaşar University*, *4*(14), 2151–2171.

Fiduccia, A., Gugliermetti, F., Pagliaro, F., Mattoni, B., Nardecchia, F., & Canu, C. (2016). Web-Based Spatial Decision Support Systems to Monitor and Manage Coastal Environments. *18th Italian National Conference on Photonic Technologies*. doi:10.1049/cp.2016.0920

Find Restaurants with Geospatial Queries. (n.d.). Retrieved from https://docs.mongodb.com/manual/tutorial/geospatial-tutorial/

Fitzpatrick, C., & Maguire, D. J. (2000). GIS in Schools: Infrastructure, Methodology and Role. In GIS: A Sourcebook for Schools. Taylor & Francis.

Fox, Eichelberger, Hughes, & Lyon. (2013). *Spatio-temporal Indexing in Non-relational Distributed Databases*. Commonwealth Computer Research, Inc.

Freire, S., & Santos, T. (2012). Advancing GeoMarketing Analyses with Improved Spatio-temporal Distribution of Population at High Resolution. In *Proceedings of the European Conference on Information Management* (pp.100-108). Academic Press.

Fung, D. S., & Remsen, A. P. (1997). Geographic Information Systems Technology for Business Applications. *Journal of Applied Business Research*, *13*(3), 17–24. doi:10.19030/jabr.v13i3.5748

Gao, H., Tang, J., Hu, X., & Liu, H. (2013). *Exploring temporal effects for location recommendation on location-based social networks*. The 7th ACM conference on Recommender systems. doi:10.1145/2507157.2507182

Gao, H., Tang, J., Hu, X., & Liu, H. (2015). Content-Aware Point of Interest Recommendation on Location-Based. *Social Network. the Twenty-Ninth AAAI Conference on Artificial Intelligence, Austin, Texas, January 25 - 30*, 1721–1727.

Gao, S., & Gao, C. (2011). Design and Implementation of Tourism Geographical Information System Based on ArcIMS. *IEEE International Conference on Computer Science and Service System*, 3916-3919.

García-Crespo, A., Chamizo, J., Rivera, I., Mencke, M., Colomo-Palacios, R., & Gómez-Berbís, J. M. (2009). SPETA: Social pervasive e-Tourism advisor. *Telematics and Informatics*, *26*(3), 306–315. doi:10.1016/j.tele.2008.11.008

García-Palomares, J. C., Gutiérrez, J., & Mínguez, C. (2015). Identification of tourist hot spots based on social networks: A comparative analysis of European metropolises using photo-sharing services and GIS. *Applied Geography (Sevenoaks, England)*, *63*, 408–417. doi:10.1016/j.apgeog.2015.08.002

Gavalas, D., & Kenteris, M. (2011). A web-based pervasive recommendation system for mobile tourist guides. *Personal and Ubiquitous Computing*, *15*(7), 759–770. doi:10.1007/s00779-011-0389-x

Gavalas, D., Konstantopoulos, C., Mastakas, K., & Pantziou, G. (2014). Mobile recommender systems in tourism. *Journal of Network and Computer Applications*, *39*, 319–333. doi:10.1016/j.jnca.2013.04.006

Ge, M., Delgado-Battenfeld, C., & Jannach, D. (2010). *Beyond accuracy: evaluating recommender systems by coverage and serendipity*. The fourth ACM conference on Recommender systems, 257-260. doi:10.1145/1864708.1864761

GeoFile. (n.d.). *Everything in the Radius with MongoDB Geospatial Queries*. Retrieved from https://www.compose.com/articles/geofile-everything-in-the-radius-with-mongodb-geospatial-queries/

Geohash and its format. (2016). Retrieved from http://geohash.org/site/tips.html

Georis-Creuseveau, J., Claramunt, C., & Gourmelon, F. (2016). A modelling framework for the study of Spatial Data Infrastructures applied to coastal management and planning. *International Journal of Geographical Information Science*, *31*(1), 122–138. doi:10.1080/13658816.2016.1188929

Geospatial Queries. (n.d.). Retrieved from https://docs.gigaspaces.com/xap110/query-geospatial.html

Geray, C. (1974). *Planlı Dönemde Köye Yönelik Çalışmalar*. Ankara: Sevinç Matbaası.

Giles, W. (2003*)*. *GIS applications in tourism planning*. GIS 340-GIS Seminar. Retrieved from https://www.scribd.com/document/72126092/GIS-Application-in-Tourism-Planning

Girardin, F., Fiore, F. D., Ratti, C., & Blat, J. (2008). Leveraging explicitly disclosed location information to understand tourist dynamics: A case study. *Journal of Location Based Services*, *2*(1), 41–56. doi:10.1080/17489720802261138

Giuliani, G., Lacroix, P., Guigoz, Y., Roncella, R., Bigagli, L., Santoro, M., ... Lehmann, A. (2016). Bringing GEOSS Services into Practice: A Capacity Building Resource on Spatial Data Infrastructures (SDI). *Transactions in GIS*.

Gkatzoflias, D., Mellios, G., & Samaras, Z. (2013). Development of a web GIS application for emissions inventory spatial allocation based on open source software tools. *Computers & Geosciences*, *52*, 21–33. doi:10.1016/j.cageo.2012.10.011

Goldenberg, J., & Levy, M. (2009). *Distance is not dead: Social interaction and geographical distance in the internet era*. arXiv preprint arXiv:0906.3202

Gomes, J. L., Jesus, G., Rogeiro, J., Oliveira, A., da Costa, R. T., & Fortunato, A. B. (2017). An Innovative Web Platform for Flood Risk Management. Advances in Network Systems, 217-231. doi:10.1007/978-3-319-44354-6_13

Gong, S. (2009). Joining case-based reasoning and item-based collaborative filtering in recommender systems. Second International Symposium on Electronic Commerce and Security (ISECS'09), 40-42. doi:10.1109/ISECS.2009.172

Gong, Y.-j., Xu, R.-t., Zhang, J., & Liu, O. (2009). *A clustering-based adaptive parameter control method for continuous ant colony optimization*. IEEE International Conference on Systems, Man and Cybernetics (SMC 2009), 1827-1832. doi:10.1109/ICSMC.2009.5346726

Goodchild, M. F. (1989). Geographic Information Systems and Market Research. *Papers and Proceedings of Applied Geography Conferences* (*vol. 12*, pp. 1-8). Academic Press.

Goodchild, M. (2010). Twenty years of progress: GIScience in 2010. *Journal of Spacial Information Science*, *1*, 3–20.

Gribb, W. (1991). Integrating visitor surveys, environmental factors, and GIS into recreation planning: Site development at Grayrocks reservoir, Wyoming. GIS/LIS'91 Proceedings, 1, 177-86.

Grimshaw, D. J. (1993). *Bringing GIS into Business Geoinformation*. London: Longman.

GSDI Cookbook. (2004). *Developing spatial data infrastructure: The SDI Cookbook, 2004*. Available from: http://www.gsdi.org/docs2004/Cookbook/ cookbookV2.0.pdf

Guan, W. W., Bol, P. K., Lewis, B. G., Bertrand, M., Berman, M. L., & Blossom, J. C. (2012). WorldMap–a spatial framework for collaborative research. *Annals of GIS*, *18*(2), 121–134. doi: 10.1080/19475683.2012.668559

Gülçubuk, B. (1997). *Kırsal Kalkınmada TKV'nin Yeri*. Ankara: Yayınlanmamış Doktora Tezi, Ankara Üniversitesi Fen Bilimleri Enstitüsü.

Gunawardana, A., & Shani, G. (2015). *Evaluating recommender systems, Recommender Systems Handbook*. Springer. doi:10.1007/978-1-4899-7637-6_8

Gupta, A., & Merchant, P. S. (2016). Automated Lane Detection by K-means Clustering: A Machine Learning Approach. *Electronic Imaging*, *14*, 1–6. doi:10.2352/ISSN.2470-1173.2016.1.VDA-502

Hadoop Support. (2017). Retrieved from https://docs.datastax.com/en/cassandra/2.1/cassandra/configuration/configHadoop.html

Hae-Kyong, K., & Ki-Joune, L. (2005). A Framework for Dynamic Updates of Map Data in Mobile Devices. *Web and Wireless Geographical Information Systems, Springer*, *3833*, 66–77. doi:10.1007/11599289_7

Haines, G. H., Leonard, S. S., & Marcus, A. (1972). Maximum Likelihood Estimation of Central-City Food Trading Areas. *JMR, Journal of Marketing Research*, *9*(2), 154–159. doi:10.2307/3149948

Hamylton, S. M., & Prosper, J. (2012). Development of a spatial data infrastructure for coastal management in the Amirante Islands,Seychelles. *International Journal of Applied Earth Observation and Geoinformation*, *19*, 24–30. doi:10.1016/j.jag.2012.04.004

Han, X., Tian, L., Yoon, M., & Lee, M. (2012, November). A big data model supporting information recommendation in social networks. In *Cloud and Green Computing (CGC), 2012 Second International Conference on* (pp. 810-813). IEEE. doi:10.1109/CGC.2012.125

Haq, F., & Medhekar, A. (2015a). Islamic Tourism Destinations in India and Pakistan. In Islamic Tourism Trends, Challenges, and Opportunities in the Global Economy (pp.190-216). IGI Global.

Haq, F., & Medhekar, A. (2016). Is Spiritual Tourism an Innovation in Tourism for India and Pakistan? Business, Economics. In Leadership, Innovation and Entrepreneurship as Driving Forces of the Global Economy. Springer.

Haq, F. (2015). A Framework to Compare Historical and Modern Islamic Spiritual Tourism: Study of Algerian and Malaysian tourism. *European Journal of Business and Management*, *3*(1), 105–118.

Haq, F., & Jackson, J. (2009). Spiritual journey to Hajj: Australian and Pakistani experience and expectations. *Journal of Management, Spirituality & Religion*, *6*(2), 141–156. doi:10.1080/14766080902815155

Haq, F., & Medhekar, A. (2013). Branding Spiritual Tourism as an Innovation for Peace between India and Pakistan. *International Journal of Social Entrepreneurship and Innovation*, *2*(5), 404–414. doi:10.1504/IJSEI.2013.059312

Haq, F., & Medhekar, A. (2015b). Spiritual Tourism between India and Pakistan: A Framework for Business Opportunities and Threats. *World Journal of Social Sciences*, *5*(2), 190–200.

Hareesh, J., & Kumari, V. V. (2015). Decision Tree Based Information System for Improving School Facilities. *International Journal of Science and Research*, *4*(10), 2013–2016.

Harper, E. (2006). *Open Source Technologies in Web-based GIS and Mapping* (Master's Thesis). Northwest Missouri State University, Maryville, MO.

Hawelka, B., Sitko, I., Beinat, E., Sobolevsky, S., Kazakopoulos, P., & Ratti, C. (2014). Geo-located Twitter as proxy for global mobility patterns. *Cartography and Geographic Information Science*, *41*(3), 260–271. doi:10.1080/15230406.2014.890072 PMID:27019645

He, J., Li, X., Liao, L., Song, D., & Cheung, W. K. (2016). Inferring a Personalized Next Point-of-Interest Recommendation Model with Latent Behavior Patterns. The 13th AAAI Conference on Artificial Intelligence, 137-143.

He, L., Yue, P., Di, L., Zhang, M., & Hu, L. (2015). Adding spatial data provenance into SDI—a service-oriented approach. *IEEE Journal of Selected Topics in Applied Earth Observations and Remote Sensing*, *8*(2), 926–936. doi:10.1109/JSTARS.2014.2340737

Herlocker, J. L., Konstan, J. A., & Riedl, J. (2000). *Explaining collaborative filtering recommendations*. The 2000 ACM conference on Computer supported cooperative work, 241-250.

Herrera, A., San Martín, H., & Hernández, J. (2015). Perceived influence on behavior of user-generated content on social network sites: An empirical application in the hotel sector. *Revista Española de Investigación de Marketing ESIC*, *19*(1), 12–23. doi:10.1016/j.reimke.2014.09.001

Hess, R. L., Rubin, R. S., & Lawrence, A. W. Jr. (2004). Geographic information systems as a marketing information system technology. *Decision Support Systems*, *38*(2), 197–212. doi:10.1016/S0167-9236(03)00102-7

Heywood, I., Cornelius, S., & Carver, S. (2006). *An Introduction To Geographical Information Systems*. Pearson, Prentice Hall. Retrieved from http://www.urisa.org/

Hinze, A., & Junmanee, S. (2005). *Travel recommendations in a mobile tourist information system*. Academic Press.

Ho, P. S. Y., & McKercher, B. (2004). Managing heritage resources as tourism products. *Asia Pacific Journal of Tourism Research*, *9*(3), 255–266. doi:10.1080/1094166042000290655

Horozov, T., Narasimhan, N., & Vasudevan, V. (2006). Using location for personalized POI recommendations in mobile environments. International symposium on Applications and the internet (SAINT 2006). doi:10.1109/SAINT.2006.55

Hsu, C., & Powers, T. (2001). *Marketing Hospitality*. Hoboken, NJ: John Wiley & Sons.

Huang, Q., Yang, C., Liu, K., Xia, J., Xu, C., Li, J., ... Li, Z. (2013). Evaluating open-source cloud computing solutions for geosciences. *Computers & Geosciences*, 5941–5952.

Hu, B., & Ester, M. (2014). *Social topic modeling for point-of-interest recommendation in location-based social networks*. IEEE International Conference on Data Mining (ICDM), 845-850. doi:10.1109/ICDM.2014.124

Huff, D. L. (1964). Estimating a Trading Area. *Journal of Marketing*, *28*(3), 34–38. doi:10.2307/1249154

Hung, C.-C., Chang, C.-W., & Peng, W.-C. (2009). *Mining trajectory profiles for discovering user communities*. The 2009 International Workshop on Location Based Social Networks, 1-8. doi:10.1145/1629890.1629892

Husain, W., & Dih, L. Y. (2012). A framework of a personalized location-based traveler recommendation system in mobile application. *International Journal of Multimedia and Ubiquitous Engineering*, *7*, 11–18.

Hussain, M., Arsalan, M. H., & Mehdi, M. R. (2006). Role of GIS in public health management in Pakistan. In *Advances in Space Technologies, 2006 International Conference on* (pp. 102-107). IEEE. doi:10.1109/ICAST.2006.313807

Idrees, M. O. I. (2015). Challenges in coastal spatial data infrastructure implementation: A review. *South African Journal of Geomatics*, *4*(4), 369–383. doi:10.4314/sajg.v4i4.2

İnan, A., & İzgi, E. (2011). *GIS (Coğrafi Bilgi Sistemi)*. Retrieved April 12, 2017, from http://cbs.ormansu.gov.tr/cob2011/wp-content/uploads/2011/05/cbs1.pdf

India Tourism Statistics at a Glance. (n.d.). Market Research Division, Ministry of Tourism, Government of India.

Instituto Nacional de Estatística. (2011). *Censos 2011 Resultados Definitivos - Região Lisboa*. INE. Available in: http://censos.ine.pt/xportal/xmain?xpid=CENSOS&xpgid=ine_censos_publicacoes

Instituto Nacional de Estatística. (2014). *Estatísticas do Turismo 2014*. INE. Available in: https://www.ine.pt/xportal/xmain?xpid=INE&xpgid=ine_publicacoes&PUBLICACOESpub_boui=139601&PUBLICACOESmodo=2&xlang=pt

International Council of Museums. (n.d.). Retrieved January 19, 2017, from http://icom.museum/the-vision/museum-definition

Intro to advanced GIS and a review of basic GIS. (n.d.). Retrieved from www.utsa.edu/lrsg/Teaching/EES6513-09/L1_Intro.ppt

Jadhav, A., & Sonar, R. (2009). An integrated rule-based and case-based reasoning approach for selection of the software packages. *Information Systems, Technology and Management*, 280-291.

Jensen, J. R., & Christensen, E. J. (1986). Solid and Hazardous Waste Disposal Site Selection Using Digital Geographical Information Systems Techniques. *The Science of the Total Environment*, *56*, 265–276. doi:10.1016/0048-9697(86)90331-1

Jian, C., Jian, Y., & Jin, H. (2005). *Automatic content-based recommendation in e-commerce*. The 2005 IEEE International Conference on e-Technology, e-Commerce and e-Service, 748-753.

Jovanivic', V., & Negus', A. (2008). The Applications of GIS and its Components in Tourism. *Yugoslav Journal of Operations Research*, *18*(2), 261–272. doi:10.2298/YJOR0802261J

Jovanović, V. (2016). The application of GIS and its components in tourism. Yugoslav Journal of Operations Research, 18(2).

Kabassi, K. (2010). Personalizing recommendations for tourists. *Telematics and Informatics*, *27*(1), 51–66. doi:10.1016/j.tele.2009.05.003

Kádár, B. (2014). Measuring tourist activities in cities using geotagged photography. *Tourism Geographies*, *16*(1), 88–104. doi:10.1080/14616688.2013.868029

Kaiser, H. F. (1958). The varimax criterion for analytic rotation in factor analysis. *Psychometrika*, *23*(3), 187–200.

Kale, S. H. (2004). Spirituality, religion, and globalization. *Journal of Macromarketing*, *24*(2), 92–107. doi:10.1177/0276146704269296

Karabıyık, A. (2007). *Çağdaş sanat müzeciliği kapsamında Türkiye'deki müzecilik hareketlerine bir bakış* (Unpublished master thesis). University of Atatürk, Erzurum, Turkey.

Karapınar, E., & Barakazı, M. (2017). Evaluation of cultural heritage tourism as a sustainable tourism: Gobeklitepe ruins. *Güncel Turizm Araştırmaları Dergisi*, *1*(1), 5–18.

Karimi, B., & Yektaei, M. H. (2015). Location Recommendation based on location-based social networks for Entertainment services. *Advances in Computer Science: an International Journal*, *4*(1), 55–64.

Kaya Bensghir, T., & Akay, A. (2007). *Yerel yönetimlerde coğrafi bilgi sistemleri Türkiye uygulamaları*. Ankara: Public Administration Institute for Turkey and Middle East Publication.

Kervankıran, İ., & Çuhadar, M. (2014). Turizm Rotalarının Oluşturulmasında Coğrafi Bilgi Sistemlerinin Önemi. III. Disiplinlerarası Turizm Araştırmaları Kongresi, 576-589.

Kim, D.-H., & Kim, M.-S. (2002). Web GIS Service Component based on Open Environment. *IEEE Geoscience and Remote Sensing Symposium, 6*, 3346-3348.

Koceski, S., & Petrevska, B. (2012). Empirical evidence of contribution to e-tourism by application of personalized tourism recommendation system. *Annals of the Alexandru Ioan Cuza University-Economics*, *59*(1), 363–374. doi:10.2478/v10316-012-0025-6

Kodge B.G., Hiremath P.S. (2012). Web based Geo-Spatial and Village Level Information Extraction System using FOSS. *Journal of Advances in Information Technology, 3*(4), 222-227. doi:10.4304/jait.3.4.222-227

Kommana, K. (2013). *Implementation of a Geoserver Applicatoin For GIS Data Distribution and Manipulation*. Department of Physical Geography and Quaternary Geology, Stockholm University. Retrieved from http://www.diva-portal.org/smash/get/diva2:640096/fulltext01.pdf

Konstan, J. A., McNee, S. M., Ziegler, C.-N., Torres, R., Kapoor, N., & Riedl, J. (2006). Lessons on applying automated recommender systems to information-seeking tasks. The 21st national conference on Artificial intelligence, Boston, Massachusetts, July 16 - 20, 1630-1633.

Koren, Y., & Bell, R. (2015). *Advances in collaborative filtering, Recommender systems handbook*. Springer. doi:10.1007/978-1-4899-7637-6_3

Köroğlu, A. & Köroğlu Ö. (2006). Kırsal Turizmin Yöre Kalkınmasındaki Rolü: Buldan Yöresi Kırsal Turizm Potansiyeli. *Buldan Sempozyumu*, 233-242.

Koswatte, S., McDougall, K., & Liu, X. (2015). SDI and crowdsourced spatial information management automation for disaster management. *Survey Review*, *47*(344), 307–315. doi:10.1179/1752270615Y.0000000008

Kotkov, D., Wang, S., & Veijalainen, J. (2016). A survey of serendipity in recommender systems. *Knowledge-Based Systems*, *111*, 180–192. doi:10.1016/j.knosys.2016.08.014

Kotler, P. (1997). *Marketing Management: Analysis, Planning, and Control* (9th ed.). Upper Saddle River, NJ: Prentice-Hall.

Kozak, N., Kozak Akoğlan, M., & Kozak, M. (2001). *Genel Turizm İlkeler- Kavramlar*. Ankara: Detay Yayıncılık.

Krek, A. (2006). Efficient Pricing of Geomarketing Internet Services: European vs. American Approach. *International Journal of Business and Management, 1*.

Krishan, G., Sharma, L. M., Yadav, B. K., & Ghosh, N. C. (2016). Analysis of Water Level Fluctuations and TDS Variations in the Groundwater at Mewat (Nuh) District, Haryana (India). *Current World Environment, 11*(2), 388–398. doi:10.12944/CWE.11.2.06

Küçükaltan, D. (1997). Trakya Ekonomisi İçin Bir Bölgesel Kalkınma Modeli: Kırsal Turizm. Ulusal Bölge Bilimi Bölge Planlama Kongresi 20-22 Ekim. İzmir: ProOfset Matbaacılık.

Kunaver, M., & Požrl, T. (2017). Diversity in recommender systems–A survey. *Knowledge-Based Systems, 123*, 154–162. doi:10.1016/j.knosys.2017.02.009

Kundu, A. (2011). *Trends and processes of urbanisation in India*. Population and Development Branch, United Nations Population Fund, UNFPA, September.

Kurata, Y., & Hara, T. (2013). *CT-planner4: Toward a more user-friendly interactive day-tour planner, Information and communication technologies in tourism 2014*. Springer.

Kushwaha, A., Chatterjee, D., & Mandal, P. (2011). *Potentials of GIS in Heritage and Tourism*. Geospatial World Forum. Retrieved on March, 8, 2017, from: http://www.geospatialworldforum.org/2011/proceeding/pdf/Avantika.pdf

Lakhshman, A., & Malik, P. (2010). Cassandra: A decentralized structured storage system. *ACM SIGOPS Operating Syst Rev., 44*(2), 35–40. doi:10.1145/1773912.1773922

Lane, B. (1994). What is Rural Tourism? *Journal of Sustainable Tourism, 2*(1-2), 7–21. doi:10.1080/09669589409510680

Larose, D. T. (2014). *Discovering knowledge in data: an introduction to data mining*. John Wiley & Sons. doi:10.1002/9781118874059

Laudon, K. C., & Laudon, J. P. (2003). *Essential of management information systems* (5th ed.). New Jersey: Prentice Hall.

Laura, J. R., Hare, T. M., Gaddis, L. R., Fergason, R. L., Skinner, J. A., Hagerty, J. J., & Archinal, B. A. (2017). Towards a Planetary Spatial Data Infrastructure. *ISPRS International Journal of Geo-Information, 6*(6), 181. doi:10.3390/ijgi6060181

Lee, K., Ganti, R. K., Srivatsa, M., & Liu, L. (2014). Efficient Spatial Query Processing for Big Data. *ACM SIGSPATIAL '14*.

Lee. (1982). On -Nearest Neighbor Voronoi Diagrams in the Plane. *IEEE Trans. Computers*.

Lee, B., & Graefe, A. (2004, October). GIS: A Tool to Locate New Park and Recreation Services. *Parks & Recreation*, 34–41.

Lee, C.-S., Chang, Y.-C., & Wang, M.-H. (2009). Ontological recommendation multi-agent for Tainan City travel. *Expert Systems with Applications*, *36*(3), 6740–6753. doi:10.1016/j.eswa.2008.08.016

Lee, M. (2014). A Recommendation System based on Big Data Technology. *International Journal (Toronto, Ont.)*, *3*(5).

Leidig, M., & Teeuw, R. (2015). Free software: A review, in the context of disaster management. *International Journal of Applied Earth Observation and Geoinformation*, *42*, 49–56. doi:10.1016/j.jag.2015.05.012

Lenka, Barik, Gupta, Ali, Rath, & Dubey. (2016). *Comparative Analysis of SpatialHadoop and GeoSpark for Geospatial Big Data Analytics*. Cornell University Library.

Lian, D., Zhao, C., Xie, X., Sun, G., Chen, E., & Rui, Y. (2014). GeoMF: joint geographical modeling and matrix factorization for point-of-interest recommendation. the 20th ACM SIGKDD international conference on Knowledge discovery and data mining, 831-840. doi:10.1145/2623330.2623638

Lian, D., Ge, Y., Zhang, F., Yuan, N. J., Xie, X., Zhou, T., & Rui, Y. (2015). *Content-aware collaborative filtering for location recommendation based on human mobility data*. IEEE International Conference on Data Mining (ICDM). doi:10.1109/ICDM.2015.69

Liao, H., Han, J., & Fang, J. (2010). Multi-dimensional Index on Hadoop Distributed File System. *Proceedings of IEEE Fifth International Conference on Networking, Architecture, and Storage*, 240-249.

Liben-Nowell, D., Novak, J., Kumar, R., Raghavan, P., & Tomkins, A. (2005). Geographic routing in social networks. *Proceedings of the National Academy of Sciences of the United States of America*, *102*(33), 11623–11628. doi:10.1073/pnas.0503018102 PMID:16081538

Li, H., Lu, J., Cai, B., & Yao, S. (2008). Study on SOA-Orient WebGIS framework. *14th IEEE International Conference on Automation and Computing*.

Lin, P. J., Kao, C. C., Lam, K. H., & Tsai, I. C. (2014). Design and implementation of a tourism system using mobile augmented reality and GIS technologies. In *Proceedings of the 2nd International Conference on Intelligent Technologies and Engineering Systems (ICITES2013)* (pp. 1093-1099). Springer. doi:10.1007/978-3-319-04573-3_133

Li, Q., Zheng, Y., Xie, X., Chen, Y., Liu, W., & Ma, W.-Y. (2008). *Mining user similarity based on location history*. The 21st national conference on Artificial intelligence, Boston, Massachusetts, July 16 - 20, 1630-1633. doi:10.1145/1463434.1463477

Liu, X., Han, J., Zhong, Y., Han, C., & He, X. (2009). Implementing WebGIS on Hadoop: A case study of improving small file I/O performance on HDFS. *Proceedings of IEEE International Conference on Cluster Computing and Workshops*, 1-8.

Liu, B., Fu, Y., Yao, Z., & Xiong, H. (2013, August). Learning geographical preferences for point-of-interest recommendation. In *Proceedings of the 19th ACM SIGKDD international conference on Knowledge discovery and data mining* (pp. 1043-1051). ACM. doi:10.1145/2487575.2487673

Liu, B., Xiong, H., Papadimitriou, S., Fu, Y., & Yao, Z. (2015). A general geographical probabilistic factor model for point of interest recommendation. *IEEE Transactions on Knowledge and Data Engineering*, *27*(5), 1167–1179. doi:10.1109/TKDE.2014.2362525

Liu, X., Liu, Y., Aberer, K., & Miao, C. (2013, October). Personalized point-of-interest recommendation by mining users' preference transition. In *Proceedings of the 22nd ACM international conference on Information & Knowledge Management* (pp. 733-738). ACM. doi:10.1145/2505515.2505639

Liu, Y., Wei, W., Sun, A., & Miao, C. (2014). *Exploiting geographical neighborhood characteristics for location recommendation*. The 23rd ACM International Conference on Information and Knowledge Management, 739-748. doi:10.1145/2661829.2662002

López, E. P. & García, F. J. C. (2006). Agrotourism, Sustainable Tourism and Ultraperipheral Areas: The Case of Canary Islands. *PASOS Revista de Turismo Y Patrimonio Cultural*, *4*(1), 85-97.

Lops, P., De Gemmis, M., & Semeraro, G. (2011). Content-based recommender systems: State of the art and trends. In Recommender systems handbook (pp. 73-105). Springer US.

Lops, P., De Gemmis, M., & Semeraro, G. (2011). *Content-based recommender systems: State of the art and trends, Recommender systems handbook*. Springer. doi:10.1007/978-0-387-85820-3_3

Lorenzi, F., Loh, S., & Abel, M. (2011). *PersonalTour: A recommender system for travel packages*. The 2011 IEEE/WIC/ACM International Conferences on Web Intelligence and Intelligent Agent Technology, 333-336.

Luberg, A., Tammet, T., & Järv, P. (2011). *Smart city: a rule-based tourist recommendation system, Information and Communication Technologies in Tourism 2011*. Springer. doi:10.1007/978-3-7091-0503-0_5

Luberichs, J., & Wachowiak, H. (2010). GIS supported segmentation study of visitors to Majorca Island. *Advances in Hospitality and Leisure*, (2), 1-37.

Lucas, J., da Silva Coelho, B., García, M., de Almeida Figueiredo, A., & Martins, C. (2011). Applying Recommender Methodologies in Tourism Sector. *Highlights in Practical Applications of Agents and Multiagent Systems*, 101-108.

Mackay, K., & Vogt, C. (2012). Information technology in everyday and vacation contexts. *Annals of Tourism Research*, *39*(3), 1380–1401. doi:10.1016/j.annals.2012.02.001

Maguire, D. (1995). Implementing Spatial Analysis and GIS Applications for Business and Service Planning. In G. Clarke & P. Longley (Eds.), *GIS for Business and Services Planning* (pp. 171–191). New York: Pearson Professional.

Mahmood, M. A., Al-Shammari, E. T., El-Bendary, N., Hassanien, A. E., & Hefny, H. A. (2013, September). Recommender system for ground-level Ozone predictions in Kuwait. In *Computer Science and Information Systems (FedCSIS), 2013 Federated Conference on* (pp. 107-110). IEEE.

Mahmood, T., & Ricci, F. (2009). *Improving recommender systems with adaptive conversational strategies*. The 20th ACM conference on Hypertext and hypermedia, 73-82. doi:10.1145/1557914.1557930

Mahmoud, S., El-Bendary, N., Mahmood, M. A., & Hassanien, A. E. (2013, December). An intelligent recommender system for drinking water quality. In *Hybrid Intelligent Systems (HIS), 2013 13th International Conference on* (pp. 285-290). IEEE. doi:10.1109/HIS.2013.6920498

Mani, M. (2014, September). Challenges & Limitation in Recommender system. *International Journal of Latest Trends in Engineering and Technology*, *4*(3), 138–142.

Mansourian, A., Rajabifard, A., Valadan Zoej, M. J., & Williamson, I. P. (2005). Using SDI and web-based system to facilitate disaster management. *International Journal of Computers & Geosciences*, *32*(3), 303–315. doi:10.1016/j.cageo.2005.06.017

Marakas, G. M. (2003). Modern data warehousing, mining, and visualization: Core concepts. In *Data mining and data visualization* (pp. 97–101). Upper Saddle River, NJ: Prentice Hall, Pearson Dedication, Inc.

Martinez, L., Rodriguez, R. M., & Espinilla, M. (2009). *Reja: a georeferenced hybrid recommender system for restaurants*. IEEE/WIC/ACM International Joint Conferences on Web Intelligence and Intelligent Agent Technologies (WI-IAT'09), 187-190.

Martin, W. E., Brown, C. V., DeHayes, D. W., Hoffer, J. A., & Perkins, W. C. (2005). *Managing information technology* (5th ed.). Prentice Hall.

Masron, T., Mohamed, B., & Marzuki, A. (2015). GIS base tourism decision support system for Langkawi Island, Kedah, Malaysia. *Theoretical and Empirical Researches in Urban Management*, *10*(2), 21.

Masthoff, J. (2004). *Group modeling: Selecting a sequence of television items to suit a group of viewers, Personalized digital television*. Springer.

Masthoff, J. (2015). *Group recommender systems: aggregation, satisfaction and group attributes, Recommender Systems Handbook*. Springer. doi:10.1007/978-1-4899-7637-6_22

Matans, V. M., Liu, Z., & Pereira, A. J. S. C. (2015). A web service and android application for the distribution of rainfall estimates and Earth observation data. *Computers & Geosciences*, *77*, 66–76. doi:10.1016/j.cageo.2015.01.011

Matheson, C. M., Rimmer, R., & Tinsley, R. (2014). Spiritual attitudes and visitor motivations at the Beltane Fire Festival, Edinburgh. *Tourism Management*, *44*, 16–33. doi:10.1016/j.tourman.2014.01.023

McAdam, D. (1999). The value and scope of geographical information systems in tourism management. *Journal of Sustainable Tourism*, *7*(1), 77–92. doi:10.1080/09669589908667327

McGehee, N. G. (2014). Volunteer tourism: Evolution, issues and futures. *Journal of Sustainable Tourism*, *22*(6), 847–854. doi:10.1080/09669582.2014.907299

McMahon, P. (1997). *Geographic Information System in the construction industry – an underused resource?* The Chartered Institute of Building, No.82.

Medhekar, A., & Haq, F. (2012). Development of Spiritual Tourism Circuits: The Case of India. *Journal of Global Business Review*, *2*(2), 212–218.

Meehan, K., Lunney, T., Curran, K., & McCaughey, A. (2013). *Context-aware intelligent recommendation system for tourism.* IEEE International Conference on Pervasive Computing and Communications Workshops (PERCOM Workshops), 328-331.

Mitchell, A. (2005). The ESRI guide to GIS analysis, Volume 2: Spatial measurements and statistics. ESRI Press. ISBN: 9781589481169

Mitchell, V.-W., & McGoldrick, P. J. (1994). The Role of Geodemographics in Segmenting and Targeting Consumer Markets: A Delphi Study. *European Journal of Marketing*, *28*(5), 54–72. doi:10.1108/03090569410062032

Mitroff, I., & Denton, E. A. (1999). *A spiritual audit of corporate America: A hard look at spirituality, religion, and values in the workplace* (Vol. 140). Jossey-Bass.

Moal-Ulvoas, G., & Taylor, V. A. (2014). The spiritual benefits of travel for senior tourists. *Journal of Consumer Behaviour*, *13*(6), 453–462. doi:10.1002/cb.1495

Mocholi, J. A., Jaen, J., Krynicki, K., Catala, A., Picón, A., & Cadenas, A. (2012). Learning semantically-annotated routes for context-aware recommendations on map navigation systems. *Applied Soft Computing*, *12*(9), 3088–3098. doi:10.1016/j.asoc.2012.05.010

Moniruzzaman, A. B., & Hossain, S. A. (2013). Nosql database: New era of databases for big data analytics—classification, characteristics and comparison. *Int J Database Theor Appl.*, *6*(4), 1–13.

Montejo-Ráez, A., Perea-Ortega, J. M., García-Cumbreras, M. Á., & Martínez-Santiago, F. (2011). Otiŭm: A web based planner for tourism and leisure. *Expert Systems with Applications*, *38*(8), 10085–10093. doi:10.1016/j.eswa.2011.02.005

Moreno, A., Sebastiá, L., & Vansteenwegen, P. (2015). *Recommender Systems in Tourism.* Academic Press.

Moreno, A., Valls, A., Isern, D., Marin, L., & Borràs, J. (2013). Sigtur/e-destination: Ontology-based personalized recommendation of tourism and leisure activities. *Engineering Applications of Artificial Intelligence*, *26*(1), 633–651. doi:10.1016/j.engappai.2012.02.014

Morris Steven, P. (2006). Geospatial Web Services and Geo archiving: New Opportunities and Challenges in Geographic Information Services. *Library Trends*, *55*(2), 285–303. doi:10.1353/lib.2006.0059

Mujabar, P. S., & Chandrasekar, N. (2013). Shoreline change analysis along the coast between Kanyakumari and Tuticorin of India using remote sensing and GIS. *Arabian Journal of Geosciences*, *6*(3), 647–664. doi:10.1007/s12517-011-0394-4

Munroe, K. (2004). Meaningful Metrics: Site Evaluation and Market Optimization Models. In The Power of Place: Advanced Customer and Location Analytics for Market Planning (pp. 127-156). San Diego, CA: Claritas.

Murakami, T., Mori, K., & Orihara, R. (2007). *Metrics for evaluating the serendipity of recommendation lists*. Annual Conference of the Japanese Society for Artificial Intelligence, 40-46.

Mustafa, A. A., Singh, M., Sahoo, R. N., Ahmed, N., Khanna, M., Sarangi, A., & Mishra, A. K. (2011). Land suitability analysis for different crops: A multi criteria decision making approach using remote sensing and GIS. *Researcher*, *3*(12), 1–24.

Musyoka, S. M., Mutyauvyu, S. M., Kiema, J. B. K., Karanja, F. N., & Siriba, D. N. (2007). Market segmentation using geographic information systems (GIS): A case study of the soft drink industry in Kenya. *Marketing Intelligence & Planning*, *25*(6), 632–642. doi:10.1108/02634500710819987

Mwange, C., Mulaku, G. C., & Siriba, D. N. (2016). Reviewing the status of national spatial data infrastructures in Africa. *Survey Review*, 1–10. doi:10.1080/00396265.2016.1259720

Nasr, M. R. (2007). *Open Source Software: The use of open source GIS software and its impact on organizations* (Doctoral dissertation). Middlesex University.

Nathan, H. S. K., Mishra, S., & Reddy, B. S. (2008). *An alternative approach to measure HDI.* Indira Gandhi Institute of Development Research (IGIDR), Working Paper, WP-2008-001.

Nicholson, R. A. (1978). *The Tarjuman al-Ashwaq: A Collection of Mystical Odes by Ibn Al-Arabi, translated from original.* London: Theosophical Publishing House.

Ning, X., Desrosiers, C., & Karypis, G. (2015). *A comprehensive survey of neighborhood-based recommendation methods, Recommender systems handbook.* Springer. doi:10.1007/978-1-4899-7637-6_2

Nişancı, R., Yıldırım, V., & Çolak, H. E. (2010). Coğrafi Bilgi Sistem Uygulamaları. *Bilim ve Teknik*, *514*, 58–63.

Niyizamwiyitira & Lundberg. (2016). Performance evaluation of trajectory queries on multiprocessor and cluster. *Proceedings of CNSA*, 145-163.

Noguera, J. M., Barranco, M. J., Segura, R. J., & Martínez, L. (2012). A mobile 3D-GIS hybrid recommender system for tourism. *Information Sciences*, *215*, 37–52. doi:10.1016/j.ins.2012.05.010

Noulas, A., Mascolo, C., & Frias-Martinez, E. (2013). *Exploiting foursquare and cellular data to infer user activity in urban environments*. IEEE 14th International Conference on Data Management (MDM), 167-176. doi:10.1109/MDM.2013.27

Nunes, I., & Marinho, L. (2013). A gaussian kernel approach for location recommendations. In Proceedings of KDMiLe-Symposium on Knowledge Discovery, Mining and Learning (Vol. 1060). Academic Press.

Observatório de Turismo de Lisboa (TOL). (2014). *Survey of tourism motivation – Lisbon City 2014*. OTL. Available in: http://www.visitlisboa.com/Observatorio.aspx

Observatório Turismo de Lisboa (TOL). (2014). *Visitor's activities and information survey in Lisbon Region 2014*. Available in: http://www.visitlisboa.com/Observatorio.aspx

OGC. (2004a). *Open GIS Web Map Server Cookbook, 2004*. Available from: http://www.opengeospatial.org/standards/wms

OGC. (2004b). *Open GIS Web Map Service Implementation Specification, Version 1.3.0, 2004*. Available from http://portal.opengeospatial.org/files/?artifact_id=4756

OGC. (2005). *OpenGIS Web Feature Service Implementation Specification, Version 1.1.0*. Retrieved from http://portal.opengeospatial.org/files/?artifact_id=8339

Okabe, M., Yanagisawa, M., Yamazaki, H., Kobayashi, K., Yoshioka, A., & Yamaguchi, T. (2008). Organizational knowledge transfer of intelligence skill using ontologies and a rule-based system. *Practical Aspects of Knowledge Management*, 207-218.

Orhun, Y. A. (2013). Spatial differentiation in the supermarket industry: The role of common information. *Quantitative Marketing and Economics*, *11*(1), 3–37. doi:10.1007/s11129-012-9123-x

Orlandini, R. (n.d.). *Geospatial Data*. Retrieved from http://researchguides.library.yorku.ca/content.php?pid=245987&sid=2176375

Özkan, E. (2007). *Türkiye'de Kırsal Kalkınma Politikaları ve Kırsal Turizm*. Ankara: Yayınlanmamış Yüksek Lisans Tezi, Ankara Üniversitesi Sosyal Bilimler Enstitüsü.

Özkasım, H. (2015). Müze ve müzecilik. In Altınsapan & Küçükhasköylü (Eds.), Müzecilik ve sergileme (pp. 2-21). Ankara: Anadolu Üniversitesi Yayınları.

Padma, G.V., Ramireddy, P.V., Tata Babu, C., Ramana Murty, M.V., & Prasada Rao, G. (2015). A Geospatial frame work for mapping of approximate cadastral sub divisions in Joniganuru village, santhipuram mandal, chittoor district, Andhra Pradesh - An Initiative. *International Journal of Engineering Sciences & Research Technology, 4*(7).

Panniello, U., Tuzhilin, A., Gorgoglione, M., Palmisano, C., & Pedone, A. (2009, October). Experimental comparison of pre-vs. post-filtering approaches in context-aware recommender systems. In *Proceedings of the third ACM conference on Recommender systems* (pp. 265-268). ACM. doi:10.1145/1639714.1639764

Parrish, J. S. (2009). *Geographic Information Systems and Spatial Analysis of Market Segmentation for Community Banks* (Unpublished master dissertation). University of North Carolina.

Patil, Y. M. (2014). GIS Enabled Tool for Effective Land Records Management. *International Journal on Recent and Innovation Trends in Computing and Communication*, *2*(6), 1627–1631.

Patra, S. S., & Barik, R. K. (2015). Dynamic Dedicated Server Allocation for Service Oriented Multi-Agent Data Intensive Architecture in Biomedical and Spatial Cloud. In *Cloud Technology: Concepts, Methodologies, Tools, and Applications* (pp. 2262–2273). IGI Global. doi:10.4018/978-1-4666-6539-2.ch107

Patwal, A. (2013). Sustainable Tourism Infrastructure Planning: A GIS Based Approach. *International Journal of Hospitality and Tourism Systems*, *6*(1).

Payton, K. F. (2010). *GIS-Based Market Analysis: A Model for Implementing a Geographic Information System as a Market Analysis Tool* (Unpublished master dissertation). The College of St. Scholastica, Duluth, MN.

Pazzani, M. J., & Billsus, D. (2007). *Content-based recommendation systems, The adaptive web*. Springer.

Peckham, S. D., & Goodall, J. L. (2013). Driving plug-and-play models with data from web services: A demonstration of interoperability between CSDMS and CUAHSI-HIS. *Computers & Geosciences*, *53*, 154–161. doi:10.1016/j.cageo.2012.04.019

Phoenix, M. (2000). *Learning with GIS, Arcuser Online*. Retrieved May 8, 2017, from http://www.esri.com/news/arcuser/0700/umbrella11.html

Plano Regional de Ordenamento do Território da área Metropolitana de Lisboa. (2009). *Diagnóstico Sectorial Sistema Urbano, Habitação e Equipamentos Constitucional. Comissão de Coordenação e Desenvolvimento Regional de Lisboa e Vale do Tejo (CCDR LVT)*. Accessed in June, 2017. Available in: http://www.ccdr-lvt.pt/pt/plano-regional-de-ordenamento-do-territorio-da-area-metropolitana-de-lisboa/54.htm

Principe, W. (1983). Toward defining spirituality. *Studies in Religion/Sciences Religiousness*, *12*(2), 127-141.

Pun-Cheng, L. S., Lai, W. W., & Chang, R. K. (2016). Exploring utility system SDI–Managerial and technical perspectives. *Tunnelling and Underground Space Technology*, *54*, 13–19. doi:10.1016/j.tust.2016.01.026

Puri, S. K., Sahay, S., & Georgiadou, Y. (2007). A Metaphor-Based Sociotechnical Perspective on Spatial Data Infrastructure Implementations: Some Lessons from India, Research and Theory in Advancing Spatial Data Infrastructure Concepts. ESRI Press.

Pussep, K., Kaune, S., Flick, J., & Steinmetz, R. (2009, March). A peer-to-peer recommender system with privacy constraints. In *Complex, Intelligent and Software Intensive Systems, 2009. CISIS'09. International Conference on* (pp. 409-414). IEEE. doi:10.1109/CISIS.2009.32

Quantum GIS User Guide, Version 1.6.0. (2010). Academic Press.

Quercia, D., Lathia, N., Calabrese, F., Di Lorenzo, G., & Crowcroft, J. (2010). *Recommending social events from mobile phone location data.* IEEE 10th International Conference on Data Mining (ICDM), 971-976. doi:10.1109/ICDM.2010.152

Raghavendran. (n.d.). *Small things of Big (Geo Spatial) Data.* Retrieved from https://www.slideshare.net/raaags/small-things-of-big-geospatial-data

Raghunathan, S., Prasad, A., Mishra, B. K., & Chang, H. (2005). Open source versus closed source: Software quality in monopoly and competitive markets. *IEEE Transactions on Systems, Man, and Cybernetics. Part A, Systems and Humans*, *35*(6), 903–918. doi:10.1109/TSMCA.2005.853493

Raghuvanshi, T. K., Belwal, R., & Solomon, N. (2006). An Approach to Develop Tourism Potential in Ethiophia Through Geographic Information System. *The Consortium Journal of Hospitality and Tourism, 11*(1).

Rajabifard, A., Feeney, M. E. F., & Williamson, I. P. (2002). Future Directions for SDI Development. *International Journal of Applied Earth Observation and Geoinformation ITC, 4*(1), 11–22.

Raju, P. L. N. (2004). *Fundamentals Of Geographical Information System.* World Meteorological Organisation.

Ravindran, A., & Jaishankar, S. (2006). *GIS Based Information System for Village Level planning.* Academic Press.

Rawat, J. S., & Kumar, M. (2015). Monitoring land use/cover change using remote sensing and GIS techniques: A case study of Hawalbagh block, district Almora, Uttarakhand, India. *The Egyptian Journal of Remote Sensing and Space Sciences*, *18*(1), 77–84. doi:10.1016/j.ejrs.2015.02.002

Rayed, C. A. (2012). Using GIS for Modeling a Spatial DSS for Industrial Pollution in Egypt. *American Journal of Geographic Information System*, *1*(4), 100–104. doi:10.5923/j.ajgis.20120104.01

Recreation Opportunity Spectrum. (1982). Retrieved July 10, 2017, from https://www.fs.usda.gov/detail/tonto/landmanagement/planning/?cid=stelprdb5412121

Reddy, B. S., & Nathan, H. S. K. (2011). *Emerging energy insecurity — the Indian dimension.* Academic Press.

Reddy, G.P., Reddy, M.N., Sontakki, B.S., & Kumar, K .V. (2011). A Framework of Participatory Geo-Spatial Information System for Micro Level Planning – A Case Study in Aquaculture. *Ind. Jn. of Agri. Econ, 66*(3).

Reference Manual.MAPWINGIS (2007). MAPWINGIS.

Reilly, W. J. (1931). *The Law of Retail Gravitation.* New York: John Wiley & Sons.

Rendle, S., Freudenthaler, C., & Schmidt-Thieme, L. (2010). *Factorizing personalized markov chains for next-basket recommendation.* The 19th international conference on World wide web, 811-820. doi:10.1145/1772690.1772773

Republic of Turkey Ministry of Culture and Tourism, General Directorate of the Cultural Assets and Museums. (n.d.). Retrieved July 05, 2017, from http://www.kulturvarliklari.gov.tr/TR,44423/dunya-miras-listesi.html

Republic of Turkey Ministry of Culture and Tourism. (n.d.). Retrieved July 04, 2017, from http://www.kultur.gov.tr/TR,96254/kultur.html

Reynolds, H. (1997). *An introduction to geographic information systems (GIS).* Retrieved July 17, 2015 from http://badpets.net/IntroGIS/

Ricci, F., Rokach, L., & Shapira, B. (2011). Introduction to recommender systems handbook. In Recommender systems handbook (pp. 1-35). springer US. doi:10.1007/978-0-387-85820-3_1

Ricci, F., Rokach, L., & Shapira, B. (2015a). *Recommender systems: introduction and challenges, Recommender systems handbook.* Springer. doi:10.1007/978-1-4899-7637-6

Ritchie, B., & Inkari, M. (2006). Host Community Attitudes Toward Tourism and Cultural Tourism Development: The Case of the Lewes District, Southern England. *International Journal of Tourism Research*, *8*(1), 27–44. doi:10.1002/jtr.545

Ruiz-Montiel, M., & Aldana-Montes, J. (2009). *Semantically enhanced recommender systems.* On the move to meaningful internet systems (OTM 2009 workshops), 604-609.

Ruotsalo, T., Haav, K., Stoyanov, A., Roche, S., Fani, E., Deliai, R., ... Hyvönen, E. (2013). SMARTMUSEUM: A mobile recommender system for the Web of Data. *Journal of Web Semantics*, *20*, 50–67. doi:10.1016/j.websem.2013.03.001

Sadoun, B., & Al-Bayari, O. (2009). A GIS system for tourism management. *IEEE International Conference on AICCSA*, 226-232.

Safoury, L., & Salah, A. (2013). Exploiting user demographic attributes for solving cold-start problem in recommender systems. *Lecture Notes on Software Engineering*, *1*(3), 303–307. doi:10.7763/LNSE.2013.V1.66

Sahebi, S., & Cohen, W. W. (2010). Community-based recommendations: a solution to the cold start problem. *Proceedings of WOODSTOCK'97.*

Salam, M. A., Mahmood, M. A., Awad, Y. M., Hazman, M., El Bendary, N., Hassanien, A. E., & Saleh, S. M. (2014). Climate recommender system for wheat cultivation in North Egyptian Sinai Peninsula. In *Proceedings of the Fifth International Conference on Innovations in Bio-Inspired Computing and Applications IBICA 2014* (pp. 121-130). Springer International Publishing. doi:10.1007/978-3-319-08156-4_13

Santiago, F. M., López, F. A., Montejo-Ráez, A., & López, A. U. (2012). GeOasis: A knowledge-based geo-referenced tourist assistant. *Expert Systems with Applications*, *39*(14), 11737–11745. doi:10.1016/j.eswa.2012.04.080

Santosh, K., Lakshmi, T. V., & Reddy, M. A. (2009). *Development Of remote Sensing and GIS Based Information System For Village Level Planning - A Case Study Of Addanki Mandal, Prakasam District, Andhra Pradesh, India.* American Geophysical Union, Spring Meeting.

Sarwat, M., Eldawy, A., Mokbel, M. F., & Riedl, J. (2013). *Plutus: leveraging location-based social networks to recommend potential customers to venues*. IEEE 14th International Conference on Mobile Data Management (MDM), 26-35.

Savage, N.S., Baranski, M., Chavez, N.E., & Höllerer, T. (2012). I'm feeling loco: A location based context aware recommendation system. *Advances in Location-Based Services*, 37-54.

Scaria, R., & Vijayan, P. K. (2012). Sustainable Rural Development With The Application Of Spatial Information Technology And Mahatma Gandhi National Rural Employment Guaranteed. *International Journal of Geomatics and Geosciences*, *2*(4), 1048–1061.

Scellato, S., Noulas, A., Lambiotte, R., & Mascolo, C. (2011). Socio-spatial properties of online location-based social networks. *ICWSM*, *11*, 329–336.

Schafer, J. B., Frankowski, D., Herlocker, J., & Sen, S. (2007). *Collaborative filtering recommender systems, The adaptive web*. Springer.

Schäffer, B., Baranski, B., & Foerster, T. (2010). Towards Spatial Data Infrastructures in the Clouds. *Spatial Thinking Lecture Notes in Geoinformation and Cartography*, *0*, 399–418. doi:10.1007/978-3-642-12326-9_21

Scripts, M. T. (n.d.). *Geohashes.* Retrieved from http://www.movable-type.co.uk/scripts/geohash.html

Sebastia, L., Garcia, I., Onaindia, E., & Guzman, C. (2009). e-Tourism: A tourist recommendation and planning application. *International Journal of Artificial Intelligence Tools*, *18*(05), 717–738. doi:10.1142/S0218213009000378

Seidel, I., Gärtner, M., Pöttler, M., Berger, H., Michael, D., & Merkl, D. (2009). *Itchy feet: a 3D e-tourism environment. In Tourism Informatics: Visual Travel Recommender Systems, Social Communities, and User Interface Design* (pp. 209–242). Hershey, PA: IGI Global.

Senaratne, H., Mobasheri, A., Ali, A. L., Capineri, C., & Haklay, M. (2017). A review of volunteered geographic information quality assessment methods. *International Journal of Geographical Information Science*, *31*(1), 139–167. doi:10.1080/13658816.2016.1189556

Sertel, E., Findik, N., Kaya, S., Seker, D. Z., & Samsunlu, A. (2008). Assessment of landscape changes in the Kizilirmak Delta, Turkey, using remotely sensed data and GIS. *Environmental Engineering Science*, *25*(3), 353–362. doi:10.1089/ees.2006.0149

Shah, S. A., & Wani, M. A. (2015). Application of Geospatial Technology for the Promotion of Tourist Industry in Srinagar City. *International Journal of u-and e-Service Science and Technology*, *8*(1), 37–50.

Shani, G., & Gunawardana, A. (2011). Evaluating recommendation systems. *Recommender systems handbook*, 257-297.

Sharma, L., & Gera, A. (2013). A survey of recommendation system: Research challenges. *International Journal of Engineering Trends and Technology*, *4*(5), 1989–1992.

Sharma, M., & Mann, S. (2013). A survey of recommender systems: Approaches and limitations. *International Journal of Innovations in Engineering and Technology*, *2*(2), 8–14.

Sharpley, R., & Sundaram, P. (2005). Tourism: A sacred journey? The case of Ashram tourism, India. *International Journal of Tourism Research*, *7*(3), 161–171. doi:10.1002/jtr.522

Shavitt, Y., Weinsberg, E., & Weinsberg, U. (2010, October). Building recommendation systems using peer-to-peer shared content. In *Proceedings of the 19th ACM international conference on Information and knowledge management* (pp. 1457-1460). ACM. doi:10.1145/1871437.1871646

Shekhar, S., & Sanjay, C. S. D. (2003). *A Tour. Prentice Hall.*

Shinde, K. A. (2007). Visiting Scared sites in India: Religious Tourism or Pilgrimage? In R. Raj & N. D. Morpeth (Eds.), *Religious Tourism and Pilgrimage Management: An international Perspective* (pp. 170–183). CAB International. doi:10.1079/9781845932251.0184

Shinde, K. A. (2016). Planning for urbanisation in religious tourism destinations: Insights from Shirdi, India. *Planning Practice and Research*, 1–20.

Shiraishi, T., Nagata, M., Shibata, N., Murata, Y., Yasumoto, K., & Ito, M. (2005). A personal navigation system with functions to compose tour schedules based on multiple conflicting criteria. *IPSJ Digital Courier*, *1*, 528–536. doi:10.2197/ipsjdc.1.528

Shyti, B., & Kushi, E. (2012). The impact of GIS application in the tourism development of Elbasan region. *The Romanian Economic Journal*, *15*(45), 189–210.

Singh, S. P., Sharma, J., & Singh, P. (2011). A GeoReferenced Information System for Tourism (GeoRIST). *International Journal of Geomatics and Geosciences*, *2*(2), 456–464.

Sitender, S., & Kumar, R. (2012). Village Information System –A Case Study of Muklan Village, Hisar, Haryana (INDIA). *International Journal of Research in Social Sciences*, *2*(2).

Smith, M. K. (2005). *Factors In the Adoption of Geographic Information Systems in Sport Marketing* (Unpublished master dissertation). California State University.

Smith, J., Mackaness, W., Kealy, A., & Williamson, I. (2004). Spatial data infrastructure requirements for mobile location based journey planning. *Transactions in GIS*, *8*(1), 23–44. doi:10.1111/j.1467-9671.2004.00166.x

Song, Y., Hu, Z., Liu, H., Shi, Y., & Tian, H. (2013, September). A novel group recommendation algorithm with collaborative filtering. In *Social Computing (SocialCom), 2013 International Conference on* (pp. 901-904). IEEE. doi:10.1109/SocialCom.2013.138

Soykan, F. (2002). *Kırsal Turizmin Sosyo-Ekonomik Etkileri ve Türkiye*. Türkiye Dağları I. Ulusal Sempozyumu, Haziran, Kastamonu.

Soykan, F. (2004). *Kırsal Alanların Turizm Potansiyelinin Saptanması ve Şirince Köyü'ne (İzmir) Uygulanması*. Bornova-İzmir: Ege Üniversitesi Edebiyat Fakültesi Yayınları.

Soykan, F. (1999). Doğal Çevre ve Kırsal Kültürle Bütünleşen Bir Turizm Türü: Kırsal Turizm. *Anatolia Turizm Araştırmaları Dergisi*, *10*(1), 67–75.

Srisuwan, P., & Srivihok, A. (2008). Personalized trip information for e-tourism recommendation system based on Bayes theorem. *Research and Practical Issues of Enterprise Information Systems*, *II*, 1271–1275. doi:10.1007/978-0-387-76312-5_53

Stanley, T. J., & Sewall, M. A. (1976). Image Inputs to a Probabilistic Model: Predicting Retail Potential. *Journal of Marketing*, *40*(3), 48–53. doi:10.2307/1249994

Star, J., & Estes, J. (1990). *Geographic Information Systems: An Introduction*. Englewood Cliffs, NJ: Prentice-Hall.

Stefanakis, E., & Prastacos, P. (2008). Development of a coastal SDI using GeoFOSS. *the Proceedings of the 11th AGILE Conference.*

Strain, L., Rajabifard, A., & Williamson, I. (2004). Spatial Data Infrastructure to facilitate coastal zone management. *Coastal Zone Asia Pacific Conference*, 1-6.

Subbu Lakshmi, E., & Yarrakula, K. (2017). Comparative analysis of digital elevation models: A case study around Madduleru River. *Indian Journal of Geo Marine Sciences*, *46*(7), 1339–1351.

Subbu Lakshmi, E., & Yarrakula, K. (2016). Monitoring Land use Land Cover Changes using Remote Sensing and GIS Techniques: A Case Study around Papagni River, Andhra Pradesh, India. *Indian Journal of Ecology*, *43*(2), 383–387.

Supak, S. K., Devine, H. A., Brothers, G. L., Rozier Rich, S., & Shen, W. (2014). An open source web-mapping system for tourism planning and marketing. *Journal of Travel & Tourism Marketing*, *31*(7), 835–853. doi:10.1080/10548408.2014.890153

Su, X., & Khoshgoftaar, T. M. (2009). A survey of collaborative filtering techniques. *Advances in Artificial Intelligence*, *2009*, 4. doi:10.1155/2009/421425

Swain, N. R., Christensen, S. D., Snow, A. D., Dolder, H., Espinoza-Dávalos, G., Goharian, E., ... Burian, S. J. (2016). A new open source platform for lowering the barrier for environmental web app development. *Environmental Modelling & Software*, *85*, 11–26. doi:10.1016/j.envsoft.2016.08.003

Swain, N. R., Latu, K., Christensen, S. D., Jones, N. L., Nelson, E. J., Ames, D. P., & Williams, G. P. (2015). A review of open source software solutions for developing water resources web applications. *Environmental Modelling & Software*, *67*, 108–117. doi:10.1016/j.envsoft.2015.01.014

Swearingen, K., & Sinha, R. (2001). Beyond algorithms: An HCI perspective on recommender systems (6th ed.). ACM SIGIR 2001 Workshop on Recommender Systems, 1-11.

Symeonidis, P., Ntempos, D., & Manolopoulos, Y. (2014). *Recommender systems for location-based social networks*. Springer Science & Business Media. doi:10.1007/978-1-4939-0286-6

T.B.M.M. (2006). *T.B.M.M. Genel Kurul Tutanağı, 22.Dönem, 3. Yasama Yılı*. Retrieved April 26, 2017, from www.tbmm.gov.tr

Tang, Yu, Aref, Mahmood, Malluhi, & Ouzzani. (2016). *In-memory Distributed Spatial Query Processing and Optimization.* Purdue technical report 2016.

Tang, Z., & Liu, T. (2016). Evaluating Internet-based public participation GIS (PPGIS) and volunteered geographic information (VGI) in environmental planning and management. *Journal of Environmental Planning and Management*, *9*(6), 1073–1090. doi:10.1080/09640568.2015.1054477

Thecla, A., & Chigozie, O. (2014). Designing and Developing a GIS Database for Tourism in Nigeria: The Case of Anambra State. *Journal of the Humanities and Social Sciences*, *19*(10), 109–120.

Thompson, J. S. (1982). *Site Selection*. New York: Lebhar-Friedman Books.

Tomlinson, R. (2003). *Thinking about GIS*. Redlands, CA: ESRI Press.

Topay, M. (2003). *Bartın-Uluyayla Peyzaj Özelliklerinin Rekreasyon-Turizm Kullanımları Açısından Değerlendirilmesi Üzerinde Bir Araştırma*. Ankara: Yayınlanmamış Doktora Tezi, Ankara Üniversitesi Fen Bilimleri Enstitüsü.

Topay, M., & Parladır, M. Ö. (2015). Isparta İli Örneğinde CBS Yardımıyla Alternatif Turizm Etkinlikleri İçin Uygunluk Analizi. *Journal of Agricultural Sciences*, *21*, 300–309.

Torre, G. M. V., & Gutiérrez, E. A. (2008). The Demand of Rural Tourism in a Natural Park in Southern Spain. In P. M. Burns & M. Novelli (Eds.), *Tourism Development: Growth, Myths and Inequalities* (pp. 113–125). London: CAB International. doi:10.1079/9781845934255.0113

Tremblay, P. (2005). GIS techniques in tourism and recreation planning: Application to wildlife tourism. In B.W. Ritchie, P. Burns, & C. Palmers (Eds.), Tourism Research Methods: Integrating theory with practice (pp. 163-178). Academic Press.

TÜİK. (2014). *Türkiye 2014 Nüfus Yoğunluğu Gridi*. Retrieved May 14, 2017, from http://www.resmiistatistik.gov.tr/?q=tr/content/18-kırkent-tanımının-revizyonu

Turkish Language Society. (n.d.). Retrieved July 04, 2017, from http://www.tdk.gov.tr/index.php?option=com_gts&arama=gts&guid=TDK.GTS.595b81c56284d9.35935636

Turkish Statistical Institute. (2016). *Kültürel miras 2015*. Retrieved January 31, 2017, from http://www.tuik.gov.tr/PreHaberBultenleri.do?id=21550

Türk, T., Kitapçı, I. O., & Dörtyol, T. (2014). The Usage of Geographical Information Systems (GIS) in the Marketing Decision Making Process: A Case Study for Determining Supermarket Locations. *Procedia: Social and Behavioral Sciences*, *148*, 227–235. doi:10.1016/j.sbspro.2014.07.038

Tussyadiah, L. (2012). An assessment of contagion on social networking sites. *Proceedings of the international conference: Information and communication technologies in tourism 2012*, 25-35. doi:10.1007/978-3-7091-1142-0_3

Tyagi, N. (2014). Web GIS application for customized tourist information system for Eastern UP, India. *Journal of Genomics*, *8*(1), 1–6.

Uçar, M., Çeken, H., & Ökten, Ş. (2017). *Kırsal Turizm ve Kırsal Kalkınma*. Ankara: Detay Yayıncılık.

UNEP. (2016). *Sustainable Tourism Gateway. Environmental Impact of Tourism*. Retrieved on August 18, 2016, from: http://www.gdrc.org/uem/eco-tour/st-documents.html

United Nations Educational, Scientific and Cultural Organization (UNESCO). (2017). Retrieved July 07, 2017, from http://whc.unesco.org/en/list

Vaccari, L., Shvaiko, P., & Marchese, M. (2009). A geo-service semantic integration in Spatial Data Infrastructure. *International Journal of Spatial Data Infrastructures Research*, *4*, 24–51.

Vansteenwegen, P., Souffriau, W., Berghe, G. V., & Van Oudheusden, D. (2011). The city trip planner: An expert system for tourists. *Expert Systems with Applications*, *38*(6), 6540–6546. doi:10.1016/j.eswa.2010.11.085

Vijay Krishna, A. (2003). *Records Management Guide* (vol. 2). Academic Press.

Vlachopoulou, M., Silleos, G., & Manthou, V. (2001). Geographic information systems in warehouse site selection decisions. *International Journal of Production Economics*, *71*(1-3), 205–212. doi:10.1016/S0925-5273(00)00119-5

Vries, F. W. T. P., Teng, P., & Metselaar, K. (1991). Systems approaches for agricultural development. *Proceedings of the International Symposium on Systems Approaches for Agricultural Development*, 2–6.

Wade, T., & Sommer, S. (2006). *A to Z GIS: An Illustrated Dictionary of Geographic Information Systems*. Redlands, CA: ESRI Press.

Wanaskar, U., Vij, S., & Mukhopadhyay, D. (2013). *A hybrid web recommendation system based on the improved association rule mining algorithm*. arXiv preprint arXiv:1311.7204

Wang, X. Z., Zhang, H. M., Zhao, J. H., Lin, Q. H., Zhou, Y. C., & Li, J. H. (2015). An Interactive Web-Based Analysis Framework for Remote Sensing Cloud Computing, *ISPRS Annals of the Photogrammetry. Remote Sensing and Spatial Information Sciences*, *4*, W2.

Wei, L.-Y., Zheng, Y., & Peng, W.-C. (2012). *Constructing popular routes from uncertain trajectories*. The 18th ACM SIGKDD international conference on Knowledge discovery and data mining, 195-203.

Wei, W. (2011). Research on the Application of Geographic Information System in Tourism Management. *International Conference on Environmental Science and Engineering- ICESE 2011*.

Wei, W. (2012). Research on the Application of Geographic Information System in Tourism Management. *Procedia Environmental Sciences*, *12*, 1104–1109. doi:10.1016/j.proenv.2012.01.394

Wen, Y., Liu, Y., Zhang, Z. J., Xiong, F., & Cao, W. (2014). Compare two community-based personalized information recommendation algorithms. *Physica A*, *398*, 199–209. doi:10.1016/j.physa.2013.12.037

What are geospatial technologies. (n.d.). Retrieved from https://www.aaas.org/content/what-are-geospatial-technologies

What are Longitudes and Latitudes. (n.d.). Retrieved from https://www.timeanddate.com/geography/longitude-latitude.html

World Bank. (2014). *Investing in the Buddhist Circuit*. International Finance Cooperation, World Bank. From: http://documents.worldbank.org/curated/en/911791487142754943/pdf/112816-WP-IN-Buddhist-Circuit-Tourism-Strategy-PUBLIC.pdf

World Tourism Organization (WTO). (2009). *Facts and Figures, The UNWTO World Tourism Barometer*. Retrieved 10 May, 2017, http://unwto.org/facts/eng/barometer.htm

World Tourism Organization (WTO). (2015). *Facts and Figures, The UNWTO World Tourism Barometer*. Retrieved 1 June, 2017, http://media.unwto.org/press-release/2015-07-08/international-tourist-arrivals-4-first-four-months-2015

WTO. (2003). *Conclusions-Seminar on Rural Tourism "Guidelines for Development"*. Retrieved April 16, 2017, from http://www.unwto.org/regional/europe/PDF/Conclusions%20_Ukraine.pdf

Xia, F., Asabere, N. Y., Ahmed, A. M., Li, J., & Kong, X. (2013). Mobile multimedia recommendation in smart communities: A survey. *IEEE Access: Practical Innovations, Open Solutions*, *1*, 606–624. doi:10.1109/ACCESS.2013.2281156

Xia, J. (2004). Library space management: A GIS proposal. *Library Hi Tech*, *22*(4), 375–382. doi:10.1108/07378830410570476

Xiaolin, Lu. (2005). An Investigation on Service Oriented Architecture for constructing Distributed WebGIS Application. *IEEE International Conference on Services Computing (SCC'05)*, *1*, 191-197.

Xiao, X., Zheng, Y., Luo, Q., & Xie, X. (2010). *Finding similar users using category-based location history*. The 18th SIGSPATIAL International Conference on Advances in Geographic Information Systems, 442-445. doi:10.1145/1869790.1869857

Yadav, L. R., & Singh, R. S. (2009). GIS Applications in Statistical System and Micro Level Planning Strengthening. *10th ESRI India User Conference, Geography in Action.*

Yağcı, Ö. (2002). *Turizmde Çeşitlendirme, Sürdürülebilirlik ve Kırsal Turizm.* Retrieved May 01, 2017, from http://www.emyo.ankara.edu.tr/iktisad/TURKONF/web/ozcan_yagci.doc

Yager, R. R. (1988). On ordered weighted averaging aggregation operators in multicriteria decisionmaking. *IEEE Transactions on Systems, Man, and Cybernetics*, *18*(1), 183–190. doi:10.1109/21.87068

Yalçın, C. (2015). *Müze assist.* Retrieved July 14, 2017, from http://www.muzeasist.com/kklavuz.pdf

Yang, C. Z., Chen, I. X., & Wu, P. J. (2008). Cross-lingual news group recommendation using cluster-based cross-training. *Computational Linguistic and Chinese Language Processing, 13*(1), 41-60.

Yang, C., Goodchild, M., Huang, Q., Nebert, D., Raskin, R., Xu, Y., ... Fay, D. (2011). Spatial cloud computing: How can the spatial sciences use and help shape cloud computing? *International Journal of Digital Earth*, *4*(4), 305–329. doi:10.1080/17538947.2011.587547

Yang, C., Huang, Q., Li, Z., Liu, K., & Hu, F. (2017). Big Data and cloud computing: Innovation opportunities and challenges. *International Journal of Digital Earth, 10*(1), 13–53. doi:10.1080/17538947.2016.1239771

Yang, C., Raskin, R., Michael, G., & Gahegan, M. (2010). Spatial cyberinfrastructure: Past, present and future. *Computers, Environment and Urban Systems*, *34*(4), 264–277. doi:10.1016/j.compenvurbsys.2010.04.001

Yang, D., Zhang, D., Yu, Z., & Wang, Z. (2013). *A sentiment-enhanced personalized location recommendation system.* The 24th ACM Conference on Hypertext and Social Media, 119-128.

Yang, Y., Tang, J., Luo, H., & Law, R. (2015). Hotel location evaluation: A combination of machine learning tools and web GIS. *International Journal of Hospitality Management*, *47*, 14–24. doi:10.1016/j.ijhm.2015.02.008

Ye, M., Yin, P., & Lee, W.-C. (2010). *Location recommendation for location-based social networks.* The 18th SIGSPATIAL international conference on advances in geographic information systems.

Ye, M., Yin, P., Lee, W.-C., & Lee, D.-L. (2011). *Exploiting geographical influence for collaborative point-of-interest recommendation.* The 34th international ACM SIGIR conference on Research and development in Information Retrieval, 325-334. doi:10.1145/2009916.2009962

Yener, D. (2017). Geographic Information Systems and Its Applications in Marketing Literature. In S. Faiz & K. Mahmoudi (Eds.), *Handbook of Research on Geographic Information Systems Applications and Advancements* (pp. 158–172). IGI Global. doi:10.4018/978-1-5225-0937-0.ch006

Yianna, F., & Poulicos, P. (2011). *GIS Contribution for the Evaluation and Planning of Tourism: A Sustainable Tourism Perspective*. Retrieved on August 18, 2016, from: http://www.gipsynoise.gr/HellasGI/KoinonikesPraktikes/papers/Farsaki.pdf

Ying, J. J.-C., Lee, W.-C., Ye, M., Chen, C.-Y., & Tseng, V. S. (2011). *User association analysis of locales on location based social networks*. The 3rd ACM SIGSPATIAL International Workshop on Location-Based Social Networks, 69-76. doi:10.1145/2063212.2063214

Ying, J. J.-C., Lu, E. H.-C., Lee, W.-C., Weng, T.-C., & Tseng, V. S. (2010). *Mining user similarity from semantic trajectories*. The 2nd ACM SIGSPATIAL International Workshop on Location Based Social Networks, 19-26.

Yin, H., Cui, B., Sun, Y., Hu, Z., & Chen, L. (2014). Lcars: A spatial item recommender system. *ACM Transactions on Information Systems*, *32*(3), 11. doi:10.1145/2629461

Yomralıoğlu, T. (2002). GIS Activities in Turkey. *Proceedings of International Symposium on GIS*, 834-840.

Yomralıoğlu, T., & Aydınoğlu, A. Ç. (2014). *Türkiye Ulusal Coğrafi Bilgi Sistemi (TUCBS) ve BB'ler İçın Coğrafi Veri Altyapısı*. Retrieved May 01, 2017, from http://arifcagdas.com/wp-content/uploads/2014/11/AYOP2014-tucbs.pdf

Yomralıoğlu, T. (2005). *Coğrafi Bilgi Sistemleri: Temel Kavramlar ve Uygulamalar*. Trabzon: Akademi Kitabevi.

Yoon, H., Zheng, Y., Xie, X., & Woo, W. (2010). *Smart itinerary recommendation based on user-generated GPS trajectories*. Springer. doi:10.1007/978-3-642-16355-5_5

Yoon, H., Zheng, Y., Xie, X., & Woo, W. (2012). Social itinerary recommendation from user-generated digital trails. *Personal and Ubiquitous Computing*, *16*(5), 469–484. doi:10.1007/s00779-011-0419-8

Yuan, Q., Cong, G., Ma, Z., Sun, A., & Thalmann, N. M. (2013). *Time-aware point-of-interest recommendation*. The 36th international ACM SIGIR conference on Research and development in information retrieval, 363-372.

Yue, P., Guo, X., Zhang, M., Jiang, L., & Zhai, X. (2016). Linked Data and SDI: The case on Web geoprocessing workflows. *ISPRS Journal of Photogrammetry and Remote Sensing*, *114*, 245–257. doi:10.1016/j.isprsjprs.2015.11.009

Yue, P., Zhou, H., Gong, J., & Hu, L. (2013). Geoprocessing in Cloud Computing platforms–a comparative analysis. *International Journal of Digital Earth*, *6*(4), 404–425. doi:10.1080/17538947.2012.748847

Zadeh, L. A. (1965). Fuzzy sets. *Information and Control*, *8*(3), 338–353. doi:10.1016/S0019-9958(65)90241-X

Zhang, Han, Liu, Wang, & Feng. (2009). Spatial Queries Evaluation with MapReduce. *Proceedings of GCC '09*.

Zhang, J.-D., & Chow, C.-Y. (2013). iGSLR: Personalized geo-social location recommendation: a kernel density estimation approach. The 21st ACM SIGSPATIAL International Conference on Advances in Geographic Information Systems, 334-343.

Zhang, J. D., & Chow, C. Y. (2013, November). iGSLR: personalized geo-social location recommendation: a kernel density estimation approach. In *Proceedings of the 21st ACM SIGSPATIAL International Conference on Advances in Geographic Information Systems* (pp. 334-343). ACM. doi:10.1145/2525314.2525339

Zhang, J. D., & Chow, C. Y. (2015). CoRe: Exploiting the personalized influence of two-dimensional geographic coordinates for location recommendations. *Information Sciences*, *293*, 163–181. doi:10.1016/j.ins.2014.09.014

Zhang, J. D., Chow, C. Y., & Li, Y. (2015). iGeoRec: A personalized and efficient geographical location recommendation framework. *IEEE Transactions on Services Computing*, *8*(5), 701–714. doi:10.1109/TSC.2014.2328341

Zhang, J.-D., & Chow, C.-Y. (2015). *Geosoca: Exploiting geographical, social and categorical correlations for point-of-interest recommendations. IEEE Transactions on Services Computing*. doi:10.1145/2766462.2767711

Zhang, J.-D., & Chow, C.-Y. (2016). TICRec: A probabilistic framework to utilize temporal influence correlations for time-aware location recommendations. The 22nd ACM SIGSPATIAL International Conference on Advances in Geographic Information Systems, 103-112 , *9*(4), 633–646. doi:10.1109/TSC.2015.2413783

Zhang, J.-D., Chow, C.-Y., & Li, Y. (2014). *Lore: Exploiting sequential influence for location recommendations*. The 38th International ACM SIGIR Conference on Research and Development in Information Retrieval, 443-452. doi:10.1145/2666310.2666400

Zhang, J., Zhan, Z., Lin, Y., Chen, N., Gong, Y., Zhong, J., ... Shi, Y. (2011). Evolutionary computation meets machine learning: A survey. *IEEE Computational Intelligence Magazine*, *6*(4), 68–75. doi:10.1109/MCI.2011.942584

Zhao, S., King, I., & Lyu, M. R. (2016a). *A survey of point-of-interest recommendation in location-based social networks*. arXiv preprint arXiv:1607.00647

Zhao, S., Zhao, T., Yang, H., Lyu, M. R., & King, I. (2016b). STELLAR: Spatial-Temporal Latent Ranking for Successive Point-of-Interest Recommendation. Academic Press 315-322.

Zhao, S., King, I., & Lyu, M. R. (2013). *Capturing geographical influence in POI recommendations*. International Conference on Neural Information Processing, 530-537.

Zhao, S., King, I., & Lyu, M. R. (2013, November). Capturing geographical influence in POI recommendations. In *International Conference on Neural Information Processing* (pp. 530-537). Springer. doi:10.1007/978-3-642-42042-9_66

Zheng, Y., & Xie, X. 2010. Learning Location Correlation from GPS trajectories. In *Proceedings of the International Conference on Mobile Data Management 2010 (MDM 2010)*. IEEE Press. doi:10.1109/MDM.2010.42

Zheng, V. W., Zheng, Y., Xie, X., & Yang, Q. (2012). Towards mobile intelligence: Learning from GPS history data for collaborative recommendation. *Artificial Intelligence*, *184*, 17–37. doi:10.1016/j.artint.2012.02.002

Zhou. (2015). *Analysis and performance test of MapReduce, Spark, Phoenix, Disco, Mars DeveloperWorks*. IBM.

Zhu, X., Ye, H., & Gong, S. (2009). *A personalized recommendation system combining case-based reasoning and user-based collaborative filtering*. Control and Decision Conference (CCDC'09), 4026-4028. doi:10.1109/CCDC.2009.5192712

About the Contributors

Somnath Chaudhuri has a M. Tech. (Fourth in the University with Distinction) in Computer Science and Engineering, B. Tech. in Computer Science and Engineering and B. Sc. in Physics, all from the University of Calcutta, India. He is working as an International faculty, in the Department of IT, Faculty of Science, Maldives National University. Before this he was working as Lecturer and Database Advisor for Royal Thimphu College, Bhutan for 4 years. His research interest lies in the domain of GIS, Web GIS, Map Mashups, Cloud Computing, Automated System, and Application of Information Technology in Business and Industry Sectors. He has some National and International publications in these domains. He is the co-editor of a book published by IGI Global, USA. Mr. Chaudhuri has worked as the project guide for some real-time projects in the government and private sectors of Bhutan. Mr. Chaudhuri has received a certificate of appreciation from the Minister of Agriculture, Royal Government of Bhutan and from the Secretary of the Department of Livestock, Royal Government of Bhutan.

Nilanjan Ray is from Kolkata, India and presently associated as an Assistant Professor of Marketing in Adamas University in School of Management, West Bengal, India. He has obtained his PhD(Mkt); M.Com (Mkt); MBA (Mkt), STC FMRM (IIT-Kgp). He has 10 years teaching experience in BBA, MBA, BCom and 6 years Research experience and guided around 56 Post Graduate students' project. Dr. Ray has contributed over 57 research papers in reputed National and International Referred, Peer Reviewed Journals, Proceedings and 6 Edited Research Hand Books from Springer, IGI-Global USA and Apple Academic Publisher CRC Press (A Taylor & Francis Group), USA. He has also associated himself as a reviewer of Tourism Management Elsevier, Journal of Service Marketing Emerald Group Publishing Limited, Journal of Business and Economics, Research Journal of Business and Management Accounting and as an Editorial Board Member of several referred Journals. He has also chaired in a technical session at IJAS Conference 2012, at Harvard University, Boston, USA. Dr. Ray is a life-member of the International

Business Studies Academia, Fellow Member of Institute of Research Engineers and Doctors Universal Association of Arts and Management Professionals (UAAMP) New York, USA.

* * *

Rahim Ali Abbaspour received his Ph.D. degree in GIS from the University of Tehran, Tehran, Iran in 2009. In 2010, he joined the School of Surveying and Geospatial Engineering, as an assistant professor. His current research interests include geospatial information science, artificial intelligence, operational research and spatial optimization, SDSS, spatio-temporal data mining and knowledge discovery, and VGI. He is a Life Member of the National Academy of Science, Iran and Iran Society of Engineering Education (ISEE). He was the Scientific chair of the 1st ISPRS International Conference on Geospatial Information Research (GIResearch 2014) held in Tehran.

Aysu Altaş, Ph.D., has been working as an Assistant Professor at Aksaray University, Faculty of Tourism, Tourism Management Department since 2016 February. Dr. Altaş has published numerous papers in several journals, conference proceedings and chapters in edited books since she got her MA. Her major is Gastronomy. She also has studies on world cuisines, food and beverage management and Ottoman Palace Cuisine.

Zahra Bahramian received B.S. degree as a top student in Civil Eng.- Surveying engineering from University of Tehran, Tehran, Iran in 2006, and M.S. degree in geospatial information system (GIS) engineering from the University of Tehran, Tehran, Iran in 2009. received her Ph.D. degree in GIS from the University of Tehran, Tehran, Iran in 2017. Her current research interests include artificial intelligence, spatial optimization, recommender systems, tour planning, temporal GIS, GIS in transportation, and disaster management.

Rabindra K. Barik is currently working as an Assistant Professor in the School of Computer Application, KIIT University, Bhubaneswar, India. He received his M.Tech and Ph.D. in Geoinformatics from Motilal Nehru National Institute of Technology, Allahabad, India. His research area includes SOA, Cloud Computing, Fog Computing, IPR and Geoinformatics. He is a member of IEEE and IAENG.

Inês Boavida-Portugal has a PhD in Geography, specialization in Geographic Information Science (Universidade de Lisboa, 2012-2016). MSc in Territorial Management, specialization in Remote Sensing and Geographical Information

Systems (Universidade Nova de Lisboa, 2010). Research interests on Geographic modelling, Complex systems, Agent-based models, Tourism development, Tourist decision making process. Assistant Professor in Spatial Planning and Modelling, Faculty of Spatial Sciences, Department of Spatial Planning and Environment, University of Groningen.

Hüseyin Çeken is an Associate Professor at Muğla Sıtkı Koçman University (Turkey) the Faculty of Tourism and Head of the Accommodation Management Department. He completed his master's and doctoral studies in the Department of Tourism Management and specializes in tourism economics. In the school he serves as a lecturer, he conducts courses in economics, tourism economics, rural tourism and rural development. He has books that the author has written about his field, numerous articles and reports. The author is married and has a daughter named Zeynep.

Probal Chaudhury works as a scientist F in Bhaba Atomic Research Centre, Mumbai, India.

Christophe Claramunt is a Professor in Computer Science and deputy director of the Naval Academy Research Institute in France. He holds a PhD in Computer Science from the University of Burgundy (France). He has been a senior lecturer in computing at the Nottingham Trent University and a senior researcher at the Swiss Federal Institute of Technology in Lausanne. His research is oriented towards theoretical and pluri-disciplinary aspects of geographical information science. He serves in the editorial boards of several international GIS journals and is regularly involved in the chairing and program committees of several international GIS conferences.

Harishchandra Dubey is interested in Audio, Speech and Language Processing, Machine Learning, Fog & Cloud Computing, Social Signal Processing. He is received Master of Science, FAU University of Erlangen-Nuremberg, Germany in 2015. He is received Bachelor of Technology in Electronics and Communication Engineering from Motilal Nehru National Institute of Technology, Allahabad, India in 2012.

Luis Encalada is a PhD Student in Geography (Geographic Information Science) at University of Lisbon, Institute of Geography and Territorial Planning, Centro de Estudos Geográficos (CEG), Lisbon, Portugal.

Subbu Esakkipandian works as a JRF in VIT University as well as doing M.Tech by Research in VIT University.

Carlos Ferreira is a Senior Researcher at University of Lisbon Institute of Geography and Spatial Planning · Centro de Estudos Geográficos (CEG), Lisbon, Portugal. PhD in Geography (Regional and Local Planning).

Anu A. Gokhale has completed twenty-five years of university teaching and is currently a professor and coordinator of the computer systems technology program at Illinois State University. She is named Fulbright Distinguished Chair in STEM at the University of Pernambuco, Brazil, 2016-17; is a Fulbright Specialist; and was a Fulbright Scholar to India in 2002. She was a Visiting Professor at Shandong University in Jinan, China during spring 2017. Dr. Gokhale was honored with the 2011 University Outstanding Researcher Award.

Farooq Haq completed his PhD in 2011 from Charles Darwin University in Australia where he was a lecturer for 3 years. Since 2012 he is lecturing in Canadian University Dubai, where he has been the Chair of Marketing Program for last 2 years. The research interests of Dr Farooq include: Marketing special products, spiritual tourism, Islamic Marketing, medical tourism and Halal Branding. He has published extensively in journal and books associated with above mentioned research areas.

Hesham A. Hefny received the B.Sc., M.Sc. and Ph.D. all in Electronics and Communication Engineering from Cairo University in 1987, 1991 and 1998 respectively. He is currently a professor of Computer Science at the Institute of Statistical Studies and Researches (ISSR), Cairo University. He is also the vice dean of graduate studies and researches of ISSR. Prof. Hefny has authored more than 160 papers in international conferences, journals and book chapters. His major research interests include: computational intelligence (neural networks – Fuzzy systems-genetic algorithms – swarm intelligence), data mining, uncertain decision making. He is a member in the following professional societies: IEEE Computer, IEEE Computational Intelligence, and IEEE System, Man and Cybernetics.

Rakesh Kumar Lenka is currently working as an Assistant Professor in the Department of Computer Science and Engineering, IIIT Bhubaneswar, India. He received his M.Tech in Computer Science and Engineering from Motilal Nehru National Institute of Technology, Allahabad, India. His research area includes Cloud Computing, Software Testing, Automata Theory and Compiler Design. He is a member of IEEE and IAENG.

Mahmoud Mahmood is an assistant professor in Cairo University at Institute of Statistical Studies and Research.

Kunal Mankodiya is an assistant professor in the Dept. of Electrical, Computer and Biomedical Engineering, University of Rhode Island, Kingston, RI, USA. He pursued his postdoctoral research at Intel Science & Technology Center affiliated with Carnegie Mellon University, Pittsburgh. He received Ph.D. degree in computer science from the University of Luebeck, Germany, with an emphasis on wearable health monitoring. He holds B.E. (Saurashtra University, India) and M.Sc. (University of Luebeck, Germany) degrees in Biomedical Engineering (BME). He has published a number of journal and conference articles and book chapters in the areas of BME, embedded computing, human-computer interaction, digital signal processing, and robotics. He has published a book on wearable health monitoring that serves as a hands-on guide to program high-end embedded processors for healthcare applications. His embedded design of a wearable ECG system based upon smart textiles has earned him the SYSTEX student award 2010, University of Ghent, Belgium. He is a member of IEEE, ACM, and Biomedical Engineering Society and serves in the professional society in various capacities, including technical program committees, grant review panels of NSF/NIH and scientific workshop organizations. He also serves as a reviewer for various journals and conference proceedings in the fields of Human-Machine Systems, Humanized Computing, Sensors, Internet- of-Things, and Pervasive Healthcare. He organizes Annual Internet-of-Things Hack-a-Thon to promote entrepreneurship and startups in the area of IoT.

Anita Medhekar is an economist. She has taught economics at undergraduate and post graduate level in India, Indonesia and Australia. Currently she is teaching at Central Queensland Rockhampton, Australia. Her research interests are in macro and micro economics, economics of the Asia and Pacific countries, development economics, international trade, tourism economics, health economics, public policy, privatization and public-private-partnerships. She has presented papers at various international conferences and has many publications to her credit as conference proceedings, books chapters and journal articles.

Sovik Mukherjee is at present a Lecturer in Economics, Department of Commerce(Evening) at Shri Shikshayatan College, Kolkata, India. He is also pursuing his research in the Department of Economics, Jadavpur University, Kolkata, India. Mr. Mukherjee has a B.A (Hons.) degree in Economics from Jadavpur University (2013) and was awarded the Ujjayini Memorial Gold Centred Silver Medal, an M.A degree in Economics from Jadavpur University (2015) and an M.Phil degree in Economics from Jadavpur University in 2017. His research interests lie in the area of public economics and taxation, applied game theory, environmental economics and issues related to economics of climate change, health economics, international economics and the econometric analysis of financial crises. He is a Life Member of the Indian Economic Association (IEA) and the Bengal Economic Association (BEA). He has completed a consultancy research project sponsored by the State Bank of India, LHO, Kolkata, India. Besides, he has served as a data analyst in a research project sponsored by the School of Media, Communication and Culture, Jadavpur University, Kolkata, India.

Jorge Rocha, geographer, born in Lisbon in 1970. He is MSc in Geographic Information Systems (2003) and in Territorial Management (2013), and PhD in Geographic Information Science (2012). Currently is an Assistant Professor of the Institute of Geography and Spatial Planning and member of the Modelling, Urban and Regional Planning, and Environmental Hazard and Risk Assessment and Management research groups of the Centre of Geographical Studies, all from the University of Lisbon. His field of expertise is Geosimulation and Geocomputation involving Artificial Neural Networks, Graphs Theory, Cellular Automata, and Multi-agent Systems. Jorge's works are quite diverse, focusing mainly on, but not exclusively, in Urban Morphology, Remote Sensing, Epidemiology, Health Geography, Smart Cities and Big Data (Geomarketing and Tourism).

Khaled Soliman is a master researcher of Institute of Statistical Studies and Research (ISSR).

Sachin Soonthodu is an Assistant Professor in the Department of Tourism, Garden City University has more than 7 years of experience in teaching both in graduate and postgraduate level. He has pursued Masters of Business Administration (Tourism) from Mangalore University and PGDCA from MAHE. He has qualified UGC- NET in 2009 and K-SET in 2015 and at present engaged in doctoral research at Karnataka Folklore University, Haveri. He is the principal investigator for the Minor Research Project funded by UGC. He also served as the visiting faculty for IATA study centers in Bangalore. Currently he is also associated with research on temples of Kodagu district of Karnataka. His key skills include Travel and Tourism Management, Tourism Geography, Indian Architecture, Airline and Cargo Management, Marketing Management and Business Administration. His passion includes research, Hindu Architecture, Travel, Photography and Blogging.

Mertcan Taşçıoğlu (PhD, Georgia Southern University) is an Assistant Professor at Istanbul Medeniyet University. His research includes ethical and social issues in consumer decision-making, international/cross-cultural consumer behavior, and green marketing.

S. Vasavi is working as a Professor in Computer Science & Engineering Department with 20 years of experience. She pursued her MS from BITS, Pilani in the year 1999 and PhD from Acharya Nagarjuna University in the year 2010. She currently holds R&D projects from UGC and ISRO-ADRIN. She published 44 papers in various Scopus indexed, Google Scholar Indexed conferences and journals. She filed two patents. She is the recipient of UGC International travel grant in the year 2015, for her visit to ICOIP 2015,USA and TEQIP grant for her visit to ICICT 2016, Thailand. She is the Keynote speaker for the IEEE international conference ICOIP conducted at Singapore in July 2017. She visited reputed universities at U.S.A, Singapore and Thailand. She also Visited Argonne National Laboratory, A multidisciplinary Science and Engineering Research Center, that address vital national challenges in Clean Energy, Environment, Technology and National Security. She is an IEEE member, life member of Computer society of India (CSI), Member Machine Intelligence Research Labs, Washington, USA. Her Research Areas are Bigdata analytics: Image object Classification. Reviewer for Scopus Indexed journals [Inderscience, Image vision and computing, Bigdata Journal] and conferences [Springer, IEEE, Elsevier]. Received Best Teacher Award, Vishitta Mahila Award, Conferred Outstanding Women in Engineering Award.

Ishrat Wahab is an Assistant Professor and Head of the Tourism Department, Garden City University, and is currently pursuing her Ph.D in Management Studies at CMR University, Bangalore. She is the managing committee member of Federation of Karnataka Chambers of Commerce & Industry - Tourism (FKCCI), and represents the university as an Allied Member of Travel Agents Association of India (TAAI). She has been instrumental in signing of MoUs between GCU and other national and international institutions as well as with MoTs and non-profit organisations. She is passionate about studies and work, and has published research papers in international journals with high Impact Factor, peer reviewed journals and other journals of repute. She is also the Chief Editor of the peer reviewed journal, Compass (ISSN 2394-0646). Her key skills include university teaching, research, event management, tourism and travel management, and creative writing. An ardent dog lover. An avid reader. Holds a keen interest in global diplomacy. A dreamer.

Kiran Yarrakula working as an Associate professor in CDMM (Centre for disaster mitigation and management) at VIT Univeraity,Vellore, Tamilnadu, India. He is expertise in Remote sensing and GIS and Hydrology. He Completed his Ph.D from IIT Kharagpur, which is one of the premier institute in India.

Dursun Yener had PhD degree on production management and marketing at Marmara University, İstanbul, Turkey. His research areas are marketing theory and consumer behaviour. He has numerous publishing in national and international journals and book. He is working at Istanbul Medeniyet University as an associate professor of marketing.

Emine Yılmaz graduated from Dokuz Eylül University (Turkey) undergraduate program in Tourism Management (100% English) and started to work as a research assistant at Muğla Sıtkı Koçman University. After completing her master's degree at the Department of Tourism Management, she started to doctorate in the same program. She still continues her PhD studies. The author is married and a fresh mother of a baby girl named Gülin.

Index

T

U

W

www.ingramcontent.com/pod-product-compliance
Lightning Source LLC
LaVergne TN
LVHW080903300926
849162LV00003B/536